Astrophysical Sources of High Energy Particles and Radiation

NATO Science Series

A Series presenting the results of scientific meetings supported under the NATO Science Programme.

The Series is published by IOS Press, Amsterdam, and Kluwer Academic Publishers in conjunction with the NATO Scientific Affairs Division

Sub-Series

I.   Life and Behavioural Sciences   IOS Press
II. Mathematics, Physics and Chemistry   Kluwer Academic Publishers
III. Computer and Systems Science   IOS Press
IV. Earth and Environmental Sciences   Kluwer Academic Publishers
V.   Science and Technology Policy   IOS Press

The NATO Science Series continues the series of books published formerly as the NATO ASI Series.

The NATO Science Programme offers support for collaboration in civil science between scientists of countries of the Euro-Atlantic Partnership Council. The types of scientific meeting generally supported are "Advanced Study Institutes" and "Advanced Research Workshops", although other types of meeting are supported from time to time. The NATO Science Series collects together the results of these meetings. The meetings are co-organized bij scientists from NATO countries and scientists from NATO's Partner countries – countries of the CIS and Central and Eastern Europe.

Advanced Study Institutes are high-level tutorial courses offering in-depth study of latest advances in a field.
Advanced Research Workshops are expert meetings aimed at critical assessment of a field, and identification of directions for future action.

As a consequence of the restructuring of the NATO Science Programme in 1999, the NATO Science Series has been re-organised and there are currently Five Sub-series as noted above. Please consult the following web sites for information on previous volumes published in the Series, as well as details of earlier Sub-series.

http://www.nato.int/science
http://www.wkap.nl
http://www.iospress.nl
http://www.wtv-books.de/nato-pco.htm

Series II: Mathematics, Physics and Chemistry – Vol. 44

Astrophysical Sources of High Energy Particles and Radiation

edited by

Maurice M. Shapiro

Department of Physics and Astronomy,
University of Maryland,
College Park, MD, U.S.A.

Todor Stanev

Bartol Research Institute,
University of Delaware,
Newark, DE, U.S.A.

and

John P. Wefel

Department of Physics and Astronomy,
Louisiana State University,
Baton Rouge, LA, U.S.A.

Kluwer Academic Publishers

Dordrecht / Boston / London

Published in cooperation with NATO Scientific Affairs Division

Proceedings of the NATO Advanced Study Institute on
and 12th Course of the International School of Cosmic Ray Astrophysics
Astrophysical Sources of High Energy Particles and Radiation
Erice, Italy
10–21 November 2000

A C.I.P. Catalogue record for this book is available from the Library of Congress.

ISBN 1-4020-0173-8 (HB)
ISBN 1-4020-0174-6 (PB)

Published by Kluwer Academic Publishers,
P.O. Box 17, 3300 AA Dordrecht, The Netherlands.

Sold and distributed in North, Central and South America
by Kluwer Academic Publishers,
101 Philip Drive, Norwell, MA 02061, U.S.A.

In all other countries, sold and distributed
by Kluwer Academic Publishers,
P.O. Box 322, 3300 AH Dordrecht, The Netherlands.

Printed on acid-free paper

This volume is dedicated

in fond memory

of two distinguished colleagues

David N. Schramm, a leader in high-energy physics and cosmology. He will be best remembered for his seminal insight into nucleosynthesis in the early universe.

John A. Simpson, an outstanding contributor to cosmic-ray research and to heliospheric physics, assuring him a significant place in the pantheon of 20th century science.

TABLE OF CONTENTS

V.   COSMIC RAY PARTICLES

VI.   FUTURE PROSPECTS

PREFACE

This volume is devoted to the NATO Advanced Study Institute which took place at the Ettore Majorana Centre in Erice, Italy from November 11 to 21, 2000. This was also the 12th Course of the International School of Cosmic-Ray Astrophysics and was combined with the International School of Particle Astrophysics. The course, "Astrophysical Sources of High-Energy Particles & Radiation", was dedicated in memory of one of its eminent charter lecturers, David N. Schramm. At the time of his untimely death in an airplane accident, Prof. Schramm had been serving as vice-president for research at the University of Chicago, while pursuing his creative endeavors in astrophysics and cosmology.

From its inception in 1977, the School has aimed to provide its student scientists with knowledge and insight that would enhance their contributions to science - - and especially to high energy astrophysics. It also provides a forum for all participants to exchange ideas and explore models which may promote progress in their fields of research. The contents of this volume reflect the multidisciplinary character of the program.

Among the highlights of this program were: new views of the X-ray sky revealed by the Chandra and XMM space observatories; detailed presentations of the roles of Supernovae and Active Galactic Nuclei in energetic processes; a new project for studies of the highest-energy cosmic rays, i.e., the Extreme Universe Space Observatory; new results from the Hubble Telescope ; theories of the nature and origin of gamma-ray bursts; the enigma of dark matter; and neutrinos and their possible oscillations. In addition, an evening session was devoted to the life and career of Prof. David N. Schramm.

Shortly before the institute convened the world of cosmic-ray physics lost one of its illustrious leaders, John A. Simpson, of the University of Chicago. A special session was held to commemorate Prof. Simpson's contributions to cosmic-ray and heliospheric science.

The twelfth course was among the best attended in the 23-year history of the School; over 100 participants came from 22 countries. J. P. Wefel of Louisiana State University and V. Ptuskin, IZMIRAN, Moscow were the NATO Co-directors primarily responsible for organizing the course, under the general supervision of the School's Director, Prof. M. M. Shapiro of the University of Maryland. Prof. T. Stanev of the Bartol Research Institute in Delaware, and Prof. P. Galeotti of the University of Torino also provided direction and guidance.

The directors are grateful to Prof. A. Zichichi, founder and director of the Majorana Centre in Erice, Italy for providing the infrastructure and support for this ASI. Ms. Maria Zaini and Ms. Fiorella Ruggio have given essential administrative support, and Mr. P. Aceto has also been helpful. Without the help of NATO's Division of Scientific Affairs, and the contributions of some generous donors, this Course would not have prospered. In particular we thank the NATO Science Committee members and acknowledge the wise support of Dr. F. Pedrazzini, director of NATO's ASI Program.

Maurice M. Shapiro

THEORY OF COSMIC RAY AND HIGH-ENERGY GAMMA-RAY PRODUCTION IN SUPERNOVA REMNANTS

E.G. BEREZHKO

Institute of Cosmophysical Research and Aeronomy,
Lenin Ave. 31, 677891 Yakutsk, Russia

1. Introduction

Considerable efforts have been made during the last years to empirically confirm the theoretical expectation that the main part of the Galactic cosmic rays (CRs) originates in supernova remnants (SNRs). Theoretically progress in the solution of this problem has been due to the development of the theory of diffusive shock acceleration (see, for example, reviews [1–3]). Although still incomplete, the theory is able to explain the main characteristics of the observed CR spectrum under several reasonable assumptions, at least up to an energy of $10^{14} \div 10^{15}$ eV. Direct information about the dominant nucleonic CR component in SNRs can only be obtained from γ-ray observations. If this nuclear component is strongly enhanced inside SNRs then through inelastic nuclear collisions, leading to pion production and subsequent decay, γ-rays will be produced.

CR acceleration in SNRs expanding in a uniform interstellar medium (ISM) [4–8], and the properties of the associated γ-ray emission [7, 8] were investigated in a number of studies (we mention here only those papers which include the effects of shock geometry and time-dependent nonlinear CR backreaction; for a review of others which deal with the test particle approximation, see for example [1–3, 9]). All of these studies are based on a two-fluid hydrodynamical approach and directly employ the assumption that the expanding supernova (SN) shock is locally plane; as dynamic variables for the CRs the pressure and the energy density are determined. Their characteristics are sometimes essentially different from the results obtained in a kinetic approach [10–12] which consistently takes the role of shock geometry and nonlinear CR backreaction into account. First of all, in kinetic theory the form of the spectrum of accelerated CRs and their maximum energy are calculated selfconsistently. In particular, the maximum particle

1

M. M. Shapiro et al. (eds.), Astrophysical Sources of High Energy Particles and Radiation, 1–17.
© 2001 *Kluwer Academic Publishers. Printed in the Netherlands.*

energy ϵ_{max}, achieved at any given evolutionary stage, is determined by geometrical factors [13], in contrast to the hydrodynamic models which in fact postulate that the value of $\epsilon_{max}(t)$ is determined by the time interval t that has passed since the explosion [4–6, 14]. Although the difference between the values of ϵ_{max} in the two cases is not very large, it critically influences the structure and evolution of the shock. For example, the shock never becomes completely modified (smoothed) by the CR backreaction [10–12, 15]. Together with the smooth precursor, the shock transition always contains a relatively strong subshock which heats the swept-up gas and leads to the injection of suprathermal gas particles into the acceleration process. In this sense diffusive shock acceleration is somewhat less efficient than predicted by hydrodynamic models. Acceleration always requires some freshly injected particles which are generated during gas heating. This prediction is in agreement with the observations that show significant gas heating in young SNRs.

A brief review of the kinetic model of CRs acceleration and subsequent γ-ray production inside SNRs is presented below.

2. Kinetic model

During the early phase of SNR evolution the hydrodynamical SN explosion energy E_{sn} is kinetic energy of the expanding shell of ejected mass. The motion of these ejecta produces a strong shock wave in the background medium, whose size R_s increases with velocity $V_s = dR_s/dt$. Diffusive propagation of energetic particles in the collisionless scattering medium allows them to traverse the shock front many times. Each two subsequent shock crossings increase the particle energy. In plane geometry this diffusive shock acceleration process [16–19] creates a power law-type CR momentum spectrum. Due to their large energy content the CRs can dynamically modify the shock structure.

In the frame of so-called kinetic model the description of CR acceleration by a spherical SNR shock wave is based on the diffusive transport equation for the CR distribution function. The gas matter is described by the gas dynamic equations which include the CR backreaction via term $-\partial P_c/\partial r$, i.e. the gradient of CR pressure. They also describe the gas heating due to the dissipation of Alfvén waves in the upstream region.

The SNR shock always includes a sufficiently strong subshock which heats the gas and plays an important dynamical role. The gas subshock, situated at $r = R_s$, is treated as a discontinuity on which all hydrodynamical quantities undergo a jump. The injection of some (small) fraction of gas particles into the acceleration process takes place at the subshock.

At present we only have some experimental (e.g. [20, 21]) and theoretical

[22–27] indications as to what value of the injection rate can be expected. A simple CR injection model, in which a small fraction η of the incoming particles is instantly injected at the gas subshock with a speed $\lambda > 1$ times the postshock gas sound speed c_{s2}, is usually used [28, 10–12, 29, 30]. In the case of protons, which are dominant ions in the cosmic plasmas, the number of particles, involved into the acceleration from each unit volume swept up by the shock, and their momentum are given by the relations:

$$N_{inj} = \eta N_1, \quad p_{inj} = \lambda m c_{s2}, \tag{1}$$

where $N = \rho/m$ is the proton number density, ρ is the gas density and m is the particle (proton) mass. The subscripts 1(2) refer to the point just ahead (behind) the subshock. Appropriate values of the injection parameters lie within the range $\eta = 10^{-4} \div 10^{-2}$, $\lambda = 2 \div 4$ consistent with the experiment [20, 21] and theoretical estimations [20–27].

It is usually assumed that the Bohm diffusion coefficient

$$\kappa(p) = \rho_B c/3, \tag{2}$$

is a good approximation for strong shocks [31], characterized by strong wave generation [18]. Here ρ_B is the gyroradius of a particle with momentum p in the magnetic field B, c is the speed of light. Note that relatively strong momentum dependence of this diffusion coefficient provides very wide range of upstream scale lengths of CR spatial distribution , so-called diffusive lengths $l = \kappa/V_s$, which typically varies from $10^{-9} R_s$ to $10^{-1} R_s$. This leads to a big difficulty in the numerical solution of the considering problem. To overcome this technical problem one can artificially suggest slowly momentum dependent CR diffusion coefficient, like $\kappa \propto p^{1/4}$, that was realized by Kang and Jones [10]. Development of the effective numerical algorithm [11,12] made it possible to investigate the process of CR acceleration and SNR evolution at arbitrary $\kappa(p)$. Note also, that at realistic Bohm diffusion coefficient $\kappa \propto p$ CR acceleration in SNR is much more effective and nonlinear effects are more pronounced compared with the case of weakly momentum dependent $\kappa(p)$.

Application of diffusive shock acceleration theory, which selfconsistently describes Alfvén wave excitation by accelerated particles, to the case of interplanetary shocks have demonstrated that Bohm limit for particle diffusion coefficient is easily reached near the front of sufficiently strong shock [32].

Alfvén wave dissipation [33] as an additional heating mechanism strongly influences the structure of a modified shock in the case of large sonic Mach number $M = V_s/c_s \gg \sqrt{M_a}$, where $M_a = V_s/c_a$ is the Alfvénic Mach number, c_s and c_a are the local sound and Alfvén speeds correspondingly, at the shock front position $r = R_s$. The wave damping substantially restricts the growth of the shock compression ratio $\sigma = \rho_2/\rho_s$ at the level $\sigma \approx M_a^{3/8}$ which, in the absence of Alfvén wave dissipation, has been found to reach extremely high values $\sigma \approx M^{3/4}$ for large Mach numbers [12, 34].

The result of a core collapse supernova, many days after the explosion, is freely expanding gas with velocity $v = r/t$. The density profile of the ejecta is described by similarity distribution, which contains high velocity power law tail $dM_{ej}/dv \propto v^{2-k}$ [35, 36], where M_{ej} is the total ejected mass. For SNRs the value of the parameter k typically lies between 7 and 12. The pressure in the expanding ejecta is negligible.

Interaction with the ambient material modifies the ejecta density distribution. The ejecta dynamics can be described in a simplified manner, assuming that the modified ejecta consist of two parts [29, 30]: a thin shell (or piston) moving with some speed V_p and a freely expanding part. The piston includes the decelerated tail of the initial ejecta distribution with initial velocities $v > R_p/t$, where R_p is the piston radius separating the ejecta and the swept-up ISM matter. The evolution of the piston is described in the framework of a simplified thin-shell approximation, in which the thickness of the shell is neglected. Behind the piston ($r < R_p$), the CR distribution is assumed to be uniform.

The high velocity tail in the ejecta distribution ensures a large value of the SNR shock speed at an early phase of evolution. It increases the CR and γ-ray production significantly compared with the case where all the ejecta propagate with a single velocity [29, 30].

The efficiency of diffusive CR penetration through the piston depends on the magnetic field structure, which is influenced by the Rayleigh-Taylor instability at the contact discontinuity ($r = R_p$) between the ejecta and the swept up medium that is contained in the region $R_p < r < R_s$.

Detailed descriptions of the model and numerical methods have been given earlier [11, 12, 29, 30].

3. CR spectrum and composition

The main fraction of the galactic volume is occupied by so-called hot and warm phases of ISM, with hydrogen number density, temperature and magnetic field values $N_H = 0.003$ cm^{-3}, $T_0 = 10^6$ K, $B_0 = 3$ μG and $N_H = 0.3$ cm^{-3}, $T_0 = 10^4$ K, $B_0 = 5$ μG respectively (e.g. [37]). The ISM temperature T_0 determines the equilibrium ionization state of elements: at $T_0 = 10^4$ K Q_0 is close to 1 for all elements, whereas at $T_0 = 10^6$ K mean ion charge number increases from $Q_0 \approx 1$ for H and He to $Q_0 \approx 10$ for heavy ions with $A \approx 100$ (e.g. [38]).

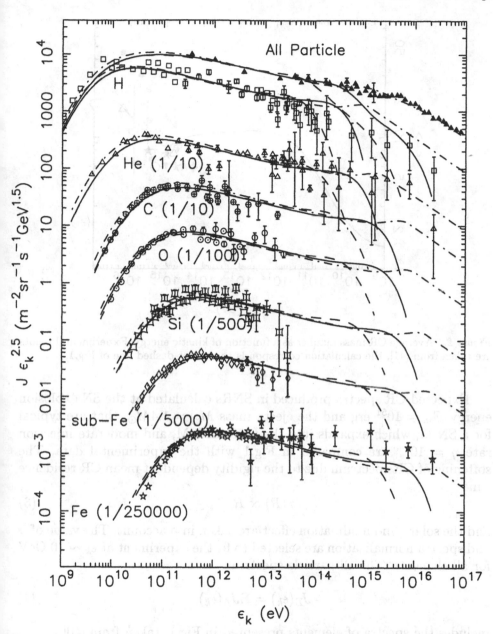

Figure 1. CR intensity near the Earth as function of the kinetic energy. Experimental points are taken from [39]. Solid (dashed) lines correspond to calculation for hot (warm) ISM with injection rate $\eta = 10^{-4}$, dot-dashed lines correspond to hot ISM with magnetic field $B_0 = 12\ \mu G$ and $\eta = 5 \times 10^{-4}$ [40].

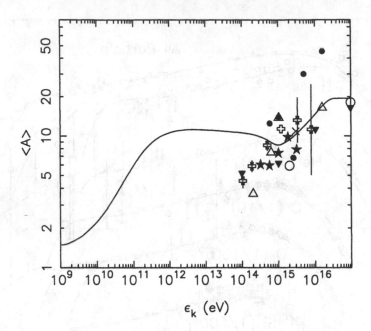

Figure 2. Average CR mass number as a function of kinetic energy. Experimental points are taken from [41], the calculation corresponds to the dot-dashed line of Fig.1.

Expected CR spectra produced in SNRs calculated at the SN explosion energy $E_{sn} = 10^{51}$ erg and the ejecta mass $M_{ej} = 1.4 M_\odot$, that are typical for a SN Ia, which expands into the uniform ISM, and moderate injection rate $\eta = 10^{-4}$ are compared in Fig.1 with the experimental data. The softening of CR spectrum due to the rigidity dependent mean CR residence time

$$\tau(R) \propto R^{-\alpha} \tag{3}$$

and the solar wind modulation effect are taken into account. The value of α and spectra normalization are selected to fit the experiment at $\epsilon_k \sim 10$ GeV for all elements. The all particle spectrum

$$J_\Sigma(\epsilon_k) = \Sigma J_A(\epsilon_k) \tag{4}$$

includes the spectra of elements presented in Fig.1, taken from [40] .

One can see that calculated spectra for all elements equally well fits the experiment at $\epsilon_k \lesssim 10^{14}$ eV for both considered ISM phases. The maximum energy in the all particle spectrum $\epsilon_{max} \approx 10^{14}$ eV and $\epsilon_{max} \approx 4 \times 10^{14}$ eV for warm and hot ISM respectively only slightly exceeds the proton maximum energy.

One can expect that the observed CR spectrum which has the only peculiarity, so-called knee at $\epsilon \approx 3 \times 10^{15}$ eV, at energies $\epsilon \gtrsim 10^{15}$ eV is produced by some reacceleration process. In this case one need to form in SNRs CR spectrum up to $\epsilon_{max} \approx 3 \times 10^{15}$ eV which is essentially higher then calculated one. To demonstrate how CR spectrum could look like at $\epsilon > \epsilon_{max}$ CR spectra calculated at $B_0 = 12 \ \mu G$ and extended towards higher energies according to the law $\epsilon^{-3.1}$ is presented in Fig.1. This rather formal procedure gives the prediction of CR composition at energies $\epsilon_k \gtrsim 10^{15}$ eV, which is expected to be sensitive to the value ϵ_{max}. The calculated mean CR atomic number

$$< A >= \Sigma A J_A(\epsilon_k)/\Sigma J_A(\epsilon_k) \qquad (5)$$

is presented in Fig.2. One can see that the expected value of $< A >$ increases from 10 to 20 in energy interval $10^{15} \div 10^{16}$ eV that is in a reasonable agreement with the existing experimental data.

To fit the data presented in Fig.1 relatively strong energy dependence of the CR residence time with $\alpha = 0.7 \div 0.8$ is required, whereas experimentally measured value $\alpha \approx 0.6$ (e.g. [37]). This discrepancy can be attributed either for existing of some dissipation process within the shock transition which makes the CR acceleration somewhat less efficient and leads to steepening of their resultant spectrum, or for a more complicated picture of CR leakage from SNRs into the galactic volume than usually assumed (see [40] for a details).

The required normalization of spectra presented in Fig.1, which is characterized by heavy elements enhancement relative to protons, is reproduced by the kinetic model due to preferential injection and acceleration of heavy elements, which at a given energy per nucleon have larger rigidity [42, 40].

Due to extremely hard CR spectrum inside SNRs, predicted by the kinetic model, an essential contribution from the single nearby SNR ($d \lesssim 1$ kpc) should be observed as a kind of bump in galactic CR spectrum at energies $\epsilon = 10^{14} \div 10^{16}$ eV, if CRs leaking from the parent SNR expands into the galactic volume more or less spherically symmetric [40]. It is not excluded that peculiarities in the galactic CR spectrum discussed by Erlykin and Walfendale [43] can be attributed to this kind of effect.

4. Gamma-rays from SNRs

A direct test, whether or not the observed Galactic CRs are indeed produced in SNRs at least up to an energy of about 10^{14} eV should be possible with observations of SNRs in high energy ($\epsilon_\gamma \sim 1$ TeV) γ-rays. If the dominant nuclear component of the CRs is strongly enhanced inside SNRs, then through hadronic collisions, leading to pion production and subsequent decay, γ-rays will be strongly produced.

Theoretical estimates of the π^0-decay γ-ray luminosity of SNRs have led to the conclusion that the expected TeV γ-ray flux from nearby SNRs in high enough ambient densities should be just detectable by present instruments [8]. The conclusions were based on a hydrodynamic approximation for the CR component and involved (very reasonable) assumptions about the CR energy spectra.

At the same time it was shown that high-energy γ-ray production is sensitive to the structure of the background ISM [30]. Since the ISM around the progenitor star can be strongly modified by the presupernova star wind the expected flux of γ-rays is presumably dependent on the SN type.

4.1. TYPE IA SUPERNOVA

The spectra of π^0-decay γ-rays, produced by shock accelerated CRs in SNRs that expand into a the uniform interstellar medium, that is typical situation for the case of SN Ia, were studied in detail in a kinetic approach [29]. We concentrate here on TeV γ-rays, measurable by the imaging Cherenkov technique.

The typical time-dependence of the expected integral flux $F_\gamma(\epsilon_\gamma)$ of γ-rays with energies ϵ_γ greater than 1 TeV from SNR Ia situated at 1 kpc distance is shown in Fig.3. Calculation was performed for a standard set of parameters: $E_{SN} = 10^{51}$ erg, $M_{ej} = 1.4 M_\odot$, $B_0 = 5$ μG, $\eta = 10^{-3}$ and hydrogen number density $N_H = 0.3$ cm^{-3} which determines the ISM density $\rho_0 = 1.4 N_H m$ (we assume 10% of helium nuclei in the ISM).

The kinetic model prediction for the peak value of the expected γ-ray flux

$$F_\gamma^{max}(1 \text{ TeV}) \approx 10^{-10} \left(\frac{N_H}{0.5 \text{ cm}^{-3}} \right) \frac{\text{photons}}{\text{cm}^2 \text{ s}} \tag{6}$$

is not very different from that obtained in the simplified models [7, 8], even though this difference is not unimportant. The main reason is that this peak value is mainly determined by the fraction of the explosion energy that is converted into CR energy, i.e. by the efficiency of CR acceleration, which is not strongly dependent on the model used.

There are more essential differences in the time variation of the predicted γ-ray fluxes. Kinetic theory revealed a much more effective CR and therefore γ-ray production during the free expansion phase of the SNR ($t \lesssim 400$ yr in the case shown in Fig.3), and a more rapid decrease of the γ-ray flux after reaching its peak value during the subsequent Sedov phase ($t \gtrsim 3000$ yr) due to the effect of different spatial distributions of the gas and the CRs inside the SNR (the so-called overlapping effect), that had not been taken into account in simplified models [7, 8].

According to Fig.3 TeV γ-ray flux $F_\gamma > 10^{-11}$ cm^{-2}s^{-1} is expected during about $t_m = 10^5$ yr of SNR evolution.It can be detected by the instrument with threshold $F_l \sim 10^{-12}$ cm^{-2}s^{-1} up to a distance $d \approx 3$ kpc.

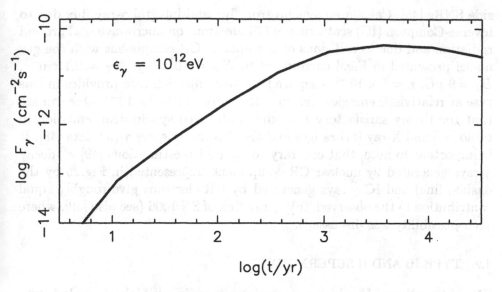

Figure 3. Integral 1 TeV γ-ray flux expected at 1 kpc distance from SNR Ia situated in a warm ISM as a function of time since the SN explosion.

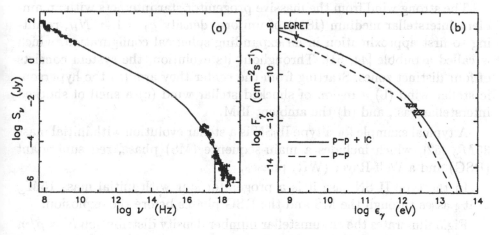

Figure 4. Synchrotron flux as a function of frequency ν (a) and integral γ-ray flux as a function of energy ϵ_γ (b) from SN 1006.

Therefore taking into account that the expected number of SNRs of this kind $N_{sn} = \nu_{sn}t_m$ is about $N_{sn} \sim 300$ (the Galactic SN Ia rate is $\nu_{sn} \approx 1/300 \text{ yr}^{-1}$ [44]) one can conclude that about 10 SNRs of this type should be observable at any given time, by the imaging Cherenkov telescopes.

Application of the kinetic model to the case of SN 1006 gives some evidence that CR nuclear component is indeed effectively generated inside SNRs [45]. Calculated synchrotron flux and integral γ-ray flux due to inverse-Compton (IC) scattering of CR electrons on microwave background radiation and due to collisions of the nuclear CR component with the gas nuclei presented in Fig.4 correspond to $E_{sn} = 10^{51}$ erg, $N_H = 0.1$ cm^{-3}, $B_0 = 9$ μG, $\eta = 5 \times 10^{-4}$. Required electron injection rate provides in this case at relativistic energies electrons to protons ratio 2×10^{-3}. One can see that the theory satisfactory reproduces observed synchrotron emission in radio [46] and X-ray [47] ranges and also fits the existing γ-ray data [48]. It is important to note, that contrary to simplified estimations [49] π^0-decay γ-rays generated by nuclear CR component (represented in Fig.2b by the dashed line) and IC γ-rays generated by CR electrons give roughly equal contribution in the observed TeV γ-ray flux of SN 1006 (see also [50], where such possibility was discussed).

4.2. TYPE IB AND II SUPERNOVA

SNe of type Ib and II, which are more numerous in our Galaxy, explode into an inhomogeneous circumstellar medium, formed by the intensive wind of their massive progenitor stars (e.g. [44, 51]).

The strong wind from the massive progenitor star interacts with an ambient interstellar medium (ISM) of uniform density $\rho_0 = 1.4mN_H$, resulting to first approximation in an expanding spherical configuration, which is called a bubble [44, 51]. Throughout its evolution, the system consists of four distinct zones. Starting from the center they are: (a) the hypersonic stellar wind (b) a region of shocked stellar wind (c) a shell of shocked interstellar gas , and (d) the ambient ISM.

A typical example for a type Ib SN is a stellar evolution with initial mass $35M_\odot$ [52], which includes a main-sequence (MS) phase, red supergiant (RSG) and a Wolf-Rayet (WR) phases.

In the type II SN case it is a progenitor star with initial mass $15M_\odot$ that passes through the MS and the RSG phases before the explosion.

Fig.5 illustrates the circumstellar number density distribution $N = \rho/m$ as a function of radial distance r for the above two types of SN situated in warm ISM with $N_H = 0.3$ cm^{-3} [30]. It corresponds to the case of a so-called modified bubble whose structure is significantly influenced by mass transport from the dense and relatively cold shell (c) into the hot region (b), and by thermal conduction in the opposite direction, these two energy fluxes balancing each other to first order. During this stage the shell (c) has collapsed into a thin isobaric shell due to radiative cooling [51]. Since the mass and heat transport between regions (b) and (c) are presumably due to turbulent motions in the bubble, it is assumed that the magnetic field grows up to the equipartition value.

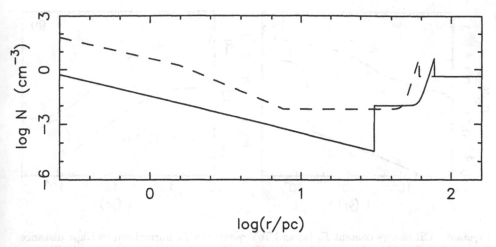

Figure 5. Circumstellar gas number density distribution as a function of radial distance for the case of a SNR Ib (full line) and SNR II (dashed line) in a warm ISM.

Fig.6 illustrates the SN shock propagation through the modified circumstellar medium for a typical set of the SN parameters: hydrodynamic explosion energy $E_{sn} = 10^{51}$ erg, ejecta mass $M_{ej} = 10M_\odot$, parameter $k = 10$ which describes the ejecta velocity distribution.

In the case of WR-star the star size $R_* = 3 \times 10^{12}$ cm, rotation rate $\Omega = 10^{-6}$ and the surface magnetic field $B_* = 50$ G, which determine the value of magnetic field in the wind region (a), are used, whereas $B_* = 1$ G, $R_* = 3 \times 10^{13}$ cm and $\Omega = 3 \times 10^{-8}$ in the case of RSG-star, progenitor of SN II.

Numerical results show that when a SN explodes into a circumstellar medium strongly modified by a wind from a massive progenitor star, then CRs are accelerated in the SNR almost as effectively as in the case of a uniform ISM: about $20 \div 50\%$ of the SN explosion energy is transformed into CRs during the active SNR evolution (see Fig.6a).

During SN shock propagation in the supersonic wind region ($t \lesssim 10^3$ yr) very soon the acceleration process reaches a quasistationary level which is characterized by a high efficiency and a correspondingly large shock modification with compression ratio $\sigma \approx 10$ [30]. Despite the fact that the shock modification is much stronger than predicted by a two-fluid hydrodynamical model [14], the shock never becomes completely smoothed by CR backreaction: a relatively strong subshock with compression ratio $\sigma_s \approx 3$ always exists [30].

Due to the relatively small mass contained in the supersonic wind region CRs absorb there only a small fraction of the explosion energy (about 1% in the case of a SN type Ib, and 10% in the case of a SN type II)

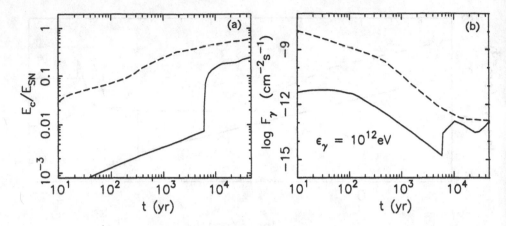

Figure 6. CR energy content E_c (a) and TeV γ-ray flux F_γ normalized to 1 kpc distance (b) as a function of time passed since the SN explosion for the same cases as in Fig.3.

and the SNR is still very far from the Sedov phase after having swept up this region (Fig.6a). Therefore we conclude, that the CRs produced in this region should not play a very significant role for the formation of the observed Galactic CR energy spectrum.

The peak value of the CR energy content in the SNR is reached when the SN shock sweeps up an amount of mass roughly equal to several times the ejected mass. This takes place during the SN shock propagation in the modified bubble. Compared with the uniform ISM case the subsequent adiabatic CR deceleration is less important in the considered case. The main amount of CRs in this case is produced when the SN shock propagates through the bubble. In this stage the dynamical scale length is much smaller than the shock size. Therefore the relative increase of the shock radius during the late evolution stage and the corresponding adiabatic effects are small.

The CR and γ-ray spectra are more variable during the SN shock evolution than in the case of a uniform ISM. At the same time the form of the resulting overall CR spectrum is rather insensitive to the parameters of the ISM as in the case of uniform ISM (see [30]). The maximum energy of the accelerated CRs reached during the SNR evolution is about 10^{14} eV for protons in all the cases considered, if the CR diffusion coefficient is as small as Bohm limit.

As one can see from Fig.6b, in the case of a SN Ib the expected TeV-energy γ-ray flux, normalized to a distance of 1 kpc, remains lower than 10^{-12} cm^{-2}s^{-1} during the entire SNR evolution except for an initial short period $t < 100$ yr when it is about 10^{-11} cm^{-2}s^{-1}. A similar situation exists at late phases of SNR evolution in the case of SN II. The expected γ-ray

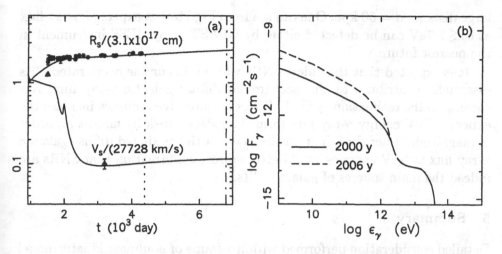

Figure 7. Shock size R_s and shock speed V_s (a) and integral γ-ray flux of SN 1987A expected at two different epochs (b): in 2000 (full line) and in 2006 (dashed line).

flux is considerably lower, at least by a factor of hundred, compared with the case of uniform ISM of the same density N_H.

In the case of a SN II during the first several hundred years t_m after the explosion, the expected TeV γ-ray flux at a distance $d = 1$ kpc exceeds the value 10^{-9} cm^{-2}s^{-1} and can be detected up to the distance $d_m = 30$ kpc with present instruments like HEGRA, Whipple or CAT. This distance is of the order of the diameter of the Galactic disk. Therefore all Galactic SNRs of this type whose number is $N_{sn} = \nu_{sn} t_m$ should be visible. But in this case we can expect at best $N_{sn} \sim 10$ such γ-ray sources at any given time.

An interesting example of type II SN represents SN 1987A in the Large Magellanic Cloud, because there are a lot of reliable observational data (e.g. see review [52]) which provide a unique opportunity to apply the existing models of CR and γ-ray production inside SNRs. In Fig.7a calculated time-dependence of SN shock size and speed [54] are compared with the experimental data [55]. During the initial period 1500 days after the explosion the SN shock propagated through the wind of BSG star which was a progenitor of SN 1987A during the last 10^4 yr before the SN event [53]. Due to low density of BSG wind the shock speed was very high $V_s \approx 30000$ k-m/s and almost constant. The essential shock deceleration from day 1500 to day 3000 indicates that the shock enters much more denser region occupied by the wind of RSG star: according to calculation the number density $N \approx 400$ cm^{-3} is required to reproduce the observed SN shock deceleration. Very high RSG wind number density leads to extremely high γ-ray luminosity on the current stage, as it is demonstrated in Fig.7b, despite the large distance $d = 50$ kpc. One can see from Fig.7b that expected γ-ray flux at $\epsilon_\gamma \lesssim 1$ TeV can be detected either by GLAST or by HESS instrument in the nearest future.

It is expected that the oldest SNRs which still confine accelerated CRs essentially contribute to the background diffuse galactic γ-ray flux. According to the estimations [56] old SNRs as unresolved sources increase the expected TeV-energy γ-ray flux from the galactic disk by almost an order of magnitude. Therefore the measurements of the predicted diffuse galactic γ-ray flux at TeV-energies would give indirect confirmation that SNRs are indeed the main sources of galactic CRs.

5. Summary

Detailed consideration performed within a frame of nonlinear kinetic model demonstrates, that the diffusive acceleration of CRs in SNRs is able to generate the observed CR spectrum up to an energy $10^{14} \div 10^{15}$ eV, if the CR diffusion coefficient is as small as Bohm limit.

Acceleration process provides more effective production of heavy elements due to the nonlinear effects inside SNRs, which expand in low temperature ISM ($T_0 \sim 10^4$ K) with low ion ionization state. Since according to the experimental (e.g. [21]) and theoretical [23] evidences heavy elements are preferentially injected into the acceleration, one can conclude that the observed CR spectrum and composition at energies $\epsilon \lesssim 10^{15}$ eV can be accounted for diffusive shock acceleration in SNRs, if they are predominantly situated in a relatively low temperature ISM.

Kinetic model predicts somewhat more hard resultant CR spectrum than it is required. This discrepancy can be attributed either for existing of some kind of dissipation process which operates within the shock transition and makes the acceleration process somewhat less efficient, or for more complicated picture of CRs leakage from the galactic volume.

Due to relatively hard CR spectrum inside SNRs measurable contribution of nearby SNRs in the galactic CR spectrum seems to be quit probable.

The typical value of the cutoff energy of the expected γ-ray flux, produced inside SNRs by CR nucleons, is about 10^{13} eV, if the CR diffusion coefficient is as small as the Bohm limit. In this respect the negative result of high-threshold arrays (e.g. [57]) in searching of γ-ray emission from Galactic SNRs is not surprising because their threshold $E_{th} \sim 50$ TeV exceeds the cutoff energy of the expected γ-ray flux. It is less obvious how to interpret the negative results of imaging atmospheric Cherenkov telescopes with thresholds less than about 1 TeV (e.g. [57]), which have up to now detected the only shell-type SNR SN 1006 in TeV γ-rays [48], whereas according to the kinetic theory $10 \div 20$ of such type of sources are expected.

Negative correlation between SN of type Ia and ISM density could be a possible explanation of this deficit in detected TeV γ-ray sources.

For core collapse SN of types II or Ib with quite massive progenitors one can in part explain this fact by the extremely low γ-ray intensity expected from such SNRs during the period of SN shock propagation through the low-density hot bubble. An alternative possibility relates to the assumption of the Bohm limit for the CR diffusion coefficient which can be too optimistic, in particular for the quasi-perpendicular geometry in wind-blown regions from rotating star.

According to the kinetic model [45] π^0-decay γ-rays generated by the nuclear CR component and IC γ-rays generated by the electronic CR component provide roughly equal contribution in the observed TeV γ-ray flux of SN 1006 and therefore SN 1006 gives some evidence that CR nuclear component is indeed produced in SNRs.

TeV γ-ray flux from SN 1987A at current evolutionary phase is about $F_\gamma \approx 7 \times 10^{-13}$ cm^{-2}s^{-1} and expected to grow at the year 2006 by a factor of two [54]. Therefore SN 1987A is a good candidate for searching high-energy γ-ray emission due to CR nuclear component in forthcoming experiments GLAST and HESS.

If SNRs produce CRs as effectively as predicted by the kinetic model then high energy diffuse galactic γ-ray flux is dominated by contribution of CRs situated inside old unresolved SNRs [56]. The measurements of the diffuse flux at TeV-energies would give an indirect test whether SNRs are indeed the main source of CRs.

6. Acknowledgments

This work has been supported in part by the Russian Foundation of Basic Research grants 99–02–16325 and 00–02–17728 and by the Federal Scientific Program 'Astronomy' (grant 1.2.3.6). The author thanks the organizing committee for the invitation to present this paper and for the sponsorship.

References

1. Drury, L.O'C. (1983) *Rep. Prog. Phys.*, **46**, 973
2. Blandford, R.D. and Eichler, D. (1987) *Phys. Rept.*, **154**, 1
3. Berezhko, E.G. and Krymsky, G.F. (1988) *Soviet Phys. Uspekhi.*, **12**, 155
4. Drury, L.O'C, Markiewicz, W.J. and Völk, H.J. (1989) *A&A*, **225**, 179
5. Markiewicz, W.J., Drury, L.O'C. and Völk, H.J. (1990) *A&A*, **236**, 487
6. Dorfi, E.A. (1990) *A&A*, **234**, 419
7. Dorfi, E.A. (1991) *A&A*, **251**, 597
8. Drury, L.O'C., Aharonian, F.A. and Völk, H.J. (1994) *A&A*, **287**, 959
9. Jones, F.C. and Ellison, D.C. (1991) *Space Sci. Rev.*, **58**, 259
10. Kang, H. and Jones, T.W.(1991) *MNRAS*, **249**, 439

16

11. Berezhko, E.G., Yelshin, V.K. and Ksenofontov, L.T. (1994) *Astropart. Phys.*, **2**, 215
12. Berezhko, E.G., Yelshin, V.K. and Ksenofontov, L.T. (1996) *JETP*, **82**, 1
13. Berezhko, E.G. (1996) *Astropart. Phys.*, **5**, 367
14. Jones, T.W. and Kang, H. (1992) *ApJ*, **396**, 575
15. Drury, L.O'C., Völk, H.J. and Berezhko, E.G. (1995) *A&A*, **299**, 222
16. Krymsky, G.F. (1977) *Soviet Phys. Dokl.*, **23**, 327
17. Axford, W.I., Leer, E. and Skadron, G., Proc. 15th ICRC 11, Plovdiv, 1977, pp.132–135
18. Bell, A.R. (1978) *MNRAS* **182**, 147 (1978); *MNRAS*, **182**, 443
19. Blandford, R.D. and Ostriker, J.P. (1978) *ApJ*, **221**, L29
20. Lee, M.A. (1982) *JGR*, **87**, 5063
21. Trattner, K.J., Möbius,E., Scholer, M. et al. (1994) *JGR*, **99**, 389
22. Quest, K.B. (1988) *JGR*, **93**, 9649
23. Trattner, K.J. and Scholer, M. (1991) *Geophys. Res. Lett.*, **18**, 1817
24. Giacalone, J., Burgess, D., Schwartz, S.J. and Ellison, D.C. (1993) *ApJ*, **402**, 550
25. Bennett, L. and Ellison, D.C. (1995) *JGR*, **100**, **A3**, 34439
26. Malkov, M.A. and Völk, H.J. (1995) *A&A*, **300**, 605
27. Malkov, M.A. and Völk, H.J. (1996) *Adv. Space Res.*, **21**, 551
28. Berezhko, E.G., Krymsky, G.F. and Turpanov, A.A., Proc. 21st ICRC 4, Adelaide, 1990, pp. 101–105
29. Berezhko, E.G. and Völk, H.J. (1997) *Astropart. Phys*, **7**, 183; (2000) *Astropart. Phys* **14**, 201
30. Berezhko, E.G. and Völk, H.J. (2000) *A&A*, **357**, 283
31. McKenzie, J.F. and Völk, H.J. (1982) *A&A*, **116**, 191
32. Berezhko, E.G., Petukhov, S.I. and Taneev, S.N., Proc.26th ICRC, 6, Salt Lake City, 1999, pp. 520–523
33. Völk, H.J., Drury, L. O'C. and McKenzie, J.F. (1984) *A&A*, **130**, 19
34. Berezhko, E.G. and Ellison, D.C. (1999) *ApJ*, **526**, 385
35. Jones, E.M., Smith B.W. and Straka, W.C. (1981) *ApJ*, **249**, 185
36. Chevalier, R.A. (1982) *ApJ*, **259**, 302
37. Berezinsky, V.S., Bulanov, S.V., Ginzburg, V.L., Dogiel V.A., Ptuskin, V.S. (1990) *Astrophysics of Cosmic Rays*, Horth-Holland Elsevier Science Publ. B.V., Amsterdam.
38. Kaplan, S.A. and Pikelner, S.B. (1970) *The Interstellar Medium*, Wiley: Chichester.
39. Shibata T. *Invited, Rapporteur & Highlight Papers*, Proc. 24th ICRC, Rome, 1995, pp. 713–754
40. Berezhko, E.G. and Ksenofontov, L.T. (1999) *JETP*, **116**, 737
41. Watson, A. *Invited, Rapporteur & Highlight Papers*, Proc. 25th ICRC, Durban, 1997, pp. 257–280
42. Ellison, D.C., Drury, L OĆ. and Meyer, J.P. (1997) *ApJ*, **487**, 197
43. Erlykin, A.D. and Walfendale, A.W. (1997) *Astropart. Phys.*, **7**, 1
44. Losinskaya, T.A. Proc 22th ICRC, 5, Dublin, 1991, pp.123.
45. Berezhko, E.G., Ksenofontov, L.T. and Petukhov, S.I. Proc. 26th ICRC 4, Salt Lake City, 1999, pp. 431–434
46. Reynolds, S.P. (1996) *ApJ*, **459**, L13
47. Hamilton, A.J.S., Sarazin, C.L., and Szymkowiak, A.E., *ApJ*, **300**, 698 (1986).
48. Tanimori, T., Hayami, Y., Kamei, S., et al. (1998) *ApJ*, **497**, L25
49. Mastichiadis, A. and De Jager, O.C. (1996) *A&A*, **311**, L5
50. Aharonian, F.A. and Atoyan, A.M. (1999) *A&A*, **351**, 330
51. Weaver, R., McCray, R., Castor J. et al.(1977) *ApJ*, **218**, 377
52. Garcia-Segura, G., Langer, N. and Mac Low, M.-M. (1996) *A&A*, **316**, 133
53. Mc Cray, R. (1993) *Ann. Rev. Astron. Astrophys.*, **31**, 175 .
54. Berezhko, E.G. and Ksenofontov, L.T. (2000) *Astron. Lett.*, **26**, 639

55. Gaensler, B.M., Manchester, R.N., Staveley-Smith, L. et al. (1997) *ApJ*, **479**, 845
56. Berezhko, E.G. and Völk, H.J., *ApJ*, **540** 923
57. Hillas, A.M. *Invited, Rapporteur & Highlight Papers*, Proc. 24th ICRC, Rome, 1995, pp.701–712

HADRONIC COLLISIONS AT VERY HIGH ENERGIES

I. Kurp

The Andrzej Sołtan Institute for Nuclear Studies
90-950, Łódź 1, Box 447, Poland.
kurp@zpk.u.lodz.pl

Abstract   Monte Carlo simulations are fundamental tools in cosmic ray data interpretation. They are based on different models of hadronic interactions
in which different values of proton–proton cross sections are used. New
results obtained recently in accelerator experiments lead to the necessity
of revising some of them. In this paper, we show some results obtained
in a framework of a so-called geometrical model. A particular proton
profile function and its energy dependence is given. The success of the
model encourages us to extrapolate calculated cross sections to much
higher energies considered in cosmic ray physics.

Introduction.

Using the Monte Carlo technique in the study of high energy cosmic
rays, we need to know the values of proton–proton cross sections. For
energies as high as considered in cosmic ray physics, we cannot measure
these quantities in accelerator experiments. The only solution available
so far is the investigation of energy dependance of proton–proton cross
sections at lower energies and then extrapolation to the high energy region. We can revise our predictions with data from extensive air shower
(EAS) experiments like Akeno or Fly's Eye which measure proton–air
nuclei absorbtive cross sections and (using some approximations) reconstruct proton–proton cross sections. This treatment of hadronic collisions is, of course, purely phenomenological, but there is still no other
possibility. We cannot use quantum chromodynamics (QCD) to predict
values in the cosmic ray energy region. However, the predictions can be
checked and maybe give some contribution to theoretical description.

19

M. M. Shapiro et al. (eds.), Astrophysical Sources of High Energy Particles and Radiation, 19–24.
© 2001 *Kluwer Academic Publishers. Printed in the Netherlands.*

1.   Proton–proton cross section parametrization.

In this paper we propose a geometrical description of the hadronic collision process. Two interacting hadrons can be treated as extended objects with some particular "hadronic matter" distributions which can depend also on interaction energy. The value of opaqueness seen by the hadrons passing each other at impact parameter b is defined by respective convolution. Using the eikonal framework, cross sections for two colliding objects can be expressed by the following formulae:

$$\sigma_{\text{tot}} = 2 \int \left[1 - Re\left(e^{i\chi(b)} \right) \right] \mathrm{d}^2 b,$$

$$\sigma_{\text{el}} = \int \left| 1 - e^{i\chi(b)} \right|^2 \mathrm{d}^2 b, \tag{1}$$

$$\sigma_{\text{inel}} = \int 1 - \left| e^{i\chi(b)} \right|^2 \mathrm{d}^2 b,$$

where χ denotes the absorbtive potential, which can be understood as the phase shift χ for the elastic scattering.

In general function χ can be complex:

$$\chi(s,b) = (\lambda(s) + i)\,\omega(s,b). \tag{2}$$

In the optical interpretation, real function ω is a particle's "opaqueness" - the matter density integrated along the collision axis.

It is a very well known experimental fact that values of cross sections increase as interaction energy increases. This implies that the integrated "opaqueness" has to increase, too. Thus, the ω function ought to become larger. Simplifying the situation, a hadron can be supposed to be getting blacker (factorization hypothesis):

$$\omega(s,b) = \omega(b)f(s), \tag{3}$$

or getting bigger (geometrical scaling):

$$\omega(s,b) = \omega(b/b_0(s)). \tag{4}$$

The best results have been obtained combining these two possibilities. So called BEL (blacker-edger-larger) behavior of ω is shown in Ref. [1] to be in a very good agreement with accelerator data at ISR energies ($\sqrt{s} \sim 50$ GeV) and at $\sqrt{s} = 540$ GeV at CERN SPS. Analytical description of ω used in Ref. [1] was of the form proposed by Henzi and Valin in Ref. [2]. It is given by the parametrization of G_{inel} which is the probability for an inelastic event to take place at b and s

$$G_{\text{inel}}(s,b) = 1 - e^{-2\,\omega(s,b)} \tag{5}$$

and has the following form:

$$G_{\text{inel}}(s,b) = P(s)\exp\{-b^2/4B(s)\}k(s,x). \tag{6}$$

Increasing values of $P(s)$ and $B(s)$ make the proton blacker and larger, respectively. $k(s,x)$ exhibits 'edge' behavior

$$k(s,x) = \sum_{n=0}^{N} \delta_{2n}(s) \left[\frac{\epsilon \exp\{1/2\}}{\sqrt{2B(s)}}x\right]^{2n}, \tag{7}$$

where

$$x = b\exp\{-(\epsilon b)^2/4B(s)\}. \tag{8}$$

(ϵ is a constant value of about $\sqrt{0.78}$). Energy dependance of these quantities was found to be a function of $y = \ln^2(s/s_0)$ ($s_0 = 100$ GeV^2):

$$P(s) = \frac{0.908 + 0.027y}{1 + 0.027y},$$

$$B(s) = 6.64 + 0.044y, \tag{9}$$

$$\delta_2(s) = 0.115 + 0.00094y, \quad \delta_4 = \delta_2^2/4.$$

In this paper the real part of χ function is taken after Ref. [1] as

$$\lambda(s) = \frac{0.077\ln(s/s_0)}{1 + 0.18\ln(s/s_0) + 0.015\ln^2(s/s_0)}. \tag{10}$$

2.   Results and conclusions.

Published in 1999 [3], very accurate data on proton–antiproton cross sections at 1.8 TeV energy allow us to extend the geometrical description of the proton. Examining all available data, we have found that a satisfactory description can be obtained by a slightly changed $B(s)$

$$B(s) = 6.65 + 0.031y. \tag{11}$$

Total and elastic cross sections data from ISR, SPS and Fermilab's Tevatron are compared with our calculations in Fig. 1. Total and elastic cross sections are only two of the bulk of the elastic scattering experiment outputs. To extend our information about the process, we can include in the analysis other quantities. One of the best measured values is the so-called elastic slope parameter B defined as

$$B = \left|\frac{\mathrm{d}}{\mathrm{d}t}\ln\left(\frac{\mathrm{d}\sigma_{\text{el}}}{\mathrm{d}t}\right)\right|_{t=0}, \tag{12}$$

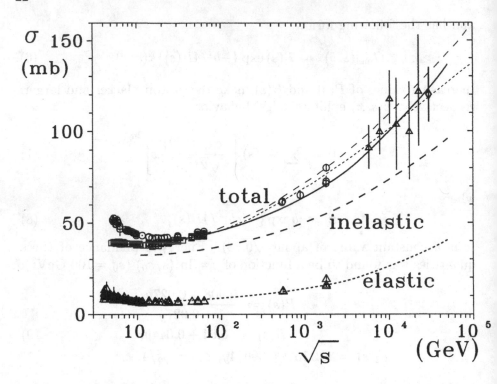

Figure 1. The cross sections of proton–(anti)proton interactions calculated with proposed parametrization. The highest energy points are from cosmic ray experiments [4]. Thin dashed lines represent other total cross section parametrizations [6] commonly used in high energy physics.

where t is the momentum transfer. From those three values (σ_{tot},σ_{el} and B) we can construct the two dimensionless ratios used in Ref. [5] for describing the character of the proton scattering:

$$X = \frac{\sigma_{el}}{\sigma_{tot}},$$
$$Y = \frac{\sigma_{tot}}{16\pi B}. \tag{13}$$

The values of X and Y for scattering of "transparent" objects are small, and they increase while particles become more "opaque". Small Y/X ratio characterizes "extended" and respectively large – "compact" protons. In Fig. 2, experimental points are compared with the theoretical predictions of our parametrization. For very low energies, scattering is getting more "transparent" with the energy increase. In the region of IS-R energies ($\sqrt{s} \approx 20$ - 60 GeV), the situation seems to be stable. Nevertheless, the SPS point at $\sqrt{s} = 546$ GeV indicates a more "opaque" and

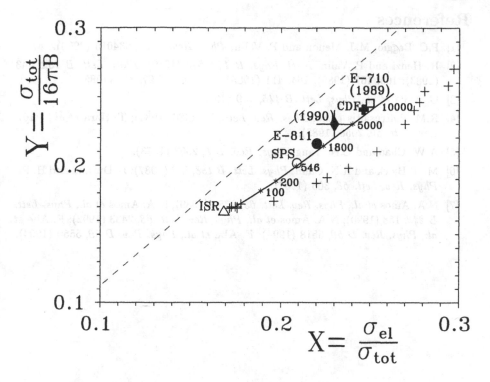

Figure 2. The Chao and Yang *XY* plot comparing all available accelerator data with our cross section parametrization (solid line). Solid circle represents the recent Tevatron data (labeled E-811). Older Tevatron results are given as E-710 (1989), (1990), and CDF. Numbers along the solid line shows the respective center of mass energy at which calculations were performed. Thin crosses represent low and very low energy data.

"compact" character of the scattering. First measurements at Fermi-lab's Tevatron energy of 1.8 TeV [7] suggested that the proton becomes "blacker" very fast (faster than expected). However, recent values of proton–(anti)proton cross sections and slope parameter for $\sqrt{s} = 1.8$ TeV [3] changed the situation. It seems that the total cross section increases with the interaction energy not so fast, and our parametrization can describe all the data including points from air shower experiments [4]. It should be remembered that the cosmic ray experiments deal with high energy protons (and nuclei) interacting with thick slab of air, and some systematic bias still could be present in the results shown in Fig. 1. Astonishing agreement with predictions encourages extrapolation to the higher energy region which is important from point of view of cosmic ray physics.

24

References

[1] P.C. Beggio, M.J. Menon and P. Valin, *Phys. Rev. D 61*, 034015 (2000).

[2] R. Henzi and P. Valin, *Nucl. Phys. B 148*, 513 (1979); *Phys. Lett. B 132*, 443 (1983); 160, 167 (1985); 164, 411 (1985); *Z. Phys. C 27*, 351 (1985).

[3] C. Avila *et al.*, *Phys. Lett. B 445*, 419 (1999).

[4] R.M. Baltrusaitis *et al.*, *Phys. Rev. Lett. 52*, 1380 (1984); T. Hara *et al.*, *Phys. Rev. Lett. 50*, 2058 (1983).

[5] A.W. Chao and C.N. Yang, *Phys. Rev. D 8*, 2063 (1973).

[6] M.M. Block and R.N. Cahn, *Phys. Lett. B 188*, 143 (1987); L. Durand and H. Pi, *Phys. Rev. Lett. 58*, 303 (1987).

[7] N.A. Amos *et al.*, *Phys. Rev. Lett. 63*, 2784 (1989); N.A. Amos *et al.*, *Phys. Lett. B 243*, 158 (1990); N.A. Amos *et al.*, *Phys. Rev. Lett. 68*, 2433 (1992); F. Abe *et al.*, *Phys. Rev. D 50*, 5518 (1994); F. Abe *et al.*, *Phys. Rev. D 50*, 5550 (1994).

ELECTRON ACCELERATION BY QUASI-PARALLEL

SHOCK WAVES

YURY UVAROV (uv@astro.ioffe.rssi.ru)
Ioffe Physical Technical Institute, St-Petersburg, Russia.

Abstract. A kinetic model of the electron acceleration by a collisionless quasi-parallel shock wave is discussed. Non-resonant interaction of electrons with instabilities induced by protons in the transition region inside the shock plays the main role in the electron heating and their injection to the first order Fermi acceleration mechanism. The electron distribution function in the vicinity of the shock front can be calculated in our model. The model is valid even in the case (important for applications) of acceleration of electrons by shock waves with high Mach numbers. The structure of such shock waves is strongly modified by the nonlinear interaction of nonthermal ions and consists from an extended predfront with smooth variation of macroscopic parameters and a viscous discontinuity in speed of moderate Mach number. Our model was applied for the case of SNR IC443 and HVC complexes M, A, C. It was found that the predicted emission produced by electrons through synchrotron, bremsstrahlung and inverse Compton mechanisms fits very well with the observational data.

1. Introduction

The multiwavelength observations appeared in recent years show that the nonthermal emission often takes place in various astrophysical objects. This is the evidence that nonthermal particles play an important role in the formation of the observable radiation spectrum in broad energy range in many objects. Thats why the mechanisms responsible for particle acceleration attract close attention. The mechanism of particle acceleration on a shock wave (also known as first order Fermi acceleration) [1, 2] is one of the most important one. It is based on simple physics but can explain the formation of power spectrum on a few orders of particle energy. Taking into account that the shock wave generation often takes place in astrophysics this mechanism is the number one candidate for explanation of the nonthermal spectrum formation in many sources.

Lets consider this process more closely. The shocks formed in rarefied cosmic plasma are collisionless. Computer simulations of the structure of collisionless shock waves that use hybrid codes, which interpret protons as particles and electrons as liquid, have made it possible to describe the main features of supercritical quasi-longitudinal shock waves [3]. The main result here is that the front of such shock wave is an extended transition region full of strong magnetic field fluctuations ($\delta B/B \sim 1$) with frequencies below the ion gyro-frequency [3]. Thermal ions which gyroradiuses in the most cases are larger than the transition region are efficiently involved in the first order Fermi acceleration. Electrons with gyroradiuses larger than the front width are also involved in this acceleration. However a nonrelativistic electron must have an energy that is m_p/m_e times higher than energy of the respective proton to be injected into the first order Fermi acceleration. So the main

25

M. M. Shapiro et al. (eds.), Astrophysical Sources of High Energy Particles and Radiation, 25–30.

problem here is injection — the acceleration mechanism that forms spectrum up to the energies from which the first order Fermi acceleration begins to work. It needs to mention that there is no way to get information about electrons from hybrid codes where they are treated as liquid. And thats why the special approach is essential to get electron energy spectrum.

To resolve the electron injection problem in [4] we suggested a kinetic model of electron acceleration and injection. Following the approach developed by Toptigin [5] we used a kinetic equation for the isotropic part of the electron distribution function N(z,p) (axis z is normal to the shock front):

$$
k(p)\frac{\partial^2 N(z,p)}{\partial z^2} - u(z)\frac{\partial N(z,p)}{\partial z} + \frac{p}{3}\frac{\partial N}{\partial p}\frac{\partial u(z)}{\partial z}
$$
$$
+ \frac{1}{p^2}\frac{\partial}{\partial p}\left(p^2 D(p)\frac{\partial N}{\partial p}\right) - \frac{1}{p^2}\frac{\partial}{\partial p}\left(p^2 F(p)N\right) = 0. \tag{1}
$$

This equation describes electron diffusion in coordinate and momentum spaces taking into account their interactions with magnetohydrodinamic (MHD) turbulence, convection and energy looses. Here $F(p) = dp/dt$ (p) — is a function of energy looses. Diffusion coefficients in coordinate $(k(p))$ and momentum $(D(p))$ spaces are depend on MHD turbulence. We used model assumptions about this turbulence based on the results of hybrid codes calculations.

The eq. (1) could be numerically solved with appropriate boundary conditions. There is no more injection problem. When we solve eq. (1) (with boundary conditions) we simultaneously take into account the injection by the second order Fermi acceleration and the first order Fermi acceleration. The resulting electron distribution function N(z,p) is valid in all the energy range. Using our model in [4] we showed that the vortical fluctuations of electric fields induced by random fluxes of ions in the supercritical shock transition region are indeed responsible for a stochastic preacceleration and heating on scales of order of the collisionless shock width. Our method applications for the modeling of radiation from the SNR IC443 and HVC complexes M, A, C are discussed in sections 2 and 3.

2. Supernova remnants

Supernova remnants are very bright sources of the nonthermal emission. The shells expanding into interstellar medium form the collisionless shock waves with very high Mach numbers (up to a few hundred). Such shocks are a very effective accelerators of all ionized particles including electrons and have a modified structure. The nonthermal ions penetrate into the incoming flow and slow it down forming the adiabatic decrease of the flow speed known as predfront. The total compression ratio in this case could be much higher than for a shock wave without predfront. The characteristic predfront size is usually several orders of magnitude greater than the characteristic size of the transition region. This shock wave structure have the following consequences for electrons: the more energetic electrons penetrate further into the incoming flow, feel more high compression and are accelerated with more efficiency. Thats why the high energy part of electron spectrum could be more flatter than the low energy part.

Our model still could be used in the case of such modified shocks. There is only need to consider adiabatic predfront and transition region individually. For transition region the

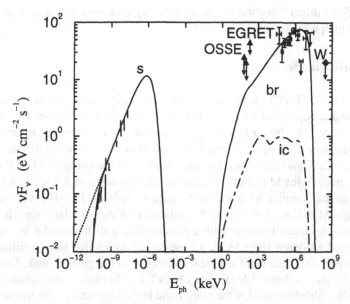

Figure 1. The calculated spectrum of the IC 443 radiation [6] in comparison with the observational data. Synchrotron emission with interstellar free-free absorption taken into account is shown by a solid line (pure synchrotron emission is shown by dotted line). Bremsstrahlung radiation of the shell with interstellar absorption (from Morrison & McCammon (1983) [7]) taken into account is shown by a solid line. The inverse Compton emission is shown by the dashed curve. The observational data points are taken from the works: Esposito et al. (1996) [8] (EGRET), Erickson & Mahoney (1985) [9] (radio), Sturner et al. (1997) [10] (OSSE) and Buckley et al. (1998) [11] (WHIPPLE).

model could be applied. And consideration of the adiabatic predfront leads to a modified boundary condition. In general our model could be applied to any SNR with regular shell structure but in particular we considered the IC 443 [6]. There are a lot of observations of this remnant on various frequencies from radio to γ-rays. From these observations the main physical parameters of IC 443 and its neighborhood could be estimated. Molecular emission that occurs from the shell of IC 443 when it interacts with dense molecular clouds is an evidence that the shock in this remnant propagates through the dense medium with $N \sim 5 - 25 \ cm^{-3}$. Such high densities make bremsstrahlung to be much more favorable in explanation of the high energy radiation from IC 443 than synchrotron [6]. This contradicts with synchrotron high energy explanation for SNR SN1006 [12] and a few other remnants with low density medium in their neighborhood. It is necessary to mention that explanation of high energy radiation up to a few GeV by bremsstrahlung needs electrons to be accelerated only up to a few dozens GeV. While synchrotron explanation needs electrons accelerated up to ~ 100 TeV. All other physical parameters used were discussed in details in [6]. Using our model we calculated an electron spectrum and then bremsstrahlung, synchrotron and inverse Compton radiation from the accelerated electrons.

On the Fig. 1 spectrum of the shell of IC443 is shown in comparison with the observational data. There is good agreement between our model results and observations. Above the preacceleration of electrons from the thermal distribution was discussed. There is a possibility of direct injection of the cosmic ray electrons to the first order Fermi acceleration which was also discussed in [6].

3. High velocity clouds

High velocity clouds (HVCs) were first discovered in 1963 in a neutral hydrogen radio survey as extended objects which radial velocities strongly differs from the gas rotation velocity in Galaxy [13]. Nevertheless high attraction for astronomers those objects in most remain mystery till now. Detected radio emission don't give the distance to them. Observations of Ca II and Si II optic lines in direction to the complex M of HVCs show that distance z to complex M from the galactic plain is in range of $1.5 \leq z \leq 4.4$ kpc [14]. This means that the complex M is within the galactic halo. In [15, 16] we assumed that HVC complexes M,A,C are falling onto the galactic disk due to tidal forces. It looks like that these complexes indeed interact with a galactic disk at distances of a few kpc.

The observed emission from M, A, C complexes looks like having diffuse nature. The γ-ray intensity expected from a typical emissivity in the galactic disk, found in [17] ($5 \cdot 10^{-25}$ ph Hatom^{-1} s^{-1}ster^{-1} MeV^{-1} at 1 MeV), is 2-3 orders of magnitude lower than needed to explain observations. On the other hand HVC interaction with matter and electromagnetic fields of the galactic disk or lower halo should produce a shock wave which is a natural particle accelerator. And energetic accelerated particles interacting with cloud are the source of the nonthermal emission. In [15, 16] we described a quantitative model that attributes the observed X-ray and γ-ray emission to bremsstrahlung from electrons accelerated due interaction of HVCs with galaxy gas. These HVCs can be expected to produce strong local disturbances and bow shocks with moderate magnetosonic Mach numbers M of $1.7 - 3$. Such Mach numbers are sufficient to consider these shock waves to be supercritical [18].

In [15, 16] we applied our model with given above parameters and calculated the distribution function of shock accelerated electrons penetrated into the HVC gas ($N_H \sim 10^{20}$ cm^{-2}), taking into account Coulomb losses and diffusion transport with a momentum-dependent mean free path $\Lambda(p) \sim p^{0.5}$. Photoelectric absorption by interstellar medium (with column density $\sim 10^{20}$ cm^{-2}) that affects the soft X-rays was taken into account using the cross sections and abundances given in [7].

Fig. 2 shows the predicted emission spectra for given above physical parameters together with the 0.75-3 MeV γ-ray intensities measured by COMPTEL. The COMPTEL intensity depends on the angular size of the source, which can't be derived from the γ-ray images because the angular resolution is insufficient. As a lower limit, the angle area of a few deg^2 could be taken and an angle area of 300 deg^2 could be taken as the maximum possible γ-ray emission area [15, 16]. In Fig. 1, the intensity levels for 3, 30 and 300 deg^2 are shown.

4. Conclusion

The model of an electron acceleration in the vicinity of the quasi-longitudinal collisionless shock wave is discussed. The model is applied for SNR IC 443 and HVC complexes A, B, C. The emission due to synchrotron, bremsstrahlung and inverse Compton mechanisms is calculated for IC 443. The bremsstrahlung radiation is calculated for the HVC complexes. It is shown that the model can explain the observable radiation from the IC 443 in the

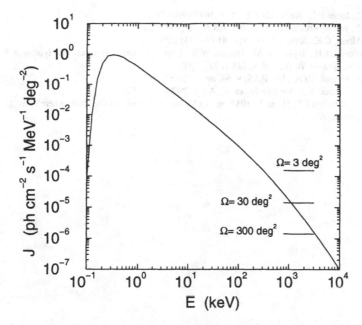

Figure 2. The spectra of bremsstrahlung radiation of electrons accelerated on a shock wave with Mach number $M = 2.7$ together with observational COMPTEL point with different assumptions about source angular area [15].

energy range from radio up to γ-rays and high energy radiation from the HVC complexes. It means that the observed spectrum of those objects could be explained by the leptonic emission only (without proton addition to radiation).

Acknowledgements

I am glad to acknowledge Prof. A.M.Bykov for fruitful joint work and for being my supervisor during the preparation of my thesis. This work was supported by INTAS 96-0390 grant and my participation in Erice school is supported in part by the organizing committee and INTAS-YSC 4216 grant.

References

1. Berezhko E.G., Krimskii G.F., Usp. Fiz. Nauk, **154** (1988) 49 [Sov. Phys. Usp., **31** (1988) 27].
2. Blandford R., Eichler E., Phys. Rep., **154** (1987) 2.
3. Quest L.B., J. Geophys. Res., **A 100** (1995) 3439.
4. Bykov A.M., Uvarov Yu.A., JETP, **v. 88**, no. 3 (1999) 465.
5. Toptigin I.N., *Cosmic Rays in Interplanetary Magnetic fields* [in Russian], Nauka, Moscow, 1983.
6. Bykov A.M., Chevalier R.A., Ellison D.C., Uvarov Yu.A., ApJ, **538** (2000) 203.
7. Morrison R., McCammon D., Astrophys. J., **270** (1983) 119.
8. Esposito J.A., Hunter S.D., Kanbach G. & Streekumar P., ApJ, **461** (1996) 820.
9. Erickson W.C., Mahoney M.J., ApJ, **290** (1985) 596.
10. Sturner S.J., Skibo J.G., Dermer C.D. & Mattox J.R., ApJ, **490** (1997) 619.
11. Buckley J.H., Akerlof C.W. et al., A&A, **329** (1998) 639.

12. Ellison D.C., Jones F.C., Reynolds S.P., ApJ, **360** (1990) 702.
13. Wakker, B.P., van Woerden, H., Ann. Rev. A&A, **35** (1997) 217.
14. Danly, L., Albert, C.E., Kuntz, K.D., ApJ, **416** (1993) L29.
15. Blom J.J, Bloemen H., Bykov A.M., Burton W.B , Dap Hartmann, Hermsen W., Iyudin A.F., Ryan J., Schonfelder V., Uvarov Yu.A., A&A, **321** (1997) 288.
16. Bykov A.M., Uvarov Yu.A., Izv. RAS, **v. 64**, no. 2 (2000) 393.
17. Strong A.W., Bennett K., Bloemen H., et al., A&A, **292** (1994) 82.
18. Kennel C.F., Edmiston J.P., Hada T., 1985, in "Collisionless Shocks in the Heliosphere" eds. B. Tsurutani et al., Washington, AGU, 1.

MAGNETOHYDRODYNAMIC WIND DRIVEN BY COSMIC RAYS IN A ROTATING GALAXY

V.N.ZIRAKASHVILI

Institute of Terrestrial Magnetism, Ionosphere and Radiowave Propagation. 142190, Troitsk, Moscow Region, Russia

D.BREITSCWERDT

Max-Planck-Institut für Extraterrestrische Physik, 85740 Garching, Germany

V.S.PTUSKIN

Institute of Terrestrial Magnetism, Ionosphere and Radiowave Propagation. 142190, Troitsk, Moscow Region, Russia

AND

H.J.VÖLK

Max-Planck-Institut für Kernphysik, Postfach 103980, D-69117 Heidelberg, Germany

We consider a galactic wind driven by cosmic rays in the rotating Galaxy, including the dynamical effects of the magnetic field. A flux tube formalism is used in order to describe the gas flow. It is further assumed that wind plasma is fully ionized and has infinite conductivity. Cosmic ray propagation is treated adiabatically. Generation of Alfvén waves due to streaming instability of cosmic rays is taking into account. We assume that nonlinear damping of waves is rapid, and their energy is turned into heating of the gas. Numerical calculations have been performed for a large set of parameters. Total mass loss rate and angular momentum loss rate of the Galaxy are obtained. Magnetic field and rotation increase total mass loss rate of the Galaxy to about 1.4 solar mass per year. Calculated total angular momentum loss rate is about 17% of the total momentum of the Galaxy per Hable time.

1. Introduction

The possibility of galactic wind in our Galaxy is in many respects similar to the solar wind and was discussed many years ago (see e.g. [5, 13, 16, 17, 10, 6]). It was found that gas temperatures several million degrees are neces-

31

M. M. Shapiro et al. (eds.), Astrophysical Sources of High Energy Particles and Radiation, 31–42.
© 2001 *Kluwer Academic Publishers. Printed in the Netherlands.*

sary for existence of thermally driven wind . Such results assume models of a hot interstellar medium suggested by McKee and Ostriker [17] or "chimney" model of Ikeuchi [19]. Now it is believed that the hot gas with density $n \approx 6 \cdot 10^{-3}$ cm^{-3} and temperature $T \approx 5 \cdot 10^5$ K heated by supernovae, and warm gas with density $n \approx 0.3$ cm^{-3} and temperature $T \approx 0.5$ to $1.0 \cdot 10^4$ K are the main components of the interstellar medium [7, 8]. In this case galactic wind models including the dynamics of cosmic rays [12, 4] become very important. In this paper we used approach suggested by Brietschwerdt, McKenzie and Völk [3] (so called flux tube formalism) for cosmic ray driven galactic wind. We applied it for the rotating Galaxy with magnetic field. Axially symmetric magnetohydrodynamic (MHD) flows were investigated in many papers (see e.g. [21, 22, 20]). In last one force balance across lines of the flow is taken into account. Further we will concentrate on effects of cosmic rays, wave dissipation, magnetic field and rotation without perpendicular balance. This means that the geometry of the flow is prescribed. Section 2 contains basic magnetohydrodynamic equations including cosmic rays. Numeric results are given in Sections 3 and 4. We will show that the inclusion of rotation in galactic wind model resumes to understand regular magnetic field structure in the halo of the Galaxy and its surroundings.

2. Basic MHD equations for galactic wind including effects of cosmic rays

Steady-state MHD equations for plasma density ρ, velocity $\mathbf{u}$ and gas pressure P_g have the form

$$\nabla(\rho \mathbf{u}) = 0 \tag{1}$$

$$\rho(\mathbf{u}\nabla)\mathbf{u} = -\nabla(P_g + P_c) + \rho\nabla\Phi - \frac{1}{4\pi}\left[\mathbf{B} \times [\nabla \times \mathbf{B}]\right] \tag{2}$$

$$\nabla\left[\rho\mathbf{u}\left(\frac{u^2}{2} + \frac{\gamma_g}{\gamma_g - 1}\frac{P_g}{\rho} - \Phi\right) + \frac{1}{4\pi}\left[\mathbf{B} \times [\mathbf{u} \times \mathbf{B}]\right]\right] = -\mathbf{u}\nabla P_c + H - \Lambda \tag{3}$$

$$\nabla \times [\mathbf{u} \times \mathbf{B}] = 0 \tag{4}$$

$$\nabla\mathbf{B} = 0 \tag{5}$$

$$\nabla_i\left(\frac{\gamma_c}{\gamma_c - 1}\left(u_i + V_{ai}\right)P_c - \frac{D_{ij}\nabla_j P_c}{\gamma_c - 1}\right) = \left(u_i + V_{ai}\right)\nabla_i P_c \tag{6}$$

Here H and Λ describe heating and energy losses of the thermal gas. The term $\mathbf{u}\nabla P_c$ in the eq. (3) describes mechanical work produced by cosmic ray pressure P_c on the volume element of the gas. The eq. (6) for the cosmic

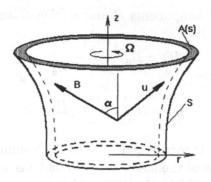

Figure 1. Magnetic flux tube geometry characterized by the surface S containing the wind stream lines. The vectors of magnetic field **B** and gas velocity **u** are tangent to this surface. The angle between the magnetic field and the meridional direction is denoted by α.

ray pressure contains diffusion tensor D_{ij} (cf. [2]). For strongly magnetized cosmic ray particles, diffusion is field alined, i.e.

$$D_{ij} = D_{\|} b_i b_j \tag{7}$$

where $D_{\|}$ is diffusion coefficient along the magnetic field. It's value is determined by the scattering on the small-scale random Alfvén waves propagating out of the Galaxy. The level of this turbulence is low and it's pressure does not directly appear in the eqs. (2) ,(3). The vector $\mathbf{V}_a = V_a \mathbf{b}$ where $V_a = B/\sqrt{4\pi\rho}$ is the Alfvén velocity, and **b** is a unit vector directed along the magnetic field out of the Galaxy.

We consider model with generation of the scattering Alfvén waves by the bulk motion of cosmic rays out of the Galaxy (cf. [2]). So, cosmic rays are scattered only on the waves which they generate itself. Generation of waves with wave vector along the regular magnetic field is the most efficient. Alfvén waves are damped due to nonlinear Landau damping [15, 1, 14, 9, 23]. So heating power equals to wave generation power (cf. [4])

$$H = -\mathbf{V}_a \nabla P_c \tag{8}$$

Due to the azimuthal symmetry and frozen magnetic field vectors of gas velocity and magnetic field lie on the same surface of rotation S. Taken another close surface containing **u** and **B** we get the volume between these two surfaces which is the flux tube where the gas streams. Let us make cross section of the flux tube perpendicular to the surface S and introduce the value $A(s)$ which is the cross section area, see Fig.1. In the case of cylindrical symmetry azimuthal field and velocity components can be expressed in terms of meridional components B and u (Weber and Davis, 1967):

$$B_\varphi = B_a \frac{\Omega}{u_a r} \frac{r^2 - r_a^2}{1 - M_a^2} \tag{9}$$

$$u_\varphi = \Omega r \frac{1 - M_a^2 r_a^2 / r^2}{1 - M_a^2} \tag{10}$$

Here $M_a = u\sqrt{4\pi\rho}/B$ is meridional Alfvén Mach number, B_a and u_a are meridional components of the magnetic field and gas velocity at the radial distance $r = r_a$, where $M_a = 1$. Hence, Ω is the angular velocity of the wind at large densities (in the disk), where Alfvén Mach number is small compared to unity. The total angular momentum per unit mass in the wind with frozen magnetic field is conserved along the surface S and is equal to Ωr_a^2 .

Due to the cylindrical symmetry $\frac{\partial}{\partial\varphi} = 0$ and , then taken into account (9), (10) the equations (1)-(3), (5), (6) have the form

$$A(s)\rho u = const \tag{11}$$

$$\rho\left(u\frac{\partial u}{\partial s} - \frac{u_r}{u}\frac{u_\varphi^2}{r}\right) = -\frac{\partial}{\partial s}\left(P_g + P_c\right) - \frac{1}{8\pi r^2}\frac{\partial}{\partial s}\left(r^2 B_\varphi^2\right) + \rho\frac{\partial\Phi}{\partial s} \tag{12}$$

$$\frac{1}{A}\frac{\partial}{\partial s}A\left[\rho u\left(\frac{u^2}{2} + \frac{u_\varphi^2}{2} + \frac{\gamma_g}{\gamma_g - 1}\frac{P_g}{\rho} - \Phi\right) - \frac{B_\varphi B_r}{4\pi}\Omega r + \frac{\gamma_c}{\gamma_c - 1}\left(u + v_a\right)P_c -\right.$$

$$\left. -\frac{D}{\gamma_c - 1}\frac{\partial P_c}{\partial s}\right] = -\Lambda \tag{13}$$

$$AB = const \tag{14}$$

$$\frac{1}{A}\frac{\partial}{\partial s}A\left(\frac{\gamma_c}{\gamma_c - 1}\left(u + v_a\right)P_c - \frac{D}{\gamma_c - 1}\frac{\partial P_c}{\partial s}\right) = \left(u + v_a\right)\frac{\partial P_c}{\partial s} \tag{15}$$

Here $v_a = B/\sqrt{4\pi\rho}$ is meridional component of the Alfvén velocity, s is coordinate in meridional direction, diffusion coefficient $D = D_\parallel \cos^2\alpha$, where α is the angle between magnetic field and meridional direction, $\cos\alpha = B/\sqrt{B^2 + B_\varphi^2}$, see Fig.1.

Using expressions for azimuthal components (9), (10) one can transform the eqs. (12), (13) to the form

$$\rho\frac{\partial}{\partial s}\left(\frac{u^2}{2} + \frac{u_\varphi^2}{2} - \Omega r u_\varphi - \Phi\right) = -\frac{\partial}{\partial s}\left(P_g + P_c\right) \tag{16}$$

$$\frac{1}{A}\frac{\partial}{\partial s}A\left[\rho u\left(\frac{u^2}{2}+\frac{u_\varphi^2}{2}-\Omega r u_\varphi+\frac{\gamma_g}{\gamma_g-1}\frac{P_g}{\rho}-\Phi\right)+\right.$$

$$\left.+\frac{\gamma_c}{\gamma_c-1}\left(u+v_a\right)P_c-\frac{D}{\gamma_c-1}\frac{\partial P_c}{\partial s}\right]=-\Lambda \qquad (17)$$

Multiplying the eq. (16) by u, subtracting it from the eq. (17), and taking into account the eq. (15) we get the equation for the gas heating

$$u\left(\frac{\partial P_c}{\partial s}-\gamma_g\frac{P_g}{\rho}\frac{\partial\rho}{\partial s}\right)=-\left(\gamma_g-1\right)\left(v_a\frac{\partial P_c}{\partial s}+\Lambda\right) \qquad (18)$$

We shall consider cosmic ray propagation adiabatically with effective adiabatic index. This assumes $D=0$ in eq. (15). Its solution is then given by

$$P_c=P_{c0}\left[\frac{u_0+v_{a0}}{u_0\frac{\rho_0}{\rho}+v_{a0}\left(\frac{\rho_0}{\rho}\right)^{1/2}}\right]^{\gamma_c} \qquad (19)$$

Here subscripts 0 correspond to the boundary values at a base level.

The equation of the gas heating (18) may be simplified for a hot gas in the halo where temperature is $T>2\cdot10^4$ K and the most important energy losses are radiative losses. One can find then that they are not very efficient and the last term in the eq. (18) may be omitted if the gas number density is less than about $n<10^{-3}$ cm^{-3}. Now, the eq. (18) may be presented as differential equation for the gas pressure as function of the gas density

$$\frac{\partial P_g}{\partial\rho}-\gamma_g\frac{P_g}{\rho}+\left(\gamma_g-1\right)\frac{v_{a0}}{u_0}\left(\frac{\rho}{\rho_0}\right)^{1/2}\frac{\partial P_c}{\partial\rho}=0 \qquad (20)$$

Here cosmic ray pressure $P_c=P_c(\rho)$ is given by the expression (19).

The eq. (20) may be integrated if the relation

$$\gamma_g=1+\frac{\gamma_c}{2} \qquad (21)$$

holds between the adiabatic indices of the thermal gas and the cosmic rays. Using (19) we find in this case

$$P_g=P_{g0}\left(\frac{\rho}{\rho_0}\right)^{1+\frac{\gamma_c}{2}}+\frac{\gamma_c^2}{\gamma_c-1}\frac{v_{a0}}{u_0}\left(\frac{\rho}{\rho_0}\right)^{1+\frac{\gamma_c}{2}}\left[1+\frac{\gamma_c+1}{2\gamma_c}\frac{v_{a0}}{u_0}-\right.$$

$$\left.-\left(\frac{u_0+v_{a0}}{u_0\left(\frac{\rho_0}{\rho}\right)^{1/2}+v_{a0}}\right)^{\gamma_c}\left(\left(\frac{\rho}{\rho_0}\right)^{1/2}+\frac{\gamma_c+1}{2\gamma_c}\frac{v_{a0}}{u_0}\right)\right] \qquad (22)$$

Note that the condition (21) is fulfilled in particular for the realistic case $\gamma_c = 4/3$, $\gamma_g = 5/3$.

With above mentioned assumptions the eq. (17) takes the form

$$\frac{u^2}{2} + \frac{u_\varphi^2}{2} - \Omega r u_\varphi + \frac{\gamma_g}{\gamma_g - 1}\frac{P_g}{\rho} - \Phi + \frac{\gamma_c}{\gamma_c - 1}\left(\frac{u + v_a}{u}\right)\frac{P_c}{\rho} = const \quad (23)$$

The gas and cosmic ray pressures P_g, P_c are determined by the expressions (22) and (19). The meridional velocity u is expressed as a function of gas density from the continuity equation. The azimuthal component of the velocity u_φ is given by the equation following from (10)

$$u_\varphi = \Omega r \frac{1 - \frac{r_a^2}{r^2}\left(\frac{u_0}{v_{a0}}\right)^2\frac{\rho_0}{\rho}}{1 - \left(\frac{u_0}{v_{a0}}\right)^2\frac{\rho_0}{\rho}} \quad (24)$$

Given $A(s)$, $r(s)$, $\Phi(s)$ the eq. (23) is transcendental equation for the gas density $\rho(s)$. For the set of initial conditions P_{c0}, P_{g0}, v_{a0}, ρ_0, Ω, r_0 the eq. (23) has two-parametric family of solutions $\rho(s, u_0, r_a)$. The topology of this family of solutions is the same as studied by Weber and Davis [21] for the solar wind, see Fig.2. There exist three critical points in a wind stream where gas velocity is equal to the velocities of slow magnetosonic, Alfvén, and fast magnetosonic waves. The solution corresponding to the galactic wind goes trough all three particular points. It fixes the value of both parameters u_0 and r_a.

3. Numeric solution

Formally, the functions $A(s)$, $r(s)$ can not be taken as prescribed and should be found from the system of the equations (1) -(6) jointly with the other variable functions. The numerical solutions of the similar equations without magnetic field and cosmic rays show that galactic wind flows have rather simple geometry. At the small heights above the galactic disk the meridional component of the gas velocity is perpendicular to the disk, and hence the cross section of flux tube $A(s) = const$ and the radial distance from the axes to the surface S $r(s) = const$. At the large distances from the disk of the Galaxy the gas velocity is almost radial, and hence $A(s) \propto s^2$, $r(s) \propto s$, see e.g. [10]. The results of the calculations of the galactic wind flow with the chosen ad hoc function $A(s)$ [4] demonstrate no considerable deviations from the proper solution of Habe and Ikeuchi [10] if appropriate asymptotic behavior of $A(s)$ was chosen.

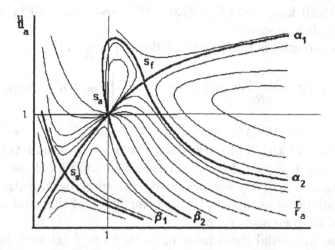

Figure 2. Family of solutions of Eq.(50). Singular points s_s, s_a, s_f are the slow magnetosonic point, the Alfvén point and the fast magnetosonic point, respectively. A galactic wind solution α_1 passes through all these points.

Here we consider the solution of our equations for the following hyperboloidal form of the surface S

$$\frac{r^2}{r_0^2} - \frac{z^2}{z_0^2 - r_0^2} = 1 \qquad (25)$$

here $z_0 = 15$ kpc is a typical scale of the Galaxy, r_0 is initial value of flux tube distance from the galactic centre, r,z are cylindrical coordinates..

We used the model of Miyamoto and Nagai [18] for the gravitation potential of our Galaxy and further include the massive dark matter halo. Gravitation potential has the form

$$\Phi = \Phi_{B,D} + \Phi_H \qquad (26)$$

Here $\Phi_{B,D}$ and Φ_H are the potentials of bulge, disk, and halo, respectively. In cylindrical coordinates the bulge and disk potentials are given by

$$\Phi_{B,D} = \sum_{i=1}^{2} \frac{GM_i}{\sqrt{r^2 + \left(a_i + \sqrt{z^2 + b_i^2}\right)^2}} \qquad (27)$$

where $G = 6.668 \cdot 10^{-8} \text{cm}^3\text{g}^{-1}\text{s}^{-2}$ denotes the gravitational constant, and the three pairs of parameters are given by $a_i = (0.0$ and $7.258)$ kpc, $b_i =$

(0.495 and 0.520) kpc, and $M_i = (2.05 \cdot 10^{10}$ and $2.547 \cdot 10^{11})M_\odot$ for the bulge and the disk, respectively.

The halo potential is taken in the following form [11]:

$$\Phi_H(x) = \Phi_0 - \frac{GM_{H0}}{R_b}\left[\ln(1+x) + \frac{1}{1+x}\right] for \quad R < 100pc \qquad (28)$$

where $M_{H0} = 1.35 \cdot 10^{11}M_\odot$, $\Phi_0 = 1.37 \cdot 10^{15} \text{cm}^2/\text{s}^2$, $R = \sqrt{r^2 + z^2}$, $x = R/R_b$, and $R_b = 13$ kpc. The dark matter halo is cut off at 100 kpc.

The eq. (23) was solved by iterations. The solution depends on two parameters u_0 and r_a. They were scanned to provide the transition through the critical points corresponding to the coincidence of the wind speed with slow and fast magnetosonic speeds.

We choose the initial gas density $n_0 = 10^{-3}$ cm^{-3} (at such low density one can neglect by the radiative energy losses) at corresponding base level at 1 kpc above the galactic disk, where all inner boundary conditions are prescribed. The flux tube is located at the position of the Sun, $r_0 = 8.5$ kpc. We choose $\gamma_c = 1.2$ and $\gamma_g = 1.6$ for the adiabatic indexes of the cosmic rays and the thermal gas, so the relation (21) is satisfied. The gas is treated as completely ionized hydrogen plasma. The initial cosmic ray pressure $P_{c0} = 2 \cdot 10^{-13}$ erg/cm^3. The initial gas pressure is taken to be equal to

$$P_{g0} = \frac{\gamma_c^2}{2(2+\gamma_c)}\frac{v_{a0}}{u_0}P_{c0} \qquad (29)$$

which marginally guarantees that the compound sound speed is not imaginary and the solution is stable, see also [4] for discussion. The meridional component of the regular magnetic field is $B = 10^{-6}$ Gs, and hence $v_{a0} = 69$ km/s. The rotational velocity $\Omega r_0 = 260$ km/s was obtained by balancing the gravitational and centrifugal accelerations at the midplane of the Galaxy. For $r_0 = 8.5$ kpc, the gravitational potential at the reference level is $\Phi_0 = 1.89 \cdot 10^{15}$ cm^2/s^2. Figure 3 shows numeric results. The numerical calculations yield an initial velocity $u_0 = 27.9$ km/s and the distances of the slow, Alfvénic and fast magnetosonic points at $z_s = 5.1$ kpc, $z_a = 6.8$ kpc, $z_f = 18.7$ kpc, respectively. The calculated initial values of the azimuthal magnetic field component and the gas temperature are $B_{\varphi0} = -0.5 \cdot 10^{-6}$ Gs and $T_0 = 4.0 \cdot 10^5$ K, respectively. The asymptotic value of the wind velocity is $u_f = 476$ km/s and is attained at very large distances, of the order 1 Mpc. We do not consider here the effect of an external intergalactic pressure which can decelerate the galactic wind flow at a smaller distance through a termination shock, supposedly around 300 kpc. The strength of magnetic field does not vary drastically up to the distances about 20 kpc from the disk. At larger distances the field is almost purely azimuthal and

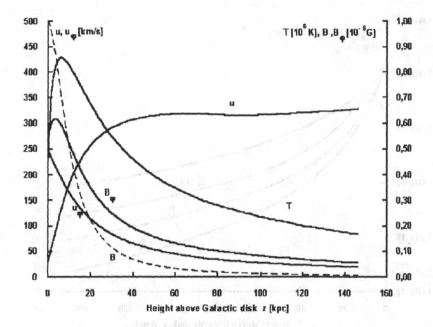

Figure 3. Variation of the meridional and azimuthal velocities u and u_ϕ, azimuthal and meridional magnetic field strength B_ϕ and B and gas temperature T with distance from the disk z. Initial cosmic ray pressure is $P_{c0} = 2 \cdot 10^{-13}$ erg/cm^3, $\rho_0 = 1.67 \cdot 10^{-27}$ cm^{-3}, $B_0 = 10^{-6}$ Gs. The resulting initial velocity is $u_0 = 27.9$ km/s, and the critical points positions are $z_s = 5.1$ kpc, $z_a = 6.8$ kpc, $z_f = 18.7$ kpc. The terminal velocity is $u_f = 476$ km/s .

falls down with distance as $1/z$. The gas temperature increases with distance up to $z = 10$ kpc and then decreases slowly because of adiabatic expansion. At $z = 75$ kpc the gas temperature is $T = 3 \cdot 10^5$ K, i.e. almost equal to the temperature $T = 4 \cdot 10^5$ K at the base level (in spite of the large expansion factor of the gas $\rho_0/\rho \approx 400$). This behavior demonstrates a very effective heating of the gas due to the damping of small scale MHD waves. Figure 4 shows the dependencies of ram pressure, cosmic ray pressure, gas pressure and magnetic tension on height z above the disk. It is easy to see that the ram pressure of the flow dominates at distances larger then 20 kpc. Also, the magnetic tension dominates over the gas and cosmic ray pressures practically everywhere (except at small heights) and therefore plays an important dynamic role.

4. The mass and angular momentum loss rates of the Galaxy

In this section we calculate the total mass and angular momentum loss rates of the Galaxy. For this purpose we have to perform the relevant galactic

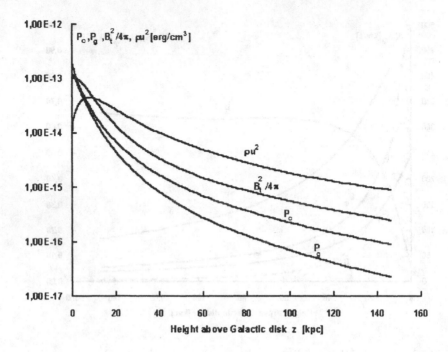

Figure 4. Variation of the dynamic pressure ρu^2, cosmic ray pressure P_c, gas pressure P_g and magnetic tension $B_t^2/4\pi$, (B_t is the total magnetic field strength) with distance z from the disk.

wind flow calculations for flux-tubes originating at different galactocentric radii and sum over the corresponding area weighted loss rates per unit flux-tube area, $\dot{m}$, is simply given by the initial wind velocity multiplied by the gas density at the reference level. The corresponding angular momentum loss rate is then given by

$$\dot{l} = \dot{m}\Omega r_a^2 \tag{30}$$

The cosmic ray pressure and the gas density at the base level 1 kpc are assumed to be constant at different galactocentric radii, i.e. $P_{c0} = 2 \cdot 10^{-13}\text{erg/cm}^3$ and $\rho_0 = 1.67 \cdot 10^{-27}$ g/cm^3 throughout. The wind flow is calculated in radial steps of 1kpc until $r_0 = 14$ kpc. The results are shown in Fig.5. Flux tubes at large galactocentric radii give the main contribution to the total mass and angular momentum loss rates $\dot{M}$ and $\dot{L}$. Their values are $\dot{M} = 1.36 \cdot M_\odot/\text{yr}$ and $\dot{L} = 0.17 \cdot L/10^{10}$ yr. About ten percent of the last value is due to magnetic torque. The total angular momentum of the Galaxy is taken to be $L = 1.7 \cdot 10^{74}\text{g}\cdot\text{cm}^2/\text{s}$ (it was estimated using the model of Miyamoto and Nagai [18], for stellar distribution in the Galactic disk).

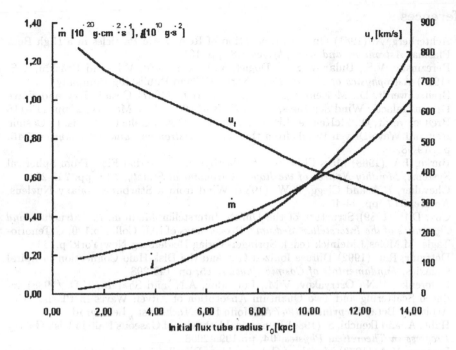

Figure 5. The dependencies of the angular momentum and mass loss rates of the Galaxy $\dot{M}$ and $\dot{l}$, and the terminal velocity of the wind u_f, on the initial flux tube radius r_0.

5. Conclusion

In this paper we consider galactic wind model, taking into account explicitly galactic rotation, together with the azimuthal magnetic field. Nonlinear wave damping is strong, which leads to significant gas heating at the expense of the cosmic ray energy flux. In the absence of radiating cooling of the thermal plasma, this results in a picture in which a late-type galaxy like ours is surrounded by a large wind halo with scale of 100 kpc, containing rarefied hot gas, chemically enriched from supernova explosions, and a strong magnetic field. The results show that galactic winds can exist under a wide range of initial conditions.

The wind is mainly driven by the cosmic rays which can separate from the gas by the Alfvénic drift (with the scattering waves) as well as by diffusion. It is also driven to some extent by the magnetic field and the thermal gas.

Finally we wish to point out that the existence of winds profoundly affects the possible modes of galactic dynamos by fixing the boundary conditions of the magnetic field in a definite form.

References

1. Achterberg, A. (1981) On the Propagation of Relativistic Particles in a High Beta Plasma, *Astronomy and Astrophysics*, **98**, pp. 161-172
2. Berezinsky, V.S., Bulanov, S.V., Dogiel, V.A., Ginzburg, V.L. and Ptuskin, V.S. (1990) *Astrophysics of Cosmic Rays*, North-Holland Publishing Company
3. Breitschwerdt, D., McKenzie, J.F. and Völk, H.J., (1987) Cosmic Ray and Wave Driven Galactic Wind Solutions, *Proceedings of 20th ICRC*, Moscow, **2**, pp. 115-118
4. Breitschwerdt, D., McKenzie, J.F. and Völk, H.J. (1991) Galactic Winds. I - Cosmic Ray and Wave-Driven Winds from the Galaxy, *Astronomy and Astrophysics*, **245**, pp. 79-98.
5. Burke, J.A. (1968) Mass Flow from Stellar Systems-I. Radial Flow From Spherical Systems, *Monthly Notices of the Royal Astronomical Society*, **140**, pp. 241-254
6. Chevalier, R.A and Clegg, A.W. (1985) Wind from a Starburst Galaxy Nucleus, *Nature*, **317**, pp. 44-45
7. Cox, D.P. (1989) Structure of the Diffuse Interstellar Medium, in *"Structure and Dynamics of the Interstellar Medium"*, Proceedings of IAU Coll. No.120, G.Tenorio-Tagle, M.Moles,J.Melnick (eds), Springer-Verlag Heidelberg New York, p.432
8. Dettmar, R.J. (1992) Diffuse Ionized Gas and the Disk-Halo Connection in Spiral Galaxies, *Fundamentals of Cosmic Physics*, **15**, pp. 143-208
9. Fedorenko, V.N., Ostryakov, V.M., Polyudov, A.N. and Shapiro, V.D. (1988) Induced Scattering and Two Quantum Absorption of Alfvén Waves in Plasma with Arbitrary Beta, *Preprint N 1267* A.F.Ioffe Phys.Tech. Inst., Leningrad
10. Habe, A. and Ikeuchi, S. (1980) Dynamical Behavior of Gaseous Halo in Disk Galaxy *Progress in Theoretical Physics*, **64**, pp.1995-2008
11. Innanen, K.A. (1973) Models of Galactic Mass Distribution, *Astrophysics and Space Science*, **22**, pp. 393-399
12. Ipavich, F.M. (1975) Galactic winds driven by cosmic rays, *Astrophysical Journal*, **196**, pp. 107-120
13. Johnson, H.E. and Axford, W.I. (1971) Galactic Winds, *Astrophysical Journal* **165**, pp. 381-390
14. Kulsrud, R.M. (1982) Plasma in astrophysics, *Physica Scripta*, **2/1**, pp. 177-181
15. Lee, M.A. and Völk, H.J. (1973) Damping and Non-Linear Wave-Particle Interactions of Alfvén-Waves in the Solar Wind, *Astrophysics and Space Science*, **24**, pp. 31-42
16. Mathews, W.G., Baker, J.C. (1971) Galactic Winds, *Astrophysical Journal*, **170**, pp.241-260
17. McKee, C.F. and Ostriker, J.P. (1977) A Theory of the Interstellar Medium - Three Components Regulated by Supernova Explosions in an Inhomogeneous Substrate, *Astrophysical Journal*, **218**, pp. 148-169
18. Miyamoto, M. and Nagai, R. (1975) Three-Dimensional Models for the Distribution of Mass in Galaxies, *Astronomical Society of Japan, Publications*, **27**, pp. 533-543
19. Norman, C.A. and Ikeuchi, S. (1989) The Disk-Halo Interaction - Superbubbles and the Structure of the Interstellar Medium, *Astrophysical Journal*, **345**, pp. 372-383
20. Sakurai, T. (1985) Magnetic Stellar Winds - A 2-D Generalization of the Weber-Davis Model, *Astronomy and Astrophysics*, **152**, pp. 121-129
21. Weber, E.J. and Davis, L.Jr. (1967) The Angular Momentum of the Solar Wind, *Astrophysical Journal*, **148**, pp. 217-228
22. Yeh, T. (1976) Mass and Angular Momentum Effluxes of Stellar Winds, *Astrophysical Journal*, **206**, pp. 768-776
23. Zirakashvili, V.N. (2000) Induced Scattering and Two-Photon Absorption of Alfven Waves with Arbitrary Propagation Angles, *Journal of Experimental and Theoretical Physics*, **90**, pp. 810-816

SUPERNOVAE, SUPERBUBBLES, NONTHERMAL EMISSION & LIGHT ELEMENT NUCLEOSYNTHESIS IN THE GALAXY

ANDREI M. BYKOV

A.F.Ioffe Institute for Physics and Technology,
194021, St.Petersburg, Russia
byk@astro.ioffe.rssi.ru

Abstract   Nonthermal particles acceleration and interactions with interstellar matter are considered with special emphasis on the supernova remnants interacting with molecular clouds, superbubbles created by massive star-formation regions, and high-velocity clouds. We discuss hard continuum and the 511 keV annihilation line emission from that objects. Multi-wavelength spectra of extended and compact sources are used to constrain the nonthermal particle generation models and hard emission spectra resulting from interactions of energetic particles with interstellar matter. We further show that galactic superbubbles are potentially important sites of light element (e.g. ^{9}Be, ^{10}B) nucleosynthesis as well as extended sources of γ-ray emission. Special attention is paid to the CR-ISM interaction regions that can be observed with *INTEGRAL* and *GLAST*.

1.   INTRODUCTION

Cosmic Rays (CR) are one of the most interesting phenomena of modern astrophysics, as they represent highly nonequilibrium states of matter, with high energy particles yet impossible to achieve in a laboratory. Thus, detailed information on their interactions with the interstellar matter (ISM) is of a fundamental value. *International Gamma-Ray Astrophysical Laboratory (INTEGRAL)* (e.g. Winkler, 1999) will be able to provide very important observational information on CR sources and interactions with the ISM.

2.   COSMIC RAYS IN SUPERNOVA REMNANTS

Supernova remnants (SNR) are considered for a long time as potential sources of CRs (e.g. Berezinskii et al., 1990; Ellison, Drury & Meyer

M. M. Shapiro et al. (eds.), Astrophysical Sources of High Energy Particles and Radiation, 43–64.
© 2001 *Kluwer Academic Publishers. Printed in the Netherlands.*

1997). An evidence for electron acceleration first came from observations of radio synchrotron radiation. The observation of nonthermal X-ray emission from SN 1006 by *ASCA* has provided some evidence for electron acceleration up to ~ 100 TeV (Koyama et al.1995; Reynolds 1998) supported by the TeV emission signature from SN 1006 reported by *CANGAROO* telescope (Tanimori et al., 1998). Featureless X-ray emission from SN1006 was interpreted as synchrotron emission of TeV regime electrons accelerated by SNR shock (e.g. Reynolds, 1998). Hard X-ray tails extending above 100 keV were observed from Cas A with *CGRO* (The et al., 1996), *RXTE* (Allen et al., 1997) *BeppoSAX* (Vink et al., 1999). X-ray image of Cas A above 4.0 keV obtained by Bleeker et al.(2001) with *XMM-Newton MOS and PN* instruments showed mostly extended continuum emission through the whole remnant of Cas A. The image morphology is somewhat different from that expected from the synchrotron interpretation of hard X-ray emission (Bleeker et al., 2001). If the emission dominated the *XMM-Newton* image of Cas A is of nonthermal origin, then bremsstrahlung and inverse Compton components generated by electrons of energy below 10 GeV might have been important. Nonthermal bremsstrahlung emission is a plausible source of hard X-rays though rather an energetically expensive one. Inverse Compton scattering of radio emitting electrons could provide an extended source of hard X-ray emission of photon index about 1.8 at 4 - 10 keV regime. However if estimated magnetic field in Cas A is above 100 μG the inverse Compton model would require a special source of IR photons to provide the observed flux. Further evidences for TeV emission from SNRs was recently reported for RXJ 1713.7-3946 by *CANGAROO* team and for Cas A by *HEGRA* (see Fegan (2001) for a review).

The barionic cosmic ray component in SNRs is more difficult to detect, but it has long been recognized that pion decays from collisions with interstellar gas could produce an observable flux of $0.1 - 1$ GeV photons. Initial estimates of γ-ray emission from supernova remnants (SNRs), concentrating on the component resulting from pion decays, were made by Chevalier (1977), Blandford & Cowie (1982), and Drury, Aharonian, & Völk (1994).

When γ-ray emission was apparently detected from SNRs by *CGRO EGRET*, the γ-ray spectrum could not be fitted by a pure pion decay spectrum and some other component was needed (Esposito et al.1996; Sturner et al.1997; Gaisser et al.1998). In addition to the pion decays, the relevant processes are bremsstrahlung emission of relativistic electrons and inverse Compton emission. Gaisser et al.(1998) modeled these processes in detail in order to fit the observed γ-ray spectra of the remnants IC443 and γ Cygni. They assumed acceleration to a power law spectrum in a shock front and determined the spectral index, the electron to proton ratio, and the upper energy cutoff. de Jager & Mastichiadis (1997) dealt with the same processes in W44, as well as inverse Compton emission. They noted that the observed radio spectrum is flatter than would be expected from shock acceleration of newly injected particles and suggested that the particles originated from a pulsar in the SNR. Baring et al.(1999) presented calculations of the broad-band emission from nonlinear shock models of shell-type SNRs. They used Sedov adiabatic shock dynamics in a homogeneous medium and a Monte Carlo simulation of the particle acceleration taking into account the nonlinear shock structure. The set of models considered by Baring et al.(1999) covers the range of shock speeds $490 \leq v_S \leq 4000$ km s^{-1} and ambient medium number densities $10^{-3} \leq n \leq 1$ cm^{-3}.

All four of the SNRs preliminary identified with $CGRO$ $EGRET$ γ-ray sources (Esposito et al., 1996) are showing some indications for SNR - molecular cloud interactions (see also Grenier, 2000).

3. SUPERNOVAE IN MOLECULAR CLOUDS

Massive stars that are the likely progenitors of core collapsed supernovae are expected to be spatially correlated with molecular clouds. An interaction of SNR with a molecular cloud is expected to manifest itself by a number of spectacular appearances in a wide range of wavelengths from radio to gamma-rays.

Chevalier (1999) discussed the evolution of SNRs in molecular clouds and concluded that many aspects of the multi-wavelength observations could be understood in a model where the remnants evolve in the interclump medium of a molecular cloud. The non-thermal multi-wavelength spectrum of a SNR interacting with a molecular cloud was studied by Bykov et al.(2000). They showed that the propagation of a radiative shock wave within a molecular cloud leads to a substantial non-thermal emission both in hard X-rays and in γ-rays. The complex structure of a molecular cloud consisting of dense massive clumps embedded in the interclump medium could result in localized sources of hard X-ray emission correlated with both bright molecular emission and an extended source of nonthermal radio and γ-ray emission.

IC443 (G189.1+3.0) is an evolved SNR of about 45 arcmin size located at an estimated distance D of about 1.5 kpc (e.g. Fesen & Kirshner, 1980). It is one of the best space laboratories to study the rich phenomena accompanying supernova interaction with a molecular cloud. This was established from the observed shock-excited molecular line emission from OH, CO and H_2 (e.g. Burton et al., 1988; van Dishoeck et al.,

1993; Cesarsky et al., 1999). Extensive study of atomic fine structure lines with *ISO LWS* and *SWS* spectrometers revealed rich spectra of these lines. The fine structure lines of oxygen [O I] 63.2 μm and silicon [Si II] 34.8 μm are the main coolants of the radiative shock propagating in the interclump medium (Reach & Rho, 2000).

Nonthermal emission from an SNR is a signature of the presence of accelerated particles that has important implications to the cosmic ray origin problem. Substantial efforts to detect γ-ray emission of cosmic rays from SNRs have been undertaken by Compton GRO. *EGRET CGRO* has detected four extended γ-ray sources (Esposito et al.1996) that are candidates to be identified with SNRs (IC443, γ-Cyg, W28, W44), though some of many unidentified *EGRET* sources might be as well related to SNRs. The remnants that are likely candidates to be γ-ray sources in *CGRO EGRET* observations show evidences for interaction with molecular gas. In the case of IC 443, the molecular line emission region is partially inside the *EGRET* γ-ray detection circle.

3.1. HARD EMISSION FROM IC443

IC 443 was a target of X-ray observations with *HEAO 1* (Petre et al., 1988), *Ginga* (Wang et al., 1992), *ROSAT* (Asaoka & Aschenbach, 1994) and *ASCA* (Keohane et al., 1997). The soft X-ray 0.2–3.1 keV surface brightness map of IC 443 from the *Einstein* Observatory (Petre et al.1988) shows bright features in the northeastern part of the remnant as well as bright soft emission from the central part of the source. The presence of nearly uniform X-ray emission from the central part of the remnant was also clearly seen by *ROSAT* (Asaoka & Aschenbach, 1994), and corresponds to the emission from hot (T$\sim 10^7$ K), low density gas interior of the shock. *ASCA GIS* observations by Keohane et al.(1997) discovered the localized character of the hard X-ray emission. They concluded that most of the 2-10 keV *GIS* photons came from an isolated emitting feature and from the southeastern elongated ridge of the hard emission. Preite-Martinez et al. (1999) and Bocchino & Bykov (2000) have reported a hard (up to 100 keV) component with BeppoSAX/PDS and two hot spots with the *MECS*. We illustrate the localized structure of IC443 hard emission in Fig. 1 showing *BeppoSAX MECS* mosaics in 4.0 - 10.5 keV energy regime taken from Bocchino & Bykov (2000).

The observed hard X-ray structure of IC443 is dominated by a few localized regions, which reflects the highly inhomogeneous complex structure of the remnant. The localized sources are likely to be represented by a pulsar wind nebula and shocked molecular clumps. With limited angular resolution of *MECS BeppoSAX* it is hard to distinguish between the

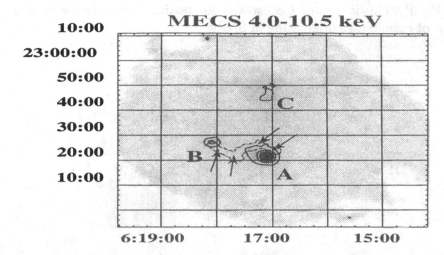

Figure 1 The image of IC 443 obtained with BeppoSAX MECS in 4.0-10.5 keV adopted from Bocchino & Bykov, (2000). Grayscale is linear and contours are 20%, 40%, 60% and 80% of the peak 2.2×10^{-3} cnt s^{-1} arcmin^{-1}. The sources are marked A, B and C, and dashed contour corresponds to the $v = 1 - 0S(1)$ molecular H line intensity of 6.6×10^{-14} erg cm^{-2} s^{-1} reported by Burton et al. (1988); arrows indicate the location of peaks with an H line intensity at least 14 times the contour levels.

two possibilities for each of the sources. However, a careful inspection of combined 0.2-2.0 keV *PSPC ROSAT* and *MECS BeppoSAX* images revealed some morphological differences between Src A and Src B (Fig.1) (Bocchino & Bykov 2000). The centre of the brighter Src A has a soft emission counterpart, while the soft counterpart of Src B has an apparent shift from the centre of the Src B. This shift might imply that Src B could be treated as a shocked clump according to the model predictions of Bykov et al.(2000). On the other hand, Src A might be treated as a pulsar wind nebula.

Modeling of an SNR interacting with a molecular cloud has been recently performed by Chevalier (1999) and Bykov et al.(2000). It has been shown that hard X-ray and γ-ray emission structure should consist of an extended shell-like feature related to a radiative shock and localized sources corresponding to shocked molecular clumps.

The extended shell of IC443 has been well studied in radiowaves since the 1960s. Relativistic electrons responsible for the observed synchrotron radio emission of the shell can be accelerated by the radiative shock and produce substantial γ-ray emission with photon energies above MeV. It has been shown by Bykov et al.(2000) that the γ-ray emission above 100 MeV observed by *EGRET* is consistent with that of relativistic bremsstrahlung of the radio emitting electrons (see the dash-dotted line on Fig.3). This is a serious constraint for testing the nucleonic component of cosmic rays originating in SNRs.

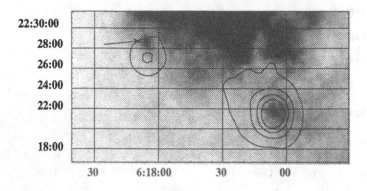

Figure 2   0.2-2.0 keV PSPC image of the region containing the two MECS hard X-ray sources. The hard X-ray contours of Fig. 1 are overlaid. The arrow shows the soft X-ray counterpart of Src B. Adopted from Bocchino & Bykov, (2000)

The shocked clumps are expected to emit hard X-ray spectra with photon spectral index about 1.5 in the 10-100 keV regime (see Fig. 3), while the photon indexes of pulsar wind nebulae are generally somewhat softer (e.g. Chevalier 2000). The measured 10 keV *MECS BeppoSAX* flux is dominated by Src A, while at higher energies the relative contribution of Src B will increase.

Thus *INTEGRAL IBIS* observations providing spatial resolution at higher energies would resolve the problem of distinguishing the sources at 30-100 keV. Such constraints on the hard emission spectrum of IC443 would be also very informative to distinguish hard emission of leptonic origin from that originating from pion decay.

3.2.   POSITRON ANNIHILATION LINE FROM SUPERNOVA REMNANTS

Another spectacular appearance of hard emission from SNRs could be the 511 keV positron annihilation line. We will consider here the case of IC443 as a generic example. There are two principal sources of positrons in a remnant.

The first one is represented by the secondary positrons produced in inelastic collisions of accelerated particles.

The second one is represented by the positrons released from decays of nucleosynthetic ^{56}Co, ^{44}Ti and ^{26}Al at different stages of the remnant

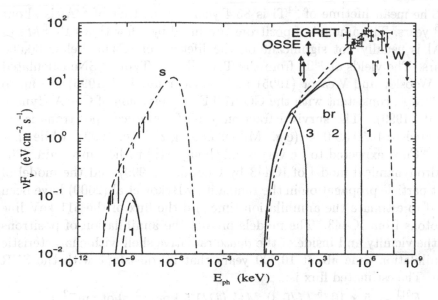

Figure 3 Broadband νF_ν spectrum for a model of nonthermal electron production by a radiative shock interacting with a IC443 from Bykov et al.(2000). The dot-dashed curves are the shell emission spectra from radiative shock of 150 km s^{-1}. The model spectra of shocked molecular clump emission are calculated for the case of a shock velocity in the clump of 100 km s^{-1}, a number density of 10^3 cm^{-3}, and a magnetic field strength of 2×10^{-4} G. The maximum energy of accelerated electrons is $E_m = 0.5$ GeV (curve 1) and $E_m = 0.05$ GeV (curve 2). The model spectrum for the case of a 30 km s^{-1} shock in a clump with number density 10^4 cm^{-3} is shown by curve 3. The internal absorption at the shocked clump (of a pc radius) is shown for EM $\approx$ 110 cm^{-6} pc (solid line radio spectrum) and EM $\approx$ 900 cm^{-6} pc (dashed line radio spectrum).

evolution. A small fraction of the positrons is likely to survive until the later stages (Chan and Lingenfelter, 1993).

The recent models of nucleosynthesis in SNe predict that a typical SN from a massive progenitor would yield about $0.1 M_\odot$ of ^{56}Ni that transforms (in 8.8 days) into ^{56}Co by an electron capture. About 19% of ^{56}Co decays (in 111 days) into ^{56}Fe through positron emitting channels.

The estimations of the positron survival fractions depend on the details of mixing of ejecta as well as on the mechanism of fast positron transport through the ejecta. The later process depends on the magnetic field strength and topology in the ejecta. The precise simulation of the processes is not yet feasible and thus we have to use some simplified models to distinguish between several limiting cases. It should be also noticed that the positron survival fractions are very sensitive to the moment when the nuclear decay positrons are released. Thus, the survival fractions are generally higher for the nuclei with longer mean lifetime.

The mean lifetime of ^{44}Ti is 85.4 years, while that of ^{26}Al is about 10^6 years. The amount of positrons produced by a few times $10^{-5} M_\odot$ of ^{26}Al is usually not significant on the lifetime of IC443, below 30,000 years. The yields of ^{44}Ti from the Type Ib and Type II SNe calculated by Woosley and Weaver (1995) and Thielemann et al.(1996) are up to $10^{-4} M_\odot$, consistent with the COMPTEL observation of Cas A (Iyudin et al. 1999). The survival fraction f of Ti-produced positrons is estimated to be 0.30-0.99 (e.g. Milne, Leising & The, 1999), while that for ^{56}Co is expected to be very smal though still rather uncertain. The hydrodynamical model of IC443 by Chevalier (1999) and the model of fast particle propagation in the remnant by Bykov et al.(2000) have been used to estimate the annihilation time and the flux of the 511 keV line photons from IC443. The models predict the annihilation of positrons in the vicinity and inside of the dense radiative shell with characteristic annihilation time about 10,000 years that is consistent with the SNR age. The estimated flux is:

$$F_\gamma^{511} \approx 5 \times 10^{-5} \ (f/0.3) \ M_{-4}^{Ti} \ (D/1.5 \ \text{kpc})^{-2} \ \text{phot cm}^{-2} \ \text{s}^{-1},$$

where M_{-4}^{Ti} is the yield of the ^{44}Ti in $10^{-4} M_\odot$ units.

The 511 keV line coming from the dense radiative shell of IC443 is expected to be narrow with FWHM about 1 keV (e.g. Guessoum, Ramaty & Lingenfelter, 1991). The line will be slightly Doppler shifted (velocity $\sim$ 100 km/s) because of the shell motion and due to the differential rotation of the Galaxy.

Thus, a 10^6 second SPI observation of IC443 with the nominal sensitivity of $2.80 \cdot 10^{-5}$ phot cm^{-2} s^{-1} at 511 keV for a 3σ detection would seriously constrain the existing models of SNRs.

Another very interesting object for a search of 511 keV line is Vela SNR ($l = 263.9$, $b = -3.3$) an extended ($\sim$ 255 arcmin) nearby SNR with estimated age about 10,000 years. If the distance to Vela SNR is $\sim$ 200 pc then $F_\gamma^{511} \sim 10^{-4}$ phot cm^{-2} s^{-1} can be expected, even though the estimated annihilation time ($\gtrsim 30,000 \ years$) might be longer than that for IC443 due to somewhat smaller density.

3.3. GAS IONIZATION BY NONTHERMAL PARTICLES

We note that a high density of energetic particles in the vicinity of a shock wave in a molecular cloud can affect the ionization and thermal properties of the gas. Rich molecular spectra have been observed from IC 443 and other molecular SNRs (Reach & Rho 2000; Rho et al.2000). It will be interesting to see whether there is a signature of the presence of nonthermal particles that can be discerned from the molecular spectra. Nonthermal particles accelerated by a MHD collisionless shock wave provide an efficient ionizing agent. We discussed above [see Fig. 3] the nonthermal emission generated by accelerated energetic electrons; one can also calculate the ionization produced by the same electrons in the molecular cloud. This might connect the high energy observations with radio, IR and optical emission from shocked atomic and molecular gas. To estimate the gas ionization due to accelerated electrons in the radiative shock structure Bykov et al., (2000) used the electron spectra calculated with the kinetic model of electron acceleration and propagation and electron ionization cross sections. In Fig. 4, we present the calculated ionization rate ζ_e and estimated ionized fraction x in the shell of a radiative shock due to accelerated electrons for the case of 150 km/s velocity shock for both models of large-scale turbulence in the postshock cooling layer considered in Bykov et al.2000.

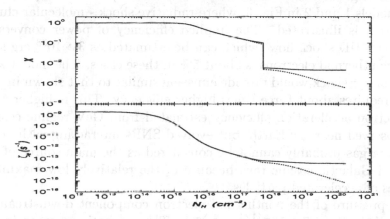

Figure 4 Postshock ionization rate due to nonthermal electrons ζ_e (lower panel) and the estimated ionization fraction x (upper panel) versus total H column density. The solid lines correspond to the model with Fermi II acceleration (see dot-dashed curve in Fig. 3) and the dotted lines to one with an inefficient Fermi II acceleration.

There should also be contributions to x from accelerated nucleons and from UV radiation which are not included in Fig. 4. Bykov et al.(2000) accounted only for radiative recombination, assuming a small molecular fraction in the radiative shell at column densities below 2×10^{20} cm^{-2}.

The solid curves in both panels of Fig. 4 correspond to the case of fully developed Alfvenic turbulence in the postshock cooling layer, which provides efficient second order Fermi acceleration making Coulomb losses relatively unimportant above $\sim$ 120 keV. This case corresponds to dash-dotted curve in Fig. 3 for emission from the shell with substantial hard X-ray emission. The dotted lines in Fig. 4 correspond to the case of a lack of large scale MHD turbulence where the Coulomb losses overcome the second order Fermi acceleration up to energies about 2 GeV. The radiative shell ionization by nonthermal electrons is sensitive to the turbulent structure of the postshock cooling layer and there is a correlation between the hard X-ray spectrum and the ionization structure of the radiative shell.

3.4. ENERGY IN NONTHERMAL ELECTRONS

The energy of the electron component for the parameter set used to compute the spectra in Fig. 3 is substantial. The power required for electron acceleration and maintenance for dot-dashed curve in Fig. 3, $\sim 1.5 \times 10^{37}$ erg s^{-1}, is high because of the strong Coulomb losses of keV electrons in the dense ionized medium, but it is consistent with the estimate of the total radiative losses from IC443, dominated by infrared emission (e.g. Reach & Rho, 2000). A similar power is required for both models 1 and 2 in Fig. 3, where radiative shock - molecular clump interaction is illustrated. The implied efficiency of power conversion from the MHD shock flow (which can be estimated as 3×10^{38} erg s^{-1}) to the nonthermal electrons is about 5% in these cases. The model with a 100 km s^{-1} shock would provide emission similar to that shown in Fig. 3, but requires about three times higher efficiency. This is higher than the electron acceleration efficiency estimated from GeV regime cosmic rays observed near the Earth, but evolved SNRs interacting with dense molecular gas probably cannot be considered as the main source of the observed Galactic cosmic rays because of the relatively low maximum energies of accelerated particles.

The pressure of the nonthermal electron component downstream of the shock is $\mathcal{P}_e \approx 7 \times 10^{-11}$ erg cm^{-3}. The magnetic pressure in the dense shell is about 1.5×10^{-10} erg cm^{-3}, which is higher than the thermal gas pressure. The uniform ($\sim 60~\mu$G) and stochastic magnetic field components dominate the total pressure in the shell.

The scenario with electron injection from preexisting cosmic rays was modeled in detail in Bykov et al.(2000). The nonthermal electron pressure in this case is $\mathcal{P}_e \approx 5.2 \times 10^{-11}$ erg cm^{-3}, similar to that for the model with injection from the thermal pool, because it is dominated by GeV particles. An advantage of this scenario is that much less power is required. It is about 4×10^{35} erg s^{-1}, which is less than 10% of that for the scenario with injection from the thermal pool. The difference is because Coulomb losses are unimportant for the high energy electrons involved in that scenario. However, relatively low energies (~ 20 MeV) of CR electron distribution flattening required to reproduce observable gamma-ray emission. The lifetime of a 20 MeV electron against ionization losses in a neutral medium of number density ~ 1 cm^{-3} is about 10^6 years. This implies that the source of MeV cosmic ray electrons should exist within about 100 pc of the molecular cloud if the diffusion coefficient is $\sim 3\times 10^{26}$ cm^2 s^{-1} at these energies. Mildly relativistic positrons from nuclei decay discussed in Sec. 3.2 can be an interesting possibility.

4.   COSMIC RAYS IN SUPERBUBBLES

Young massive stars formation is known to be spatially and temporally correlated with OB associations. Massive star formation occurs in massive molecular clouds (e.g. Blitz, 1993). The most massive O stars begin to explode as core collapsed supernovae about a million years after the formation of an OB association, creating a superbubble (SB) filled with hot tenuous plasma with supersonic turbulence. Bright X-ray emission has been observed from the hot gas in superbubbles in the Large Magellanic Cloud (LMC). An attempt to identify primary SNR shocks in SB interior has been made by Chen et al.(2000). They used Hubble Space Telescope WFPC2 emission-line (H_α and $[SII]$ lines) images of three SBs in the LMC to identify SNR shocks inside the superbubbles. Such strong and moderate strength SNR shocks could be attributed to filamentary nebular morphology seen in some SBs. From the other hand numerous weak shocks expected in the hot tenuous superbubble interiors are not producing optical signatures and can hardly be observed with such a technique.

There are some HI, IR, radio and X-ray evidences for the presence of several supershells and SBs within a local kpc from the Sun. The most impressive local supershell GSH 238+00+09 with the mass $\sim 2.7 \times 10^6$ $M_\odot$ and radius ~ 220 pc at the distance ~ 0.8 kpc has been found by Heiles (1998). Heiles estimated the energy required to produce such a shell as 3.4×10^{52} erg, that implies some 30 supernovae to be involved. The kinematic age of the superbubble can be estimated about 10 Myr. Assuming the standard Salpeter's stellar initial mass function (IMF) and using nucleosynthetic yields of exploded massive stars e.g. (Woosley & Weaver, 1995), one may estimate the metallicity $\gtrsim 2Z_\odot$ for hot SB interior at the current stage. The hot SB gas metallicity as high as $\gtrsim 10Z_\odot$ could be expected at earlier stages of the SB evolution. Being supplied with kinetic energy from extremely powerful sources as core collapsed supernovae and winds of massive early type stars, SBs should be very plausible sites of nonthermal particle acceleration (Bykov & Fleishman, 1992; Bykov, 1995; Parizot, 1998; Higdon et al., 1998). The

SBs in the local vicinity must be taken into account in the cosmic ray (CR) propagation modeling.

The structures of velocity, density and magnetic fields in a SB are rather complicated due to discrete nature of energy and momentum sources which is important during the first few million years as well as due to the interactions of the parent molecular cloud with winds and shocks. Direct observational data on the MHD motions of hot tenuous gas inside the SB are rather scarce yet. The shock turbulence formation inside the SB should occur due to multiple interactions of the shocks with the clouds following the models suggested by Bykov (1988). Simulations of 3D global dynamics of SBs in the interstellar medium (ISM) with account of the effect of the ISM magnetic field and ISM stratification have been performed by Tomisaka, (1998). Korpi et al., (1999) simulated the 3D dynamics of a SB accounting for inhomogeneous ISM structure and large scale ISM turbulence. These simulations assume continuous momentum supply from the OB star winds and supernovae as mechanical luminosity and do not resolve supersonic MHD turbulent motions inside the SB that are important for nonthermal particle production. Shocks and MHD turbulent motions inside a SB can efficiently transfer their energy to CRs because the timescale of particle acceleration in a SB is below Myr and the efficiency of energy conversion could be $\gtrsim$ 30 % at least during the first 3 Myr of the SB evolution (Bykov, 1999). Recent global models of SB evolution are based on nonrelativistic one-component perfect gas law inside the SB (e.g. Korpi et al.1999). The effect of CR acceleration inside a SB would make the gas specific heat ratio to be closer to 4/3 and provide effective energy leakage from SB interiors due to escaping of fast particles even before the radiative stage. These effects could be important for simulations of the global dynamics of SBs.

At the early stage of a SB evolution the elemental abundances inside a SB filled with a rarefied hot gas can differ strongly from the standard cosmic abundances due to ejection of matter enriched with some heavy elements from SNe and stellar winds of massive stars of (WR and OB type).

We can indicate the two most important injection processes in that model (Bykov, 1995).

(i) Creation of suprathermal nuclei by collisionless shock waves within a hot bubble. The injection produced directly by a collisionless shock depends on the rigidity. One may expect that the injection of O, C, Ne, Mg, Si nuclei (as well as of other nuclei with $A/Q = 2$) in a hot plasma of the bubble has the same efficiency as the injection of α-particles in a hydrogen-helium plasma.

(ii) A very important process of superthermal nuclei injection might be associated with fast moving knots and filaments very highly enriched with oxygen and other star burning products due to explosions of massive stars observed in some SN remnants like CAS A (Chevalier & Kirshner, 1978), Puppis A etc. These filaments moving with typical velocities of about 1000 - 5000 km s^{-1} are the sources of injected niclei of relatively low ionisation stages. Note that a neutral atom which is evaporated from a metal rich knot or a filament (even of a low velocity) and then ionized within the bubble will be picked up by supersonically moving magnetized plasma and injected into the acceleration.

Both injection processes are expected to inject metal-rich nonthermal component. The second process (ii) should dominate at the early stages of SB evolution during the first few million years and might contribute substantially later on. Since the particle acceleration time in a SB is about a few times 10^5 years (see Fig.1) one may expect to have a source of nonthermal nuclei with greatly enhanced fluxes of metals during such a period. Recent measurements of ^{59}Ni and ^{59}Co abundances in galactic cosmic rays by *CRIS* onboard the *ACE* mission indicated a long cosmic ray acceleration period of $\sim 10^5$ years after the nucleosynthesis (Wiedenbeck et al., 1999), which is in a good agreement with the SB model discussed.

It is important to note that the efficiency of SB energy (MHD) conversion to nonthermal component is about a factor of 1.5–4 higher during the earliest stage of SB evolution (the first 3-10 Myr depending on the SB scale). The energy injection from supernovae explosions into a SB is roughly time independent for about 5×10^7 years for the standard IMF (e.g. McCray & Kafatos, 1987), but an account of massive star winds contribution would increase the energy injection rate at the earliest stages. Thus the energy converted to CRs at the stage of WR stars and most massive SN explosions could be comparable to that of the CRs accelerated after the first 10 Myrs in the SB model. This results in the enhancement of the abundances of the elements produced by the most massive SNe and the WR stars in the accelerated nonthermal component. The ^{22}Ne, ^{12}C -rich CR component (e.g. Cassé & Paul, 1982; Maeder & Meynet, 1993; Meyer et al., 1997) could be naturally accounted for in the SB model. The injection mechanism (ii) - due to local ionization of the neutral atoms evaporating from knots and filaments ejected by a SN- can explain observed A/Q enhancement (see e.g. Meyer et al., 1997, for detailed analysis of observations), because of the low ionisation stage of the fast moving atoms evaporated from the metal rich knots.The quantitative prediction of CR abundances expected in the SB model depends on the details of the structure of SN ejecta. Complex kinetics of material mixing and condensation in a highly nonequilibrium SNR condition is not well established at the moment to model the structure of SN ejecta. A growing body of high resolution observations of SNRs with *ISO, Chandra, XMM* as well as high-quality optical data (e.g. Blair et al., 2000) is indicating complex structure of SN fast-moving debris of nuclear-processed material.

Later on an extended SB should be mostly a source of the standard cosmic rays (Axford, 1992; Bykov & Fleishman, 1992; Higdon, et al.1998) with the nuclei injection processes discussed above. The SB thermal plasma composition is close to the standard one at that stage, with possible excess of ^{22}Ne, etc. in the nonthermal component accelerated at the previous stage. The ISM dust grains are also contributed like in the scenario of galactic CRs acceleration by isolated SNR shocks developed by Meyer et al.(1997) and Ellison et al.(1997).

4.1. NONTHERMAL PROCESSES IN SUPERSHELLS

SBs might manifest themselves as a class of galactic objects with greatly enhanced fluxes of nonthermal nuclei with a non-standard composition. At a certain stage of their evolution, depending also upon the environmental conditions (which are different for early galaxies) SBs could be treated as sources of low-energy nonthermal nuclei. Nucleosynthesis and spallation reactions due to interactions of accelerated nonthermal nuclei with the ambient medium would be an efficient source of light elements (Cassé et al., 1995; Bykov, 1995; Ramaty et al., 1996, 2000; Duncan et al., 1997; Vangioni-Flam et al., 1998; Parizot & Drury, 1999; Fields et al., 2000; Parizot, 2000; Vangioni-Flam & Cassé, 2000). These reactions can drastically change the isotope composition in the supershell surrounding a SB which makes them responsible for variations of some isotope ratios observed in the ISM (e.g. Bykov, 1995). Due to diffusion of the nonthermal nuclei in the dense shell the scale of the variations could be as small as a few parsecs. Then the "starformation wave" is able to reflect the abundance variations in the next generation of stars. It is important that being supplied with a source of violent MHD motions from SNe explosions in a SB a supershell should have MHD turbulence during at least 30 Myr. The reacceleration of nuclei inside a dense supershell is an important effect to compensate strong Coulomb losses. Modeling of light element production in such supershells with account of reacceleration effect has been performed by Bykov et al.(1999). In Fig. 6 we presented a model result for production of ^{9}Be, ^{10}B, ^{11}B as a function of the supershell depth. Diffusive propagation of fast nuclei in the shell is described by models with coefficients $\kappa_0 = 3\times 10^{25}$ cm^2 s^{-1} (solid line) and $\kappa_0 = 3\times 10^{27}$ cm^2 s^{-1} (dotted line) at GeV/nucl energy. These models account for the cases of strong and moderate scattering rates of nuclei by MHD waves, respectively. Using the light element productions in supershells illustrated in Fig.6 and taking into account the SB statistics one may conclude that the superbubble model can explain the observed abundances of some light elements (^{9}Be, ^{10}B) while an extra source of ^{11}B is required. It may come from neutrino spallation of C in supernovae that formed the SB (see e.g. Woosley & Weaver, 1995).

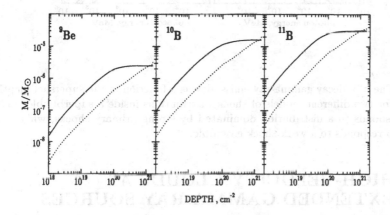

Figure 5 The light element deposition as a function of the supershell depth calculated for two different diffusion regimes (Bykov et al., 1999) (see text). The supershell surrounds a SB of R = 220pc.

Nuclear interaction lines are a natural test for observational diagnostic of the nonthermal nuclei component. Having in mind the efficiency of power conversion from the nonthermal nuclei to γ-ray lines to be typically below one percent, the source has to be rather nearby to be observed with the current instruments like *INTEGRAL*. The integrated line luminisity at 0.7 - 8 MeV regime is time dependent and it is typically below 10^{33} erg s^{-1} for the standard SB model. Because of the temporal evolution of nonthermal particle spectra and composition in a SB one may expect the γ-ray line spectrum to be dominated by broad lines at the early stage of the evolution while narrow lines should dominate the late evolution stages. Gamma-ray emission from pion decays in the supershell is also time dependent. We illustrated in the Fig.7 the gamma-ray emissivity due to pion decays calculated for different depths inside the shell for two different models of MHD shock distribution inside a SB. The emission can be detected with *GLAST* from sources at the distances within a few kpc.

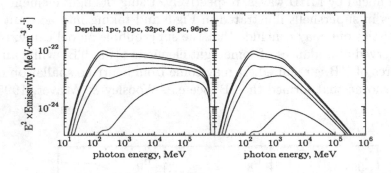

Figure 6 The π^0 - decay gamma-ray emissivity as a function of the supershell depth calculated for two different models of shock distributions inside a superbubble. Left panel corresponds to a distribution dominated by strong primary shocks, while the right one corresponds to a weak shock ensemble.

5. HIGH-VELOCITY CLOUDS AS EXTENDED GAMMA-RAY SOURCES

The interaction of High-Velocity Clouds (HVCs) with the matter and electromagnetic fields of the lower galactic halo (we consider Complex M to be located between 1.5 and 4.4 kpc from the galactic plane) seems a natural source of non-thermal emission.

For the adopted temperature $T \sim (2-4) \times 10^5$ K, a number density $n \sim 3 \times 10^{-3}$ cm^{-3}, and a magnetic field strength $B \sim 1\mu$G, it follows that the adiabatic sound velocity is $\sim 55 - 75$ km s^{-1} and the local Alfvén velocity is ~ 35 km s^{-1}. In such an environment, HVCs with velocities like that of the M IV cloud (~ 150 km s^{-1}) can be expected to produce strong local disturbances and bow shocks with moderate magnetosonic Mach numbers M $\sim$ 2–3 for both quasi-parallel and quasi-perpendicular waves. This is sufficient to consider the shock waves to be supercritical. The basic characteristics of supercritical collisionless shock waves are fairly well understood from numerical simulations with hybrid codes, which treat the protons as particles and the electrons as a fluid. The spectra of accelerated electrons cannot be obtained from a pure hybrid code, so we proposed a method to calculate the electron distribution function in the vicinity of a quasi-parallel shock with moderate Mach number (Bykov & Uvarov, 1999). It was shown that the vortical fluctuations of electric fields induced by random fluxes of ions in the supercritical shock transition region are responsible for a stochastic pre-acceleration and heating on scales of the order of the collisionless shock width.

Blom et al.(1997) described a quantitative model that attributes the observed X-ray and γ-ray emission from Complex M to bremsstrahlung from locally accelerated electrons interacting with the HVC gas using plausible parameter values.

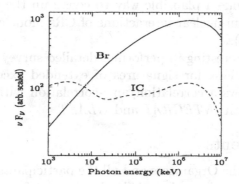

Figure 7 Typical νF_ν gamma-ray flux (arbitrarily scaled) expected from nonthermal electrons accelerated by a large scale shock interacting with galactic HI disc. Bremsstrahlung emission is indicated by a solid line, while inverse Compton is a dashed line.

They concluded that the γ-radiation observed by *GRO COMPTEL* from the sky region between the HVC Complexes M and A may be largely due to non-thermal emission arising from HVC interactions with the ambient matter. Supporting arguments were:

- The γ-ray emission seen from the general direction of HVC complexes cannot be attributed to a single source and is probably partly of diffuse nature. This interpretation is supported by simulations and model fitting.

- Enhanced X-ray emission is seen by ROSAT in the same sky area. The shock acceleration model (see Blom et al.(1997)) predicts non-thermal X-ray and γ-ray fluxes in agreement with the *ROSAT* and *GRO COMPTEL* detections. At least a part of the observed γ-ray emission is indeed non-variable, which is expected in the framework of shock model.

The positional anti-correlation of the HVC complexes A,M and C with the γ-ray excess was pointed out by Blom et al.(1997). We also modelled nonthermal emission due to posible interaction of HVCs with the galactic disc. The resulting spectrum (Fig. 7) dominated by electron bremsstrahlung and inverse Compton scattering is very hard and peak at GeV regime with an energy flux about 10 eV cm^{-2} s^{-1} from the inner radian. Some excesses of pion decay emission may also be produced, however, in the event of shock acceleration of protons due to HVC - disc interactions. These results, in comparison with the recent measurements of hard diffuse emission from the inner radian (e.g. Strong & Mattox, 1996; Bloemen et al.1997; Valinia & Marshall, 1998; Kinzer, Purcell & Kurfess, 1999) provide a plausible way to constrain the diffuse component originating from nuclear interactions of CRs modeled in details by Strong et al., (2000).

It will be very interesting to perform a detailed survey of the galactic diffuse emission to look for signatures of extended excess gamma-ray emission which shows a correlation/anti-correlation with high velocity cloud complexes with *INTEGRAL* and *GLAST*.

Acknowledgments

I am grateful to the Organizers and all the participants for a most enjoyable school. The author acknowledged a support from the organizing commetee. This work was supported by INTAS/ESA 99-1627 grant.

References

Allen, G. E. et al.(1997) Evidence of X-Ray Synchrotron Emission from Electrons Accelerated to 40 TeV in the Supernova Remnant Cassiopeia A *Astrophys. J.* **487**, L97-L100.

Asaoka I., & Aschenbach B. (1994) An X-ray study of IC443 and the discovery of a new supernova remnant by ROSAT *Astron. Astrophys* **284**, 573-582.

Axford, W. I. (1992) 'Particle Acceleration on Galactic Scales', In: *Particle Acceleration in Cosmic Plasmas*, eds.G. P. Zank, T.K.Gaisser, AIP Conf. Proc. **264**, 45-56.

Baring M.G., Ellison D.C., Reynolds S.P., et al. (1999) Radio to Gamma-ray emission from shell-type supernova remnants: predictions from nonlinear shock acceleration models, *Astrophys. J.* ,**513**, 311-338.

Berezinskii, V. S.; Bulanov, S. V., Dogiel, V. A., Ginzburg, V.L., Ptuskin, V. S. (1990) Astrophysics of cosmic rays, North-Holland (Amsterdam).

Blair, W. P. et al.(2000) *HST* Observations of Oxygen-Rich Supernova Remnants in the Magellanic Clouds II *Astrophys. J.*, **537**, 667-689.

Bleeker, J.A.M. et. al. (2001) Cassiopeia A: On the origin of the hard X-ray continuum and the implication of the observed ion Oviii Ly-alpha/Ly-beta distribution, *Astron. Astrophys*, **365**, L225-L230.

Blitz, L. (1993) Giant Molecular Clouds In: *Protostar & Planets III*, E.H. Levy, J.I.Lunine, Tucson: Univ. of Arizona, 125-161.

Bloemen, H. et al.(1997) COMPTEL spectral study of the inner Galaxy. In: *AIP Conf. Proc.* **v.410**, 1074-1079.

Blom, J. J. et al.(1997) COMPTEL detection of low-energy gamma rays from HVC Complex M and A regions? *Astron. Astrophys*, **321**, 288-292.

Bocchino, F., Bykov, A. M. (2000) Hard X-ray emission from IC443: evidence for a shocked molecular clump? *Astron. Astrophys*, **362**, L29-L32.

Burton M.G., Geballe T., Brand P.W.J.L., Webster, A.S. (1988), Shocked molecular hydrogen in supernova remnant iC 443 *MNRAS*, **231**, 617-634.

Bykov, A.M. (1988) A Model for the Generation of Interstellar Turbulence, *Sov. Astron. Lett.* **14**, 60-63.

Bykov, A. M. (1995) Nucleosynthesis from nonthermal particles, *Space Sci. Rev.* **74**, 397-406.

Bykov, A. M. (1999) Nonthermal Particles in Star Forming Regions, In: ASP Conf. Series, **171**, 146-153.

Bykov A.M, Chevalier R.A, Ellison D.C, Uvarov, Yu.A. (2000) Nonthermal emission from a supernova remnant in a molecular cloud, *Astrophys. J*, **538**, 203-216.

Bykov, A. M., Fleishman, G. D. (1992) On non-thermal particle generation in superbubbles, *MNRAS* **255**, 269-275.

Bykov, A. M, Gustov, M. Yu., Petrenko, M. V. (1999) Energetic-Nuclei Acceleration and Interactions in the Early Galaxy, In: *Astronomy with Radioactivities*, eds. R.Diehl, D.Hartmann, MPE Report **274**, 241-248.

Bykov, A. M., Toptygin, I. N. (1987) Effect of Shocks on Interstellar Turbulence and Cosmic-Ray Dynamics, *Astrophys. Space Sci.*, **138**, 341-354.

Bykov A.M, Uvarov, Yu. A. (1999) Electron kinetics in collisionless shock waves, *JETP*, **88**, 465-475.

Cassé, M., Lehoucq, R., & Vangioni-Flam, E. (1995) Production and Evolution of Light Elements in Active Star-Forming Regions, *Nature* **373**, 318-321.

Cassé, M., Paul, J. (1982) On the stellar origin of the ^{22}Ne excess in cosmic rays, *Astrophys. J.* **258**, 860-863.

Cesarsky D., Cox P., Pineau de Forets G., et al. (1999) ISOCAM spectro-imaging of the H_2 rotational lines in the supernova remnant IC443, *Astron. Astrophys*, **348**, 945-949.

Chan, K., Lingenfelter, R.E. (1993) Positrons from supernovae, *Astrophys. J.* **405**, 614-636.

Chen, C. H, Chu, Y.H., Gruendl, R. A., Points, S. D. (2000) HST WFPC-2 Imaging of Shocks in Superbubbles, *Astron. J.* **119** , 1317-1324.

Chevalier, R. A. (1977) The evolution of supernova remnants. V - Cosmic rays in the dense shell, *Astrophys. J.*, **213**, 52-57.

Chevalier, R. A. (1999) Supernova remnants in molecular clouds, *Astrophys. J.* **511**, 798-811.

Chevalier, R. A. (2000) A model for the X-ray luminocity of pulsar nebulae, *Astrophys. J.* **539**, L45-L48.

Chevalier, R. A., Kirshner, R. P. (1978) Spectra of Cassiopeia A. II - Interpretation *Astrophys. J.* **219**, 931-941.

de Jager O., Mastichiadis, A. (1997) A relativistic bremsstrahlung/inverse Compton origin for 2EG J1857 +0118 associated with supernova remnant W44, *Astrophys. J.* **482**, 874-880.

Duncan, D. et. al. (1997) The Evolution of Galactic Boron and the Production Site of the Light Elements, *Astrophys. J.* **488**, 338-349.

Drury, L.O'C., Aharonian, F.A, Völk, H.J. (1994) The gamma-ray visibility of supernova remnants. A test of cosmic ray origin, *Astron. Astrophys*, **287**, 959-971.

Ellison, D. C., Drury, L.O'C., & Meyer, J. P. (1997) Galactic Cosmic Rays from Supernova Remnants- II, *Astrophys. J.* **487**, 197-217.

Esposito, J.A. et. al. (1996) EGRET Observations of Radio-bright Supernova Remnants, *Astrophys. J.* **461**, 820-827.

Fegan, S. (2001) TeV observations of SNRs and unidentified sources, *Astro-ph/0102324*.

Fields, B. D., Olive, K. A., Vangioni-Flam, E., Cassé, M. (2000) Testing Spallation Processes With Beryllium and Boron, *Astrophys. J.* **540**, 930-945.

Gaisser, T. K., Protheroe, R. J., Stanev, T. (1998) Gamma-Ray Production in Supernova Remnants *Astrophys. J.* **492**, 219-234.

Grenier, I. A. (2000) EGRET Unidentified Sources, *In: AIP Conf. Proc.* **515**, 261-275.

Guessoum, N., Ramaty, R, Lingenfelter, R.E. (1991) Positron annihilation in the interstellar medium, *Astrophys. J.* **378**, 170-180.

Iyudin, A.F. et al. (1999) COMPTEL All-Sky Survey in ^{44}Ti Line Emission, *Astrophysical Letters and Communications*, **38**, 383-386.

Heiles, C. (1998), Whence the local bubble, *Astrophys. J.* **498**, 698-703.

Higdon, J. C., Lingenfelter, R. E., Ramaty, R. (1998), Cosmic Ray Acceleration from Supernova Ejecta in Superbubbles, *Astrophys. J.* **509**, L33-L36.

Keohane J.W., Petre R., Gotthelf, et al. (1997) A Possible Site of Cosmic Ray Acceleration in the Supernova Remnant IC 443, *Astrophys. J.* **484**, 350-362.

Kinzer, R. L., Purcell, W. R., Kurfess, J.D. (1999) Gamma-Ray Emission from the Inner Galactic Ridge *Astrophys. J.*, **515**, 215-225.

Korpi, M. J., Brandenburg, A., Shukurov, A. & Tuominen, I. (1999) Evolution of a superbubble in a turbulent, multi-phased and magnetized ISM, *Astron. Astrophys*, **350**, 230-239.

Koyama, K., Petre, R., Gotthelf, E.V et. al. (1995) Evidence for Shock Acceleration of High-Energy Electrons in the Supernova Remnant SN 1006, *Nature* **378**, 255-258.

McCray, R., Kafatos, M. (1987) Supershells and Propagating Star Formation, *Astrophys. J.* **317**, 190-196.

Maeder, A., Meynet, G. (1993) Isotopic anomalies in cosmic rays and the metallicity gradient in the Galaxy, *Astron. Astrophys.* **278**, 406-414.

Meyer, J. P., Drury, L.O'C., Ellison, D. C. (1997) Galactic Cosmic Rays from Supernova Remnants- I, *Astrophys. J.* **487**, 182-196.

Milne, P., Leising, M. D., The, L. S. (1999) Galactic positron production from supernovae, In: *Astronomy with Radioactivities*, eds. R.Diehl, D.Hartmann, MPE Report **274**, 189-196.

Parizot, E. (1998) The Orion Gamma-ray emission and the Orion-Eridanus bubble, *Astron. Astrophys.*, **331**, 726-736.

Parizot, E. (2000) Superbubbles & the Galactic evolution of Li,Be,B, *Astron. Astrophys.* **362**, 786-7798.

Parizot, E., Drury, L. (1999) Superbubbles as the Source of ^{6}Li, Be and B in the Early Galaxy, *Astron. Astrophys.* **349**, 673-684.

Petre R., Szymkowiak A.E., Seward F.D., et al. (1988) A comprehensive study of the X-ray structure and spectrum of IC 443, *Astrophys. J.*, **335**, 215-238.

Preite-Martinez, A., Feroci, M., Strom, R.G., Mineo, T. (1999) Hard X-ray emission from IC443: the BeppoSAX view, *In: AIP Conf. Proc.* **510**, 73-76.

Ramaty, R., Kozlovsky, B., & Lingenfelter, R. E. (1996) Light Isotops, Extinct Radioisotopes and Gamma-Ray Lines from Low-Energy Cosmic-Ray Interactions, *Astrophys. J.* **456**, 525-540.

Ramaty, R., Scully, S.T., Lingenfelter, R. E. & Kozlovsky, B. (2000) Light-Element Evolution & Cosmic-Ray Energetics, *Astrophys. J.* **534**, 747-756.

Reach, W. T., Rho, J. (2000) Infrared spectroscopy of molecular supernova remnants, *Astrophys. J.* **544**, 843-858.

Reynolds, S. P. (1998) Models of Synchrotron X-Rays from Shell Supernova Remnants, *Astrophys. J.* **493**, 375 - 386.

Seo, E. S., Ptuskin, V. S. (1994) Stochastic reacceleration of cosmic rays in the interstellar medium, *Astrophys. J.* **431**, 705-714.

Strong, A. W. & Mattox, J.R. (1996) Gradient model analysis of EGRET diffuse Galactic γ-ray emission, *Astron. Astrophys.* **308**, L21-L24.

Strong, A. W., Moskalenko, I. V., Reimer, O. (2000) Diffuse Continuum Gamma Rays from the Galaxy, *Astrophys. J.* **537**, 763-784.

Sturner S.J., Skibo J., Dermer, C., Mattox, J.R. (1997) Temporal Evolution of Nonthermal Spectra from Supernova Remnants, *Astrophys. J.*, **490**, 617 - 632.

Tanimori, T. et al.(1998) Discovery of TeV Gamma Rays from SN 1006, *Astrophys. J.*, **497**, L25-L28.

The, L. S. et. al. (1996) CGRO/OSSE observations of the Cassiopeia A SNR, *Astron. Astrophys. Suppl.* **120**, 357 -360.

Thielemann, F.-K., Nomoto, K., Hashimoto, M. (1996) Core-Collapse Supernovae and Their Ejecta *Astrophys. J.* **460**, 408-436.

Valinia, A., Marshall, F. E. (1998) RXTE Measurement of the Diffuse X-Ray Emission from the Galactic Ridge, *Astrophys. J.* **505**, 134 - 147.

van Dishoeck E.F., Jansen D., & Phillips T. (1993) Submillimeter observations of the shocked molecular gas associated with the supernova remnant IC 443, *Astron. Astrophys.* **279**, 541 - 566.

Vangioni-Flam, E., Cassé, M. (2000) LiBeB Production and Associated Astrophysical Sites, *Astro-ph/0001474*.

Vangioni-Flam, E., Ramaty, R., Olive, K. & Cassé, M. (1998) Testing the primary origin of Be and B in the early galaxy, *Astron. Astrophys.* **337**, 714-720.

Vink, J. et al.(1999) A comparison of the X-ray line and continuum morphology of Cassiopeia A *Astron. Astrophys.*, **344**, 289-294.

Wang Z.R., Asaoka I., Hayakawa S., Koyama, K. (1992) Hard X-rays from the supernova remnant IC 443, *PASJ*, **44**, 303-308.

Wiedenbeck, M. E. et. al. (1999) Constraints on the time delay between nucleosynthesis and Cosmic-Ray Acceleration, *Astrophys. J.* **523**, L61-L64.

Winkler, C. (1999) INTEGRAL: The current status. *Astroph. Lett. & Comm.* **39**, 309 -316.

Woosley, S. E., Weaver, T. A. (1995) The evolution and Explosion of Massive Stars II, *Astrophys. J. Suppl.* **101**, 181-235.

COSMIC RAY INTERACTIONS IN THE GALACTIC CENTER REGION

P.L. BIERMANN[1,2], S. MARKOFF[1], W. RHODE[3], E.-S. SEO[4]

[1] *Max Planck Institute for Radioastronomy,*
D-53121 Bonn, Germany, and
[2] *Department of Physics and Astronomy,*
University of Bonn, Bonn, Germany
[3] *Physics Dept., Univ. of Wuppertal,*
D-42097 Wuppertal, Germany
[4] *Institute for Physical Science and Technology, Univ. of Maryland,*
College Park, MD 20742, USA
plbiermann@mpifr-bonn.mpg.de
http://www.mpifr-bonn.mpg.de/div/theory

Abstract. EGRET data on the Gamma ray emission from the inner Galaxy have shown a rather flat spectrum. This spectrum extends to about 50 GeV in photon energy. It is usually assumed that these gamma-rays arise from the interactions of cosmic ray nuclei with ambient matter. Cosmic Ray particles have been observed up to $3\,10^{20}$ eV, with many arguments suggesting, that up to about $3\,10^{18}$ eV they are of Galactic origin. Cosmic ray particles get injected by their sources, presumably supernova explosions in the Galaxy. Their injected spectrum is steepened by diffusive losses from the Galaxy to yield the observed spectrum. As cosmic ray particles roam around in the Galactic disk, and finally depart, they encounter molecular clouds and through p-p collisions produce gamma rays from pion decay. The flux and spectrum of these gamma rays is then a clear signature of cosmic rays throughout the Galaxy. Star formation activity peaks in the central region of the Galaxy, around the Galactic Center. Looking then at the gamma ray spectrum of the central region of our Galaxy yields clues as to where the cosmic ray particles interact, and with what spectrum. We find a best fit for a powerlaw spectrum of cosmic rays with a spectrum of 2.34, rather close to the suggested injection spectrum for supernovae which explode into their own winds. This suggests that most cosmic ray interaction happens near the sources of injection; it has already been shown elsewhere that this is consistent with the spectrum of cosmic ray nuclei

M. M. Shapiro et al. (eds.), Astrophysical Sources of High Energy Particles and Radiation, 65–79.
© 2001 *Kluwer Academic Publishers. Printed in the Netherlands.*

derived from spallation. We point out several consequences of such a picture: i) cosmic ray heating and ionization should be strong in the Galactic Center region, ii) the knee of the cosmic ray spectrum should reappear in the gamma-spectrum, detectable with future instruments and iii) the neutrino emission should be a fair bit stronger at TeV energies than heretofore expected, and iv) Neutrons offer a chance to scan the Galaxy spatially; we go through this latter possibility in some detail.

1. Introduction

Cosmic rays were discovered early in the 20th century (Hess, 1912; Kohlhörster, 1913). The spectrum of Galactic cosmic ray particles extends to probably $3\,10^{18}$ eV; the various contributions have been reviewed extensively by (Wiebel-Sooth & Biermann, 1999), and basic fits to the data for the various chemical elements have been given in (Wiebel-Sooth $et\ al.$, 1998). These cosmic ray particles interact with interstellar matter, and so spallate to produce the secondary nuclei (see, e.g., (Garcia-Munoz $et\ al.$, 1987)), as well as the gamma ray emission above GeV photon energies (see, e.g., (Stecker, 1971)).

There is a recent AGASA paper (Hayashida $et\ al.$, 1998) suggesting that neutrons near 10^{18} eV are seen from both the Galactic Center region as well as the Cygnus region; this is a correlation in arrival direction. There is now confirmation from the SUGAR-array data (Bellido $et\ al.$, 2000). The data suggest neutrons, because protons at such an energy cannot get through the Galaxy on a straight line path with its magnetic field, since at that energy the Larmor radius is about 1 kpc. On the other hand, one might wonder, whether protons in their meanderings through the Galaxy just fortuitously have a small excess coming from those two directions; with the present statistics this is hard to check. The data cannot also easily be explained as gamma ray photons, arising from pion-decay following p-p collisons, since then the correlation would be stronger at lower energy even, where nothing is seen. Of course, neutrons are just one more channel in p-p collisions.

It is worth remembering, that the Galactic Center region, as well as the Cygnus region are prime candidate regions for star formation and supernova activity in the Galaxy, as clearly shown in radio, far-infrared and gamma ray data. Neutrons at EeV energies also can get here within their life time from the Galactic Center, but not at much lower energy; at higher energy they should be very little production of neutrons from nuclear collisions, since almost all plausible sources of Galactic cosmic rays cut off at around $3\,10^{18}$ eV. This leaves neutrons as the most probable origin of these events.

If this result is accepted with the interpretation as neutrons, then it is a consistency argument that cosmic rays are indeed Galactic to 10^{18} eV. However, it is also immediately obvious, that the flux of cosmic ray particles required to produce so many neutrons as suggested by the AGASA data needs to be rather high relative to the flux observed at Earth.

The first discussions of modern shock acceleration to explain cosmic ray energies stem from (Fermi, 1949; Fermi, 1954). Cosmic Ray Physics has beend extensively reviewed and developed in (Shklovskii, 1953; Ginzburg, 1953a; Ginzburg, 1953b; Hillas, 1984; Blandford & Eichler, 1987; Jones & Ellison, 1991; Drury, 1983; Ginzburg, 1993; Zatsepin, 1995; Ginzburg, 1996). Spallation and abundances have been reviewed by (Reeves, 1974; Reeves, 1994). Important books on cosmic rays are (Ginzburg & Syrovatskij, 1963; Hayakawa, 1969; Ginzburg & Syrovatskij, 1969; Berezinsky et al., 1990; Gaisser, 1990).

By now historical, but still challenging and inspiring compilations on cosmic rays and stochastic processes useful for understanding cosmic ray transport are (Heisenberg, 1953; Flügge, 1961; Rosen, 1969; Wax, 1954).

For the transport of cosmic rays we require an understanding of turbulence: The turbulence spectrum in the interstellar medium and its understanding can also be traced back quite a bit (Prandtl, 1925; Karman & Howarth, 1938) and (Kolmogorov, 1941a; Kolmogorov, 1941b; Kolmogorov, 1941c; Obukhov, 1941), with more recent work described in (Heisenberg, 1948; Kraichnan, 1965; McIvor, 1977; Achterberg, 1979; Matthaeus & Zhou, 1989; Biskamp & Müller, 2000), and some reviews in (Sagdeev, 1979; Rickett, 1990; Goldstein et al., 1995). Much of todays thinking goes back to the papers of the 1940ies.

The interstellar medium of relevance here is the hot phase (Snowden et al., 1997; Kaneda et al., 1997; Valinia & Marshall, 1998; Pietz et al., 1998).

Finally, we need an understanding of Galactic magnetic fields, which were most recently reviewed by (Kronberg, 1994; Beck et al., 1996; Beck, 2000).

The presently available evidence supports the point of view, that there is a thick disk of hot gas, magnetic fields and cosmic rays, with a vertical exponential scale of about 2 kpc, and a radial exponential scale of about 5 kpc.

In this paper we wish to review first the interpretations of the cosmic ray particles in the Galactic Center region, and then show that a simple concept may be sufficient to explain the data from EGRET, which show a rather flat gamma ray spectrum. This is a severe test for any theory of cosmic ray origin. This then leads to considerable ionization and heating by low energy cosmic rays in the Galactic Center region. Related results have been reported by us at the Hirschegg conference (Biermann, 1998b), the

Tucson Galactic Center conference (Rhode *et al.*, 1998), at the 26th ICRC meeting (Markoff *et al.*, 1999; Rhode *et al.*, 1999), and in CRIX (Biermann *et al.*, 2001).

2. Origin of high energy cosmic rays

Our Galactic Center harbors a black hole, which probably went through many activity episodes during its growth. Therefore we want to ask first whether this activity could possibly explain high energy cosmic rays, and as a consequence gamma rays.

(Biermann & Strittmatter, 1987) have shown that radio galaxy hot spots can accelerate protons to about 10^{21} eV. Scaling this result with the power of the underlying source and using the jet/disk-symbiosis picture developed by Falcke et al. ((Falcke & Biermann, 1995; Falcke & Biermann, 1999) and other papers) we obtain for the maximum proton energy

$$E_{p,max} = 6.7 \, 10^{20} \, Q_{jet,46}^{1/2} \, eV \tag{1}$$

where $Q_{jet,46}$ is the power of the jet in units of 10^{46} erg/s. The most extreme inferred jet luminosity is about $3 \, 10^{47}$ erg/s, and so energies up to

$$E_{p,max} = 4. \, 10^{21} \, eV \tag{2}$$

appear possible (Biermann, 1998a). Therefore radio galaxies and their various counterparts such as compact radio quasars (Farrar & Biermann, 1998) are clearly a suitable source for high energy cosmic rays. The jet-disk symbiosis does seem to work down to stellar size black holes (Falcke & Biermann, 1999), and so we may be permitted to use it for lower powers of a proposed source.

The jet power of our Galactic Center, assuming that the compact radio source does signify the existence of a jet, is

$$Q_{jet} = 5 \, 10^{38} \, erg/s \tag{3}$$

and so $E_{p,max} = 1.5 \, 10^{17}$ eV. Therefore the activity of our central black hole is insufficient to produce neutrons at 10^{18} eV, and so is unlikely to help to explain the data.

We can expand upon this type of argument and ask, whether there could have been any time in the past, when our Galactic Center black hole could have been sufficiently powerful to do this: Its mass is fairly reliably estimated to be $2.6 \, 10^6$ solar masses and so its maximum power available for the jet is about $3 \, 10^{43}$ erg/s, using 10 % of the Eddington limit as a guideline. This produces a maximum energy of particles accelerated in the jet of $3 \, 10^{19}$ erg/s; however, this is a spatial limit, and close to a powerful activity center the maximum particle energy is limited by losses to much lower energies (Biermann & Strittmatter, 1987). Therefore its cannot be ruled out completely, that our Galactic Center at some time in the past could have produced particles directly near 10^{18} eV, but it is rather unlikely. Within the travel time difference of neutrons at 10^{18} eV compared to light, basically a minute fraction of 20,000 years (Biermann, 1997b), this is clearly completely impossible.

The Galactic Center region does harbor many interesting binary systems, some of which are referred to as mini-quasars; however, there again, their power is just not sufficient to explain particles near 10^{18} eV.

There is a new hot disk model, where weakly relativistic protons produce various secondaries in their interaction in the disk (Mahadevan, 1998), but this model also cannot explain any particles at 10^{18} eV.

Therefore we propose to explore in the following the acticity and cosmic ray injection properties of supernovae focussing on those supernovae that explode into their own stellar winds (Biermann, 1997a).

3. Galactic Cosmic Rays

In a series of papers Biermann et al. (from 1993, (Biermann, 1993a; Biermann & Cassinelli, 1993; Biermann & Strom, 1993; Stanev et al., 1993; Biermann et al., 1995; Wiebel-Sooth et al., 1998; Biermann et al., 2001; Rachen & Biermann, 1993; Rachen et al., 1993)) have proposed that cosmic rays get injected from three sites predominantly:

− Supernovae that explode into the interstellar medium, *Interstellar medium supernovae, or ISM-SN.*
− Supernovae that explode into their own stellar wind, *wind-supernovae, or wind-SN.*
− Radio galaxies and compact radio quasars.

The predictions of this model have been described and subjected to various tests in some reviews (Biermann, 1993c; Biermann, 1995; Biermann, 1997a; Biermann, 1998b; Biermann, 1997b; Wiebel-Sooth & Biermann, 1999) and we summarize here briefly:

The cosmic ray particles which interact the most derive from the wind-supernovae. This happens since massive stars explode close to their birthplace, where the original material is still around from which they formed (Tinsley & Biermann, 1974). Their source spectrum has been predicted to be $E^{-2.33-0.02\pm0.02}$ below the knee at $5\,10^{15}$ eV particle energy. For particles above the knee the corresponding prediction is $E^{-2.74-0.07\pm0.07}$. This peculiar way of writing the expected theoretical error range signifies an asymmetric error distribution, extending here from 2.33 to 2.37

in the first case; this is due to the finite wind velocity range as compared to the Supernova shock velocity. The bend (to explain the knee) has been predicted to be at 600 Z TeV, where Z is the charge of the chemical element nucleus under consideration, and the cutoff is near 100 Z PeV. Because the energy of the bend depends on the charge, the element abundance gets heavier at the knee, as noted already by (Peters, 1959; Peters, 1961). This energy to charge ratio is uncertain by probably a factor up to 3.

To understand the concept of wind-supernovae we must remember that massive stars have a range of wind properties, resulting in different wind shells:

- First, stars above 8 solar masses, but below about 15 solar masses explode as supernovae, but do so directly into the interstellar medium. This is the classical case.
- Then stars above 15 but below about 25 solar masses explode into their wind, and that wind may be powerful enough to sweep up interstellar material into a shell mixed with shocked wind material, but the shell is still rather thin. The material in this shell is enriched in Helium from the nuclear reactions inside the star.
- Finally, above about 25 solar masses the winds get very powerful, leading to Wolf-Rayet stars, and is heavily enriched. Then the wind-shell may be rather thick.

For the more massive stars it is likely that there are many clouds still left in the surrounding region from the phase of star formation; therefore the supernova shock meets first the wind, then the shocked wind material, then the shocked interstellar medium material, and finally the debris from the star forming phase.

The diffusion through the Galaxy is with a wavefield that derives from a Kolmogorov spectrum of interstellar turbulence, and so the cosmic ray spectrum is steepened by 1/3 (see, e.g., (Biermann, 1995)), to yield $E^{-2.67-0.02\pm0.02}$ below the knee.

This result can be directly compared with the data for Helium through Iron, which give a best fit of $E^{-2.64\pm0.04}$ (Wiebel-Sooth *et al.*, 1998).

Many of the most important checks will come from isotopic abundace ratios (Connell & Simpson, 1997; Connell, 1998; Connell *et al.*, 1998; Simpson & Connell, 1998). The relevance of WR-stars has been recognized for some time, e.g. (Prantzos *et al.*, 1987; Yanagita *et al.*, 1990).

However, where the interaction really happens is not clear. There are many possible points of view on this question, but two conceptually simple notions are documented in the literature:

3.1. CR-INTERACTION IN THE ISM

First, there is the standard model of cosmic ray acceleration in the interstellar medium (Drury, 1983); the latest development of this model has been discussed as regards the chemical abundances by (Ramaty et al., 1997; Ellison et al., 1997; Meyer et al., 1997). The transport of cosmic rays in this context has been described by (e.g., (Garcia-Munoz et al., 1977; Garcia-Munoz et al., 1987)), and it adopts the point of view that the average cosmic rays interact with the average interstellar matter. In such a picture the gamma rays should have a spectrum that nicely fits the average cosmic ray spectrum, near $E^{-2.7}$. In such a picture the secondary to primary ratio of spallation products such as Bor derived from Carbon spallation gives the spectrum of interstellar irregularities with an implied energy dependence of the leakage time scale as $E^{-0.6}$. One problem with this argument is that there is little evidence for such a spectrum of irregularities (Biermann, 1995), but it does give a good fit. This latter point could be remedied with reacceleration as proposed by (Seo & Ptuskin, 1994), and then could be consistent with an interstellar spectrum of irregularities as derived from a Kolmogorov-type turbulence.

The standard model has been explored to fit the gamma ray spectrum in two papers, using the best EGRET data (Hunter et al., 1997; Mori, 1997). The standard model fails by a wide margin.

The failure is due to the spectrum. The observed spectrum is just too flat.

There are several ways out of this conundrum.

First, one might argue that the Monte-Carlo codes used to predict the gamma rays are not good enough. This is what (Mori, 1997) has tried. The uncertainties in the Monte-Carlos are not sufficient to explain the flat spectrum.

Second, one might argue, that pion decay does not explain the data. This has been tried by (Pohl et al., 1997). They suggest that the spectrum can be partially derived from Inverse Compton off a population of energetic electrons produced by pulsars created in supernova exlosions. In the progressive leakage of the electrons as a function of energy and time from injection the observed spectrum can be matched. What speaks for such an interpretation is the large latitude extent of the gamma emission. If the new AGASA data are correctly interpreted with arising from energetic neutrons, then cosmic ray nucleon interaction is about as high as can possibly be, and so it is difficult to see how to avoid gamma ray production from pion decay being a strong contributor.

Conversely, a success of an alternative model that also explains other data would be very helpful, and this is what we have tried.

3.2. NEAR-SOURCE CR-INTERACTION

Second, there is the notion that most cosmic ray interaction happens near the source (Biermann, 1998b). In such a picture the spallation leading to secondary nuclei production happens in the shell around the stellar wind and in the surrounding high density region, when the supernova induced shock smashes through that material. This leads to an energy dependence for the local leakage time scale of $E^{-0.55}$ (Biermann, 1998b); however, this happens only when the shell is thick enough to allow diffusive interaction to be dominant over convective losses. This latter process is likely to dominate for the more abundant, but thinner shells around slightly lower mass stars. Therefore, in such a picture we expect that the more common stars in the range 15 to 25 solar masses would produce most gamma rays. And as a corollary we expect that the gamma rays should correspond to the injection spectrum.

Therefore we have adopted a simple powerlaw model for the Bremsstrahlung and inverse Compton contribution and then fitted the data with two main parameters in mind, quite in the spirit of (Gaisser *et al.*, 1998):

First, and most importantly, the power law spectrum of the cosmic rays; this law is given as a strict power law in momentum from many MeV to many GeV and beyond in energy.

Second, a smaller point of relevance is the content of heavy elements. Since the injected spectrum is a power law in particle momentum, and not of energy, there is change in spectrum near threshold, when transforming to and from the center of mass system for the collision calculation. Near the pion production threshold the cosmic ray particle energy is not too far from its rest mass, so that subtle effects due to mass become relevant.

We have tried three different Monte-Carlo codes from CERN and Fermi-Lab to do this analysis, and we have adopted for this work the code FLUKA (specifically the version GEANT3.21/FLUKA from the CERN library). This code could readily be adapted to include the subtle effects of Helium for instance.

The data fits show a clear minimum in the χ^2-distribution at spectral index 2.34, with the lowest minimum for the highest Helium abundance used, 50 %. This minimum is at a quite acceptable level of χ^2_{red} of 1.5.

The next step will be to check how in high energy we can push this model of cosmic ray interaction near the source; there are severe limits now on the inner Galaxy from the CASA-MIA experiment (Glasmacher *et al.*, 1999a; Glasmacher *et al.*, 1999b; Ong, 1998), and there are checks from observed supernova remnants (Gaisser *et al.*, 1998; Protheroe & Stanev, 1997).

4. Consequences

First of all, what the fit demonstrates is that there is a spectrum for the cosmic rays that fits their gamma ray emission. This spectrum is consistent with a power law, and is in fact quite close to the original prediction of a source spectrum for wind-supernovae.

Second, it shows that source related interaction may be worth pursuing in detail. What has not yet been done here, is a fit to the detailed isotope abundances (see the recent discussion of this point by (Westphal *et al.*, 1998)). Together with the earlier work (Biermann, 1998b; Biermann *et al.*, 2001) this means that there is a viable proposal how to explain a) the cosmic ray spectrum itself, b) the gamma ray spectrum, and c) the spectrum of spallation secondaries.

There is two recent confirmations already of these predictions, and that is a) the new finding by (Jones *et al.*, 2000), that the best data fit for the energy dependence of the ratio of Boron to Carbon has an exponent of 0.54, rather close to the prediction of 5/9, dating to 1997 (Biermann, 1998b); and b) that the source spectrum should be close to an index of 2.35 (Ptuskin *et al.*, 1999), rather close to the prediction of 1993 (Biermann, 1993c).

If this scenario could be confirmed in more detail, there are some consequences also for the stars with strong winds, long before they explode as supernovae:

- Wolf-Rayet and OB stars have shock waves running through their winds.
- These shocks accelerate electrons and produce observed radio emission (Biermann & Cassinelli, 1993).
- These shocks accelerate also protons, resulting in a steep pion decay spectrum; this spectrum is steep because the Alfvénic Machnumber of these shocks is low (Nath & Biermann, 1994b).
- These shocks also accelerate nuclei, which can give rise in collisions to spallation products in an excited nuclear state, then explaining gamma ray lines (Nath & Biermann, 1994b) from active regions of star formation.

To summarize the essential idea again, before we consider some important consequences:

- The supernova shock races through the wind.
- The shock accelerates particles.
- Cosmic ray injection of Helium, and some of the heavier elements, in addition to a small amount of Hydrogen is produced from the shocks in winds of stars of a zero age sequence mass between about 15 and 25 solar masses.

– Cosmic ray injection of elements such as Carbon and Oxygen and most heavier elements originates from this injection, for stars of a zero-age sequence mass above about 25 solar masses.

4.1. COSMIC RAY HEATING AND IONIZATION

Once outside their site of origin the protons (and other nuclei) at energies below about 50 MeV use up their ionization and heating power near their origin (Nath & Biermann, 1994a). Therefore the cosmic ray induced ionization and heating should be high in the Galactic Center region. This may be a tell-tale sign of recent supernova activity, since it is well known, that most recent supernovae, never detected, should have been in the Galactic Center region.

4.2. A GAMMA-RAY KNEE

In the standard model of interaction with the average cosmic ray spectrum the knee present in the cosmic rays observed should produce a corresponding knee in the gamma spectrum, and a steeper spectrum beyond the knee.

The knee of the cosmic ray spectrum should reappear in the gamma-spectrum also in the model introduced here, but at much higher flux: Since we suggest here, that freshly injected cosmic rays do most of their interaction near their site of origin, the knee should be visible in the gamma-ray spectrum. We predict the gamma ray spectrum to correspond to the injection spectrum of $E^{-2.33-0.02\pm0.02}$ up to the knee, corresponding to the knee at about 0.3 PeV/charge unit in particles, and thereafter to correspond to the steeper spectrum predicted for cosmic rays beyond the knee, of $E^{-2.64-0.07\pm0.07}$; the gamma ray spectrum should extend to much higher energies, and finally cut off at the photon energies corresponding to a particle energy of near 100 PeV/nucleon (Biermann, 1993a; Biermann, 1993c; Biermann, 1997a).

4.3. NEUTRINO EMISSION

The neutrino emission should be a fair bit stronger at TeV energies than heretofore expected (Berezinsky et al., 1994a; Berezinsky et al., 1994b), because the interaction is predicted here to arise from a spectrum 1/3 flatter than the average spectrum, and so at TeV energies, the emission from the Galaxy is predicted to be about an order of magnitude stronger than in previous models.

4.4. NEUTRONS: SCANNING THE GALAXY

Neutrons offer a chance to scan the Galaxy spatially. Since neutrons decay after about 1000 seconds, the different energies offer the possibility to scan the cosmic ray interaction, with EeV probing distances around 10 kpc, and lower energies distances correspondingly smaller. This means that we need first to obtain cosmic ray fluxes with sufficiently good statistics to observe the anisotropies.

The second step is to realize that local magnetic field structure should also produce such anisotropies, which we need to take into account. The magnetic field should produce a local characteristic for protons and other charged nuclei from streaming in the magnetic field (Clay, 2001). This should produce a pronounced asymmetry considering a given direction and the direction directly opposite to it. Such an asymmetry should scale with energy of the charged particles, since at any given energy the particles trace out the magnetic field structure at the corresponding mean free path for scattering. We note that the mean free path for cosmic rays can be estimated from the residence time in the disk to be

$$\lambda = 1 \, (\frac{E}{EeV})^{1/3} \, \text{kpc} \qquad (4)$$

in the context of the model presented here, where we take all scatterings of cosmic rays in the interstellar medium to be described well by a Kolmogorov spectrum of turbulence. This estimate also assures us that even at EeV energies, cosmic rays should still be approximately isotropic. This then requires the following steps:

- First we get the sky distribution of cosmic ray fluxes at energies near 10^{16} to 10^{19} eV with outmost accuracy.
- We determine an isotropic flux as a baseline.
- Then we unfold this distribution into multipole components, to obtain the proton and nuclei fluxes corresponding to the local magnetic field structure. A first step in such a direction was taken by (Clay, 2001).
- We check this unfolding procedure with the electron synchrotron emissivity, from radio maps of the sky.
- We check again with a Rotation Measure map of the sky, similar to what (Simard-Normandin & Kronberg, 1979) have done, using distant polarized radio sources.
- We try to separate gradient effects from streaming of protons from source components of neutrons; the gradient effects as discussed by (Clay, 2001) should have a very much broader spatial signature than source components of any hypothetical neutron component.

- We subtract multipole components derived from gradient effects of streaming and isotropic flux from the total sky distribution of cosmic ray fluxes, at all energies considered.
- We estimate the sky distribution of hypothetical neutron fluxes.
- We check with the gamma ray emissivity on the sky, again at different energies; this step is an important consistency check, but will require very good photon statistics at high energy.
- We check with anti-protons and positron fluxes for self-consistency.
- We build a simple model connecting the source related fluxes, dominant for gamma ray emission and neutron production in our model, in order to combine deduced neutron fluxes with observed charged particle flux. This step will indicate immediately, if the entire model approach fails, contradicting the model presented here.
- We improve the estimates of the isotropic flux, and of the multipole components, and iterate the steps outlines above, until we have converged to the limit of the statistics.
- We then may have a map of the Galaxy, outlining the sites of cosmic ray interaction in three-domensional space.

5. Conclusions

We have introduced a model in which most cosmic rays interact near their sources, developed over the last five years (Biermann, 1998b; Biermann *et al.*, 2001).

We also propose a way to scan the galaxy spatially, using high energy neutrons. This requires an airshower array, which covers the energy range from 10^{16} eV through 10^{19} eV, leading to a spatial three-dimensional map of the activity of cosmic rays in the Galaxy. Such an airshower array requires lower energy than HIRES and AUGER, but very much higher flux accuracy, and so is a highly accurate development beyond AGASA.

6. Acknowledgments

We acknowledge numerous discussions on these matters with many friends and colleagues, especially John Bieging, Andrey Bykov, Jim Connell, Alina Donea, Tom Gaisser, Stan Hunter, Hyesung Kang, Bob Kinzer, Norbert Langer, Karl Mannheim, Jim Matthews, Hinrich Meyer, Rene Ong, Buford Price, Ray Protheroe, Vladimir Ptuskin, Jörg Rachen, Dongsu Ryu, Todor Stanev, Samvel Ter-Antonyan, and Dave Thompson. Work on these topics with PLB has been supported by grants from NATO, the BMBF-DESY, the EU, the DFG and other sources.

References

Achterberg, A.: 1979 *Astron. & Astroph.* **76**, 276 - 286

Beck, R., Brandenburg, A., Moss, D., Shukurov, A., & Sokoloff, D., Galactic Magnetism: Recent Developments and perspectives, 1996, *Ann. Rev. Astron. & Astrophys.* **34**, 155 - 206.

Beck, R., in *Astrophysics of galactic Cosmic Rays*, Eds. Diehl *et al.*, *Space Science Reviews*, Kluwer (in press), 2000

Bellido, J. A., Clay, R. W., Dawson, B. R , Johnston-Hollit, M., astro-ph/0009039 (2000).

Berezinsky, V.S., *et al.*: 1990 *Astrophysics of Cosmic Rays*, North-Holland, Amsterdam

Berezinsky, V.S., Gaisser, T.K., Halzen, F. & Stanev, T., 1994, *Astropart. Phys.* **1**, 281–287

Berezinsky, V.S., Gaisser, T.K., Halzen, F. & Stanev, T., 1994, *Astropart. Phys.* **2**, 101-101

Biermann, P.L. & Strittmatter, P.A., *Astrophys. J.*, **322**, 643, (1987).

Biermann, P.L.: 1993a *Astron. & Astrophys.* **271**, 649 - 661 (paper CR I), astro-ph/9301008.

Biermann, P.L., and Cassinelli, J.P.: 1993 *Astron. & Astrophys.* **277**, 691 - 706 (paper CR II), astro-ph/9305003.

Biermann, P.L., and Strom, R.G.: 1993 *Astron. & Astrophys.* **275**, 659 - 669 (paper CR III), astro-ph/9303013.

Biermann, P.L.: 1993c in *23rd ICRC, Invited, Rapporteur and Highlight Papers*, ed. D.A. Leahy *et al.* (World Scientific, Singapore, 1993), 45 - 83.

Biermann; in the Proc. Gamow Jubilee Seminar, St. Peterburg, Sept. 1994, *Space Science Rev.*=A074, 385, 1995; astro-ph/9501003

Biermann, P.L., Strom, R.G., Falcke. H., *Astron. & Astroph.* **302**, 429, 1995 (paper CR V); astro-ph/9508102

Biermann, P.L.: 1997a in *Cosmic Winds and the Heliosphere*, ed. J.R. Jokipii *et al.* (University of Arizona Press, Tucson), p. 887 - 957; astro-ph/9501030.

Biermann, P.L.: 1997b *J. of Physics G* **23**, 1 - 27

Biermann, P.L., 1998a, in "Workshop on observing giant cosmic ray air showers from $> 10^{20}$ eV particles from space", Eds. J.F. Krizmanic *et al.*, AIP conf. proc. No. 433, p. 22 - 36

Biermann, P.L.: 1998b, in Proc. *Nuclear Astrophysics*, Hirschegg, GSI, Darmstadt, p. 211 - 222

Biermann, P.L., Langer, N., Seo, E.-S., Stanev, T., 2001, *Astron. & Astroph.* (in press), (paper CR IX)

Biskamp, D., Müller, W.-Ch., 2000, *Phys. of Plasma* **7**, 4889 - 4900

Blandford, R.D., Eichler, R.D.: 1987 *Phys. Rep.* **154**, 1 - 75.

Clay, R.W., preprint, 2001

Connell, J.J., Simpson, J.A.: 1997 *Astrophys. J. Lett.* **475**, L61 - L64

Connell, J.J.: 1998 *Astroph. J. Lett.* **501**, L59 - L62

Connell, J.J., DuVernois, M.A., Simpson, J.A.: 1998 *Astrophys. J. Lett.* **509**, L97 - L100

Drury, L.O'C: 1983 *Rep. Prog. Phys.* **46**, 973 - 1027.

Ellison, D.C., Drury, L. O'C., Meyer, J.-P.: 1997 *Astrophys. J.* **487**, 197

Falcke, H., Biermann, P.L., *Astron. & Astroph.* **293**, 665 - 682, 1995; astro-ph/9411096

Falcke, H., Biermann, P.L., *Astron. & Astroph.* **342**, 49 - 56, 1999, astro-ph/9810226

Farrar, G.R., Biermann, P.L., *Phys. Rev. Letters* **81**, 3579 - 3582, 1998; astro-ph/9806242

Fermi, E.: 1949 *Phys. Rev.* 2nd ser., **75**, no. 8, 1169 - 1174.

Fermi, E.: 1954 *Astrophys.J.* **119**, 1 - 6.

Flügge, S., 1961, Editor, Encyclopedia of Physics, *Cosmic Rays I*, Springer, Berlin

Gaisser, T.K.: 1990 *Cosmic Rays and Particle Physics*, Cambridge Univ. Press

Gaisser, T.K., Protheroe, R.J., Stanev, T.: 1998 *Astrophys. J.* **492**, 219, also in astro-ph

Garcia-Munoz, M., Mason, G.M., Simpson, J.A.: 1977 *Astrophys.J.* **217**, 859 - 877

Garcia-Munoz, M., Simpson, J.A., Guzik, T.G., Wefel, J.P., Margolis, S.H.: 1987 *Astro-*

phys.J. Suppl. **64**, 269 - 304
Ginzburg, V.L.: 1953a *Usp. Fiz. Nauk* **51**, 343 - . See also 1956 *Nuovo Cimento* **3**, 38 - .
Ginzburg, V.L.: 1953b *Dokl. Akad. Nauk SSSR* **92**, 1133 - 1136 (NSF-Transl. 230).
Ginzburg, V.L. & Syrovatskii, S.I., *The origin of cosmic rays*, Pergamon Press, Oxford 1964, Russian edition 1963
Ginzburg, V.L., Syrovatskij, S.I.: 1969 *The Origin of Cosmic Rays*, Gordon and Breach, New York
Ginzburg, V.L.: 1993 *Phys. Usp.* **36**, 587 - 591.
Ginzburg, V.L.: 1996 *Uspekhi Fizicheskikh Nauk* **166**, 169 - 183
Glasmacher, M.A.K., *et al.*: 1999a, *Astropart. Phys.* **10**, 291 - 302
Glasmacher, M.A.K., *et al.*: 1999b, *Astropart. Phys.* **12**, 1 - 17
Goldstein, M.L., Roberts, D.A., Matthaeus, W.H.: 1995 *Ann. Rev. Astron. & Astroph.* **33**, 283 - 325
Hayakawa, S.: 1969 *Cosmic Ray Physics*, Wiley-Interscience, New York
Hayashida *et al.* 1998 preprint astro-ph/9807045
Heisenberg, W. 1948 Z. Physik 124, 628
Heisenberg, W., 1953, Editor, *Kosmische Strahlung*, Springer, Berlin
Hess, V.F.: 1912 *Phys. Z.* **13**, 1084.
Hillas, A.M.: 1984 *Ann. Rev. Astron. Astrophys.* **22**, 425 - 444.
Hunter, S.D. *et al.*: 1997 *Astrophys. J.* **481**, 205 - 240
Jones, F.C., Ellison, D.C. : 1991 *Space Science Rev.* **58**, 259 - 346.
Jones, F.C., Lukasiak, A., Ptuskin, V., Webber, W., *Astrophys. J.* (submitted) 2000, astro-ph/0007293
Kaneda, H., Makishima, K., Yamauchi, S., Koyama, K., Matsuzaki, K., Yamasaki, N. Y., Complex Spectra of the Galactic Ridge X-Rays Observed with ASCA, 1997, *Astrophys. J.* **491**, 638.
Karman, Th. de, Howarth, L.: 1938 *Proc. of the Royal Soc. of London* **164**, 192 - 214.
Kohlhörster, W.: 1913 *Phys. Z.* **14**, 1153.
Kolmogorov, A.N. 1941a *Dokl. Akad. Nauk SSSR* **30**, 299 - 303
Kolmogorov, A.N. 1941b *Dokl. Akad. Nauk SSSR* **31**, 538 - 541
Kolmogorov, A.N. 1941c *Dokl. Akad. Nauk SSSR* **32**, 19 - 21
Kraichnan, R.H.: 1965 *Phys. Fl.* **8**, 1385.
Kronberg, P.P., Extragalactic magnetic fields, 1994, *Rep. Prog. Phys.*, **57**, 325 - 382.
McIvor, I. 1977 *Month. Not. Roy. Astr. Soc.* **178**, 85 - 99
Mahadevan, R., 1998 *Nature* **394**, 651 - 653
Markoff, S., Rhode, W., Enßlin, T.A., Donea, A.C., Biermann, P.L., in Proc. 26th ICRC (Salt Lake city), vol. 4, p. 411 - 414, 1999
Matthaeus, W.H., Zhou, Y.: 1989 *Phys. Fluids* **B1**, 1929 - 1931
Meyer, J.-P., Drury, L. O'C., Ellison, D. C.: 1997 *ApJ* **487**, 182
Mori, M.: 1997 *Astrophys. J.* **478**, 225 - 232
Nath, B.B., Biermann, P.L., *Month. Not. Roy. Astr. Soc.* **267**, 447 - 451, 1994a; astro-ph/9311048
Nath, B.B., Biermann, P.L., *Month. Not. Roy. Astr. Soc.* Letters **270**, L33 - L36, 1994b; astro-ph/9407001
Obukhov, A.M. 1941 *Dokl. Akad. Nauk SSSR* **32**, 22 - 24
Ong, R., 1998 *Physics Reports* **305**, 93 - 202
Peters, B., 1959 *Nuovo Cimento Suppl.* XIV, ser. X, p. 436 - 456.
Peters, B. 1961 *Nuovo Cimento* XXII, p. 800 - 819
Pietz, J., Kerp, J., Kalberla, P. M. W., Burton, W. B., Hartmann, D., Mebold, U., *Astron. & Astroph.* **332**, 55 - 70 (1998).
Pohl, M. *et al.*: 1997 *Astrophys. J.* **491**, 159 - 164
Prandtl, L.: 1925 *Zeitschrift angew. Math. und Mech.* **5**, 136 - 139.
Prantzos, N., Arnould, M., Arcoragi, J.-P.: 1987 *Astrophys. J.* **315**, 209
Protheroe, R.J., Stanev, T.: 1997 at the ICRC meeting in Durban, South Africa, OG 3.4.3

Ptuskin, V.S., Lukasiak, A., Jones, F.C., Webber, W.R., 1999, in Proc. 26th ICRC (Salt Lake City), vol. 4, 291 - 294

Rachen, J.P. & Biermann, P.L., *Astron. & Astroph.* **272**, 161, 1993; astro-ph/9301010

=A0Rachen, J.P., Stanev, T., Biermann, P.L., *Astron. & Astroph.* **273**, 377, 1993; astro-ph/9302005

Ramaty, R., Kozlovsky, B., Lingenfelter, R.E., Reeves, H.: 1997 *Astrophys. J.* **488**, 730 - 748

Reeves, H.: 1974 *Ann. Rev. Astron., & Astroph.* **12**, 437 - 469.

Reeves, H.: 1994 *Rev. of Modern Physics* **66**, 193 - 216

Rickett, B.J.: 1990 *Ann. Rev. Astron. & Astroph.* **28**, 561 - 605.

Rhode, W., Enßlin, T.A., Biermann, P.L., 1998, at the Galactic Center conference, Eds. H. Falcke et al., p.

Rhode, W., Markoff, S., Enßlin, T.A., Donea, A.C., Biermann, P.L., in Proc. 26th ICRC (Salt Lake City), vol. 4, p. 415 - 418, 1999

Rosen, S., 1969, Editor, *Selected papers on cosmic ray origin theories*, Dover, New York

Sagdeev, R.Z.: 1979 *Rev. Mod. Phys.* **51**, 1 - 20

Seo, E.S., & Ptuskin, V.S., 1994, *Astrophys. J.* **431**, 705–714,

Shklovskii, I.S.: 1953 *Dokl. Akad. Nauk, SSSR* **91**, no. 3, 475 - 478 (Lib. of Congress Transl. RT-1495).

Simard-Normandin, M., & Kronberg, P.P., *Nature* **279**, 115 (1979).

Simpson, J.A., Connell, J.J.: 1998 *Astroph. J. Lett.* **497**, L85 - L88

Snowden, S.L., *et al.*, ROSAT Survey Diffuse X-Ray Background Maps. II., 1997, *Astrophys. J.* **485**, 125.

Stanev, T., Biermann, P.L., & Gaisser, T.K.: 1993, *Astron. & Astrophys.* **274**, 902 - 908 (paper CR IV), astro-ph/9303006.

Stecker, F.W.: 1971 *Cosmic Gamma Rays*, NASA SP-249

Thum, C., Morris, D., 1999, *Astron. & Astroph.* **344**, 923

Tinsley, B.M. & Biermann, B.M., *Publ.Astron.Soc.Pacific* **86**, 791, 1974

Valinia, A., Marshall, F. E., RXTE Measurement of the Diffuse X-Ray Emission from the Galactic Ridge: Implications for the Energetics of the Interstellar Medium, 1998, *Astrophys. J.* **505**, 134 - 147.

Wax, N., 1954, Editor, *Selected papers on noise and stochastic processes*, Dover, New York

Westphal, A.J., *et al.*, 1998 *Nature* **396**, 50 - 52

Wiebel-Sooth, B., Biermann, P.L., Meyer, H.: 1998 *Astron. & Astroph.* **330**, 389 - 398 (paper CRVII), astro-ph/9709253

Wiebel-Sooth, B., Biermann, P.L., in Landolt-Börnstein, vol. VI/3c, Springer Publ. Comp. (1999), p. 37 - 90.

Yanagita, S., Nomoto, K., Hayakawa, S.: 1990 Proc. 21st ICRC, Adelaide, Australia, Ed. R.J. Protheroe,(Univ. of Adelaide, 1990), 4, 44

Zatsepin, V.I.: 1995 *J. of Phys. G* **21**, L31 - L34

DARK MATTER HALO OF OUR GALAXY

RAMANATH COWSIK
Indian Institute of Astrophysics, Bangalore 560 034, India
and
Tata Institute of Fundamental Research, Mumbai 400 005, India

The fascination and the challenge for the search for the constituents of dark matter stem from the connection it bears with astronomy, astrophysics and cosmology on the one hand and with nuclear and particle physics on the other. Dark matter particles are the only relicts that still servive to-day from epochs prior to the primordial helium synthesis which took place at about 400s after the big bang. The first clues pointing to the presence of hidden mass in astronomical systems date-back to the 1920s and 1930s. Kepteyn (1922), Jeans (1922) and Oort (1930) after determining the mass density in the solar neighbourhood had noted that the visible stars contributed only a part to this measured value[1]. Zwicky (1933) measured the velocity dispersion of galaxies in the Coma-cluster (though based on observations of a small number of galaxies) and noted that a substantial part of the mass in the cluster has to be in some unseen form to account for the large observed value of the velocity dispersion[2]. This work was followed up by a few other workers and finally in 1972 based on observations of the velocities of hundreds of galaxies in the cluster, Rood et al., confirmed that atleast 75 % of the mass of the cluster was in some unseen form[3]. As the astronomical observations proceeded with increased vigor, very many ideas to explain the hidden mass were proposed. Among these the most exotic one (for that time) was presented by Cowsik & Mc Clelland (1972, 1973)[4,5] who suggested that weakly interacting particles with finite rest-mass, generated in the early epochs of a hot big bang universe, would thermodynamically decouple from radiation and matter, and will evolve without substantial annihilation, to form a relict background of particles, which will gravitationally dominate over the normal baryonic matter, trigger the formation of galaxies, and thus generally form halos of invisible dark matter around galactic systems. This set of ideas has been developed and made more sophisticated by the work of many others and forms to-day the basic paradigm for the study of dark matter in the Universe.

M. M. Shapiro et al. (eds.), Astrophysical Sources of High Energy Particles and Radiation, 81–91.
© 2001 *Kluwer Academic Publishers. Printed in the Netherlands.*

In a parallel development the observations of the rotation curves of galaxies showed that they were flat upto large galactocentric distances, often far beyond the optical "edge" of the galaxies[6-11]. With such data on hand the existence of dark matter in spiral galaxies and its dominance at large galactocentric distances in the from of a halo extending to several hundred kiloparsecs was generally accepted[12-13]. These observations gave support to the seminal paper written some years earlier by Kahn and Woltjer (1959) who had derived quite a high mass to our Galaxy based on the fact that M 31 and the Galaxy were approaching each other indicating that there was enough mass in the pair to reverse the Hubble flow at the present epoch[14]. Early observations of the rotation curve of our Galaxy and the motions of stars in the solar neighbourhood[15-16] supported the view that the Galaxy as well possessed a halo of dark matter, just like the external galaxies.

Let us now quickly review the decade of development of the general dark matter problem: What is the concentration of matter and energy in the Universe and what are their constituent components ? This is generally discussed as a fraction of the critical density ρ_{crit} and is called the Ω parameter:

$$\rho_{crit} = \frac{3H_o^2}{8\pi G} \approx 1.9 \times 10^{-29} h^2 gcm^{-3} \tag{1}$$

$$\Omega_o = \Sigma \rho_i / \rho_{crit} \equiv \Sigma \Omega_i \tag{2}$$

In the above equations h $\equiv$ H$_o$/100 km s^{-1} $\approx$ 0.65 according to the preponderance of current analysis of the astronomical data, so that $\rho_{crit} \approx 8 \times 10^{30} gcm^{-3} \approx 6$ keV cm^{-3}, and ρ_i are the energy density of various components. The best current numbers of the various components are listed in Table 1.

1. Mass-models of the Galaxy

Important insights into the structure of the Galaxy have come from the study of mass models which allows one to make use of the insights gained from the astronomical observations of external galaxies to the interpretation of the data on our own Galaxy which is limited oweing to the specific location of the solar system in the galactic plane, observation by dust and so on. Of particular interest here is the fact that from these models it is possible to calculate the gravitational field of the Galaxy and thence the rotation curve representing the circular velocity as a function of galactocentric distance in such a field. An example of such a curve with the contribution of the various components labelled as 'Bulge + Spheroid', 'Disk'

TABLE 1. The energy density of various components of the Universe and the method of their determination

Component	ρ_i	Method
Black-body photons of $\sim 2.7^\circ$K	0.25 eV cm^{-3}	direct observation[17]
Neutrinos	$\geq$ 20 eV cm^{-3}	atmospheric neutrino oscillation experiment [18]
Stars	$\sim$ 18 eV cm^{-3}	mean luminosity density in the Universe
Clusters of galaxies	$\geq$ 1.2 keV cm^{-3}	Luminosity density combined with $<M/L>_{clusters}$ (R< 500 kpc) [19, 20]
Total baryons	$\sim$ 270 eV cm^{-3}	deuterium observations and primordial nucleosynthesis[21]
Λ_{vacuum}	$\sim$ 3 keV cm^{-3}	SNe Ia observations and expansion rate of the Universe [22, 23]
Cold dark matter	$\sim$ 1.5 keV cm^{-3}	structure formation theories [24-26]
ρ_{tot}	$\sim$ 1 keV cm^{-3} $<\rho \ll$ 6 keV cm^{-3}	peculiar velocity field [27,28,29]
	$\sim$ 4 keV cm^{-3}	measurements of anisotropy of the micowave background and theory [17]

and 'Corona' is reproduced in the Fig.1, taken from Maarten Schmidt [30]. The components of the astronomically luminous matter such as the disk, bulge etc can be constrained by the observations. In contrast the corona is made up mostly of non-luminous matter and its density distribution can be ascertained only through indirect means such as those adopted by Oort, Bahcall, Maarten Schmidt and others[31].

Now let us turn to a discussion the mass-model presented in Fig.1 with particular focus on the corona whose presence has been inferred only through its dynamical effects. Today there is no dearth of candidates to populate the halo of the galaxy: (a) MACHOS or massive - compact - halo - objects which are low-mass white dwarfs, brown dwarfs and smaller compact forms of normal matter [32], cold dark matter particles like neutralinos

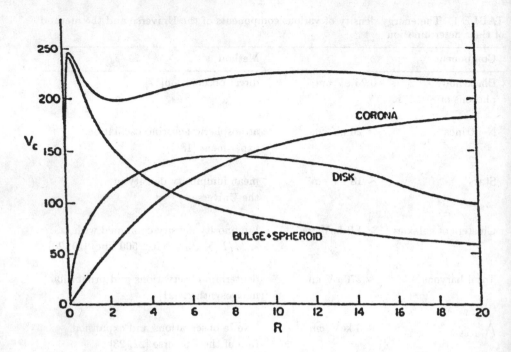

Figure 1. Mass model of the Galaxy to fit the rotation curve [21].

or axions [33] and even neutrinos with m $\sim$ 10 eV if the halo is extensive enough to envelope the local group [34]. Now, irrespective of the nature of the constituents we can pose the question: how should the velocity of the constituent particles be distributed in order that they would spread out in space as we want them and yield the correct rotation curve of the galaxy? In answering the question we should take care to include the effects of gravitation of the various components like disk, bulge etc., and also the contribution of the dark matter particles themselves. Since the mass within r is given by $m(r) \sim r.v_c^2$, notice from Fig. 1, the contributions of the halo component to the mass of the galaxy increases as r^3 upto a galactocentric distance of $\sim$ 5 kpc. Beyond this distance the contribution to v_c^2 by the corona progressively flattens off. What does this tell us about the velocity-distribution of the constituent particles? If the velocities are too low then the particles will tend to orbit closer to the galactic centre and would concentrate there. If on the other hand they had very high velocities then they would essentially have uniform density upto very large distances, leading to $m(r) \sim r^3$. These ideas are quantified in the next section.

2. Qualitative estimate of the velocity dispersion of halo particles

Before we develop the mathematical model to obtain the self-consistent density distributions, note that the scale l, length for the distribution of a gas of temperature T in a regions of total density ρ is given by

$$
l = \left(\frac{kT}{4\pi G \rho} \right)^{1/2} = < v^2 >^{1/2} (12\pi G \rho)^{-1/2} \approx 0.24 < v^2 >_{100}^{1/2} \rho_1^{-1/2} \; kpc
$$

$$(3)$$

Here $< v^2 >_{100}^{1/2}$ is the r.m.s. velocity in units of 100 km s^{-1} and ρ_1 is the total density in units of 1 M$_\odot$ pc^{-3}. With $< v^2 >_{100}^{1/2} \sim 5$ and $\rho_1 \sim 0.1$ we can get $l \approx 3.8$ kpc as suggested by the mass - models.

3. A self-consistent dynamical model for the halo

In developing a detailed dynamical model let us focus specifically on density the dark-matter component but assume as known the spatial density distributions of the various components of normal matter as known and given for example by a spheroidal bulge and an axisymmetric disk [35-37]; the density distribution of these two components are given by

$$
\rho_s(r) = \rho_s(0) \left(1 + \frac{r^2}{\alpha^2} \right)^{-3/2}
$$

$$(4)$$

$$\rho_d(r) = \frac{\Sigma}{2h} e^{-(R-R_o)/R_d} \, e^{-|z|/h} \tag{5}$$

Here $r = (R^2 + z^2)^{1/2}$, z =height above the central plane of the disk, Σ = surface density of the disk at the solar position in the Galaxy and the parameters take the values $\alpha = 0.103$ kpc, $R_d = 3.5$ kpc $h = 0.3$ kpc, $\rho_s(0) = 7 \times 10^{-4} M_\odot pc^{-3}$ and Σ is in the range 40 - 80 $M_\odot pc^{-2}$.

The phase space structure of the dark matter component is described by two possible distribution functions: (a) isothermal and (b) reduced isothermal or King model. These are given by the equations given below:

3.1. ISOTHERMAL:

$$f \equiv f(E) = \rho_o(2\pi\sigma^2)^{-\frac{3}{2}} \, exp\left[-\frac{\phi + \frac{1}{2}v^2}{\sigma^2}\right] \tag{6}$$

Here, $E = \phi + \frac{1}{2}v^2$ is the total energy (per unit mass) of a particle of velocity v in a gravitational potential ϕ. The potential is normalized to be $\phi = 0$ at $r = 0$ and the velocity dispersion is constant everywhere, $< v^2 >= 3\sigma^2$ Since E is a constant of motion, f is stationary ρ_o is the density of dark matter at $r = 0$.

3.2. KING:

$$f \equiv f(\epsilon) = \rho_o(2\pi\sigma^2)^{-\frac{3}{2}} \left(e^{\epsilon/\sigma^2} - 1\right) \, for \; \epsilon > o \tag{7}$$

$$o, \, for \; \epsilon \leq o$$

In the King model, the halo has a finite extent. The "relative energy" $\epsilon = -E + \phi_o$ with $\phi_0 = $ constant, is chosen in such a way that f vanishes for $r \geq r_k$. The density of dark matter is now simply obtained by integrating the DF over the velocity space

$$\rho_{dm} = \int f d^3 v \tag{8}$$

The two distribution functions chosen above will describe halos which are "pressure supported" without any rotation and are the forms most extensively adopted in the study of a variety of astrophysical problems.

The next step is to write the Poisson equation for the potential ϕ, assuming azimuthal symmetry

$$\frac{1}{r^2}\frac{\partial}{\partial r}\left(r^2\frac{\partial\phi}{\partial r}\right) + \frac{1}{sin\theta}\frac{\partial}{\partial\theta}\left(sin\theta\frac{\partial\phi}{\partial\theta}\right) = 4\pi G(\rho_{dm} + \rho_d + \rho_s) \qquad (9)$$

Notice the nonlinear dependence of ρ_{dm} on ϕ. The solution of the non-linear Poisson equation is accomplished through numerical techniques described elsewhere. The numerical code is verified against analytical solutions which are obtained easily both for very small r and for large r. Once $\phi(r,\theta)$ is thus determined it is a simple matter to obtain from it ρ_{DM} from Eqns.(1-3) and also derive the circular velocity as a function of R, $v_c = \sqrt{R\,|\,d\phi/dR\,|}$. Notice that the solutions for the total potential ϕ are a family of functions dependent on the parameters ρ_o and σ, and of course of ϕ_o for the King models. Comparison with the available data on the rotation curve of our Galaxy upto galactocentric distances of $\sim$ 20 kpc (measured with decreasing precision with increasing distance) [38] is shown in Fig.2. The best fit is obtained for the parameter set[39].

$$v_{rms}(R \sim 8.5 \ kpc) \geq 600 \ kms^{-1} \qquad (10)$$

$$\rho_{dm}(R \sim 8.5 \ kpc) \sim 0.3 \ GeV cm^{-3} \qquad (11)$$

These values of v_{rms} and ρ_{dm} at 8.5 kpc are essentially the same for the two distribution functions that we have studied. The two distribution functions indeed yield different predictions at large distances. The isothermal distribution yields v_c (R $\to \infty$) $\approx \sqrt{2/3}v_{rms}$ as expected. The King-model yields v_{rms} which for $r \geq 100$ kpc decreases steadily with distance to zero at the King radius R_k; also for $R > R_k$ the rotation curve becomes Keplarian with $v_c \sim R^{-1/2}$. Thus the numbers quoted for ρ_{dm} and v_{rms} in Eqns. (7 and 8) for the solar neighbourhood may be incorporated into the analysis of laboratory experiments that attempt to measure nuclear recoils of DM particles. We have performed such an analysis and find that the cross-section bounds for the scattering by DM particles is lowered considerably, especially for $m_{dm} < 80$ GeV; also the expected sidereal and annual assymetries are suppressed considerably. At this juncture we should note that the results presented here are valid for any dark matter halo which supports itself through pressure gradients against the force of gravity due to normal matter of the Galaxy and its own self gravity. Rotating halos, where part of the support may come from centrifugal forces, may well yield different results.

As noted earlier, the rotation curve of the Galaxy is poorly determined beyond the solar circle; indeed there are no direct observations beyond $\sim$ 18 kpc. However, a knowledge of v_c at larger distances is crucial for

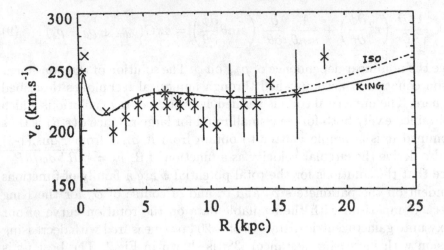

Figure 2. Dynamical model with ρ_{dm} (kpc) = 0.3 GeV cm^{-3} and $<\upsilon^2>^{1/2} = 515$ km s^{-1} to fit the observed rotation curve of the Galaxy.

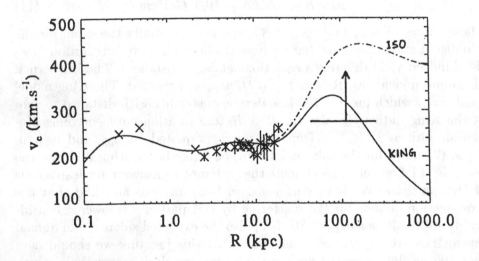

Figure 3. Extenstion of Fig. 2 to include the lower-bound on $<\upsilon_c^2>^{1/2}$ derived from data on dwarf spheroidals.

assessing the size of the galactic halo and the mass associated with our Galaxy - both visible and dark. To this end we now turn our attention to a theorem derived by Lynden-Bell and Frenk [40] based on the virial theorem: They start with the identity

$$\frac{1}{2}\frac{d^2 r^2}{dt^2} - v^2 = r.\nabla\phi = -v_c^2 \tag{12}$$

and averaging eq. 12 over a system of test particles (and noting $\frac{d}{dt}(\sum r^2)$ vanishes for a virialised system) they obtain

$$< v^2 >=< v_c^2 > \tag{13}$$

Thus by measuring the $< v^2 >$ for any system of test particles we may derive the value of $< v_c >$. As emphasised by Lynden Bell and Lynden Bell [41], the validity theorem does not require that the potential should arise only due to the ensemble of test particles under consideration. However, the formula in Eq.13 is valid only when the averaging is done over an entire ensemble rather than on a subset of particles contained in any radial interval. When one applies these considerations to a subset of test particles beyond any particular radius then the Eqn. 13 yields a lower bound on $v_c(r)$.

Now we are faced with another difficulty. The test particles we have at hand especially at large galactocentric distances ≥ 50 kpc are the dwarf-spheroidals for which we are able to measure only the radial velocities v_r rather than $< v^2 >$ that appears in Eq.13. We may take

$$< v^2 >= 3 < v_r^2 > \tag{14}$$

for an isotrophic distribution; however, the distribution of velocities of dwarf spheroidals is expected to be highly anisotropic with $< v_r^2 >$ suppressed considerably below the value expected for an isotropic distribution. This is because the orbits with relatively large value of $| v_r |$ will yield orbits which will bring the dwarf spheroidals to closer proximity of our Galaxy, where they will be tidally disrupted. Even when one supposes that, at the time of galaxy formation, the dwarf-spheroidals had an isotropic distribution, those with large v_r would have been disrupted, leaving behind a distribution that is depleted at large v_r.

(1) With the farthest member of this population being at $r_x \approx 300$ kpc we may expect that the DF for the system of dwarf spheroidals has a King-form with $r_k = r_x$; this will ensure that v_r vanishes at r_x as required and $v_r(max)$ will slowly increase at shorter distances $r < r_x$.

(2) Secondly, there is an inner radius $r_n \approx 50$ kpc below which dwarf spheroidals are not found. This probably represents tidal disruption of such

candidates during the early phases of the evolution of our local group. The radial velocities have to vanish again at r_n and increase only slowly beyond.

Indeed it is possible to derive bounds on the value of $v^2_{r_{max}}(r)$ in the range $r_n < r < r_x$ from straight forward kinematics. Taking these considerations into account the eqn. 14 is modified to

$$< v^2 >= \mu < v^2_r > \tag{15}$$

with μ taking values considerably greater than 3 (for $r_n = 50$ kpc and $r_k = 250$ kpc, $\mu \approx 7.5$). This when taken together with the observed radial velocities of the dwarf spheroidals yields $< v^2 >^{1/2} \approx 275$ kms^{-1} at R $\sim$ 100 kpc. In fig.3 we show the theoretical rotation curve upto very large galactocentric distance for the parameter set: $\rho_{dm}(R_\odot) \approx 0.3$ $GeV, cm^{-3}, <$ $v^2 >^{1/2} \approx 550\, km\, s^{-1}$ and $r_k \approx 300$ kpc and find a good fit over the entire region including the lower bound on $< v^2_c >^{1/2}$ of 275 km s^{-1} derived using Lynden-Bells formalism. These sizes and dispersions are surprisingly close to the DM distribution in rich clusters. Thus the results presented here will bear significantly on the theories of structure formation as well.

Acknowledgement

I acknowledge the help of Dr P.Bhattacharjee, in preparation of this manuscript.

References

1. Oort, J.H. (1932) BAN, 6, 249.
2. Zwicky, F.(1933) Helvetica Physica Acta, 6, 10.
3. Rood, H.J., Page, T.L., Kinter, E.C., and King, I.R. (1972) Ap.J., 172, 627.
4. Cowsik, R. and McClelland, J. (1972) Phys. Rev. Lett., 29, 669.
5. Cowsik, R. and McClelland, J.,(1993) Ap.J., 180, 7.
6. Roberts, M.S., (1974)in Stars & Stellar Systems, (ed. Sandage, A.,)
7. Vaucouleurs, G.de. (1964) Astrophys. Lett., 4, 17.
8. Arp, H., and Bertola, F. (1969) ibid 4, 23.
9. Shostaki, G.S., and Rogstad, D.H. (1973) A&A., 24, 411.
10. Seielstad, G.A., and Wright, M.C.H., (1973) Ap.J., 184, 343.
11. Rogstad, D.H. et. al. (1973) A&A., 22, 111.
12. Einasto, J. et al.,(1974) Nature, 250, 309.
13. Ostriker, J.P. et al., (1974) Ap.J. Lett., 193, L1.
14. Kahn, F.D., and Woltjer,L., (1959) Ap.J., 130, 705.
15. Rubin, V.C., et al. (1980) Ap.J., 238, 471; (1982) Ap.J., 261, 439 ; (1985) Ap.J., 289, 81; Faber, S.M. and Gallagher, J.S.,(1979) ARAA, 17, 135.
16. Bahcall, J.N., et al. (1983) Ap.J., 265, 730; (1984) Ap.J., 276, 169.
17. Brandis, P.de. et al. (2000) Nature 404, 955.
18. Fukunda, Y. et al. (1998)(Super-Kamiokande collaboration) Phys. Rev. Lett., 81, (4562); R.Cowsik, (1998) Current Science k75, 558.
19. Carlberg, R. et al. (1997) Ap. J. 478, 462.
20. Evrard,A.E., Mehr, J.J., Febriant, D.G. and Gek ,M.J.,(1993) Ap.J. Lett. 419, 69 (and refs. therein).
21. Burles, S. and Tytler,D. (1998) Ap.J., 507, 732; (1998) Ap.J., 499,699.
22. Reiss, A., et. al. (1998) A. J. 116, 1009.

23. Perlmutter, S. et. al. (1999) Astro-phy 98/12133.
24. White, S.D.M. et. al. (1983) ApJ Lett.274, L1.
25. Gawiser, E., and Silk,J. (1998) Science 280, 1405.
26. Gates, E.I., Turner,M.S. (1994) Phys. Rev. Lett, 72, 2520.
27. Deckel, A. (1994) A R AA 32, 371.
28. Sigath et al. (1998) Ap.J 495, 516.
29. Willick, J. and Strauss, M. (1998) ApJ 507, 64.
30. Schmidt,M., (1985) *in The Milky Way Galaxy*, IAY Symp. 106,75, (eds H.van Woerden et al)
31. Zeritsky,D., Third Stromlo Symposium, ASP Conference Series 165, 34, 199 (eds: B.K. Gibson et. al.)
32. Stubbs, C.W., *ibid* p.503
33. Turner, M.S., *ibid* p.431
34. Cowsik, R. and Ghosh, P. (1987) Ap.J. 317, 26.
35. Binney, J. and Tremarne, S. (1987)*Galactic Dynamics*, (Princeton Univ. Press)
36. Caldwell, J.A.R., and Ostriker, J.P. (1981) Ap.J, 219, 18.
37. Kuijken, K. and Gilmore,G. (1989) M.N.R.A.S., 239, 571.
38. Fich, M. and Tremaine, (1991) ARAA, 29, 409.
39. Cowsik,R., Ratnam, C. and Bhattacharjee,P. (1997) Phys. Rev. Lett., 28 2262.
40. Lynden-Bell, D. and Frenk,C.S., (1981) The Observatory 101, 200.
41. Lynden-Bell, D., and Lynden-Bell, R.M. (1995) M.N.R.A.S., 275, 429.

A NEW MODEL FOR THE THERMAL X-RAY COMPOSITES AND THE PROTON ORIGIN γ-RAYS FROM SUPERNOVA REMNANTS

O. PETRUK

Institute for Applied Problems in Mechanics and Mathematics
3-b Naukova St., 79053 Lviv, Ukraine

Abstract Recent nonthermal X-ray and γ-ray observations, attributed to electron emission processes, for the first time give an experimental confirmation that electrons are accelerated on SNR shocks up to the energy $\sim 10^{14}$ eV. We have no direct observational confirmations about proton acceleration by SNR. Different models of γ-emission from SNRs predict different emission mechanisms as dominating. Only π^o decays created in proton-nucleon interactions allow us to look inside the CR nuclear component acceleration processes. A new model for the thermal X-ray composites strongly suggest that thermal X-ray peak inside the radio shell of SNR tells us about entering of one part of SNR shock into a denser medium compared with other parts of the shell. This makes a TXCs promising sites for γ-ray generation via π^o decays. Detailed consideration of SNR-cloud interaction allows to increase an expected proton induced γ-ray flux from SNR at least on an order of magnitude, that allows to adjust the theoretical π^o decay γ-luminosities with observed fluxes at least for a few SNRs even for low density ($n_o \sim 10^1 \div 10^2$ cm^{-3}) cloud.

1 Introduction

Since SNRs possess high enough energy, their shocks are belived to be a major contributor of both electron and nuclear components of Galactic CRs up to energies 10^{15} eV [5]. SNRs emit in all wavelengths. Their radio emission gives clear evidence about relativistic electrons accelerated on the shock front. Observed nonthermal X-ray emission from several SNRs (SN 1006 [16], Cas A [3], Tycho [4], G347.3-0.5 [17], IC443 [15], G266.2-1.2 [29]) is thought to be also synchrotron, from the shock accelereted electrons with a much higher energy ($\sim 10^{13}$ eV). Observed synchrotron X-ray [16] and TeV γ-ray [33] emission from SN 1006 as well as from G347.3-0.5 [28, 19], by attributing the emission to the synchrotron radiation, gives firm confirmations that electrons are accelerated on the SNR shocks

M. M. Shapiro et al. (eds.), Astrophysical Sources of High Energy Particles and Radiation, 93–100.

up to energies $\sim 10^{14}$ eV. The theory of acceleration of CRs on the shock front via the first order Fermi mechanism [5] predicts that the number of electrons N_e involved in the acceleration process is much smaller than the number of protons N_p: $N_e(\varepsilon)/N_p(\varepsilon) = (m_e/m_p)^{(\alpha-1)/2} \simeq 10^{-2}$ (α is the spectral index of accelerated particles, ε is the energy of particles). Thus, if there are electrons accelereted to such high energies, we should expect that protons with the same high energy have to reveal themselves in observations, too. We have to look for the objects which might emit γ-rays mainly via neutral pion decays.

2 Observations of the γ-rays from SNRs

The visibility of SNRs in Gev and TeV γ-rays was treated in [8, 1]. The main conclusion from the theory is: if SNRs are mainly responsible for the galactic CRs, it may be difficult to observe SNRs by EGRET, but should be possible to detect compact (with angular size $\sim 0.25'$) remnants from a distance less than several kpc in TeV band, with Cherenkov telescopes. Unfortunately, as first results on TeV γ-observations [12] as well as most of the next (G78.2+2.1, W28, IC 443, γ-Cygni, W44, W63, Tycho; see short reveiw in [25]) have given a negative result. Till now TeV γ-rays were observed from only two SNRs.

SN 1006 is the first shell-like SNR detected as a TeV source [33]. Whereas authors in [33] attributed this emission to synchrotron, the emission is tried to be explained by other mechanisms, too. Recent discussion on the origin of TeV γ-rays form SN 1006 [2] shows that a proton origin of γ-rays is possible if distance to the remnant is about 1 kpc and there is a significant compression of gas. Nevertheless other emission mechanisms may not be excluded (see, e. g. [21]). The observations of G347.3-0.5 reported recently [19] also reveal a TeV γ-flux. Authors attributed this emission to inverse Compton scattering of CMBR by shock accelerated electrons and estimate π^o induced contribution as too low, but did not exclude it because of possibility of interactions with a cloud [28].

The first [9], second [34] and the third [11] EGRET catalogue of the high-energy γ-ray sources among the general categories (solar flare, pulsars, γ-ray bursts, radio galaxy, active galactic nuclei and blazars), include lists of sources for which no identification with objects at other wavelengths is yet surely found. The analysis of possible associations of the SNR positions with error circles of the unidentified γ-sources was first performed in [31]. Analysis of unidentified sources from the third EGRET catalog [23] shows 22 such possible assosiations with probability of being of chance less than 10^{-5}. Some other unidentified EGRET sources might be related to yet undetected SNR [7]. Of course, a part of these remnants have or may have their compact stellar remnants [6], pulsars, which can be responsible for the γ-ray emission. There are 225 known SNRs in our Galaxy [10]. The short history of studies on association of EGRET sources with SNRs may be found in [23].

Up to now, there is no clear observational confirmation that nuclear component of CRs is generated by SNRs. There is only one observation reported as confirmation of this thought [7]. The nature of γ-emission in all observations is very questionable.

There are several emission processes in competition in analysis of observed SNR γ-ray spectra. It depends on the conditions in emission site and on the way to the observer which one will dominate. It is well known that we should have a high number density of target nuclei in order to make the π^o decay mechanism dominate in a model. Therefore, γ-rays from proton-proton collisions are expected from the SNRs which are located near the dense interstellar material and reveal evidence about interaction with it.

3 Thermal X-ray composites. A new model in 3 dimensions

To look for signatures of proton acceleration in SNR, it is interesting to consider a mixed-morphology class of SNRs which is known also as thermal X-ray composites (TXCs). These are remnants with a thermal X-ray centrally filled morphology within the radio-brightened limb. Such remnants were first reviewed in [26]. The authors of [22] argued that they create a separate morphological class of SNR. It is interesting that most of these SNRs reveal observational evidence about the interaction with nearby molecular clouds. There are 7 remnants in the analysis of assosiation of the 2EG sources with SNRs [32]. Four of them (W28, W44, MSH 11-61A, IC443) belong to the class mentioned and interact with clouds.

The authors of [22] have emphasized two prominent morphological distinctions of TXCs: a) the X-ray emission is thermal; the distribution of X-ray surface brightness is centrally peaked and fills the area within the radio shell, b) the emission arises primarily from the swept-up ISM material, not from ejecta. Besides similar morphology, the sample of SNRs also has similar physical properties [22, 20]. Namely, i) the same or higher central density compared to the edge, ii) complex interior optical nebulosity, iii) higher emission measure in the region of X-ray peak localisation, v) temperature profiles are close to uniform. Seven objects from the list of 11 TXCs interact with molecular clouds [22]. Thus, ambient media in the regions of their location are nonuniform and cause a nonsphericity of SNRs.

Two physical models have been presented to explain TXC (see review in [22, 20]). These models are used to obtain the centrally filled X-ray morphology within the framework of one-dimensional (1-D) hydrodynamic approaches. When we proceed to 2-D or 3-D hydrodynamical models, we note that a simple projection effect may cause the shell-like SNR to fall into another morphology class, namely, to become TXC [14]. The main feature of such a SNR is the thermal X-rays emitted from the swept-up gas and peaked in the internal part of the projection. Densities over the surface of a nonspherical SNR may essentially differ in various regions. If the ambient density distribution provides a high density in one of the regions across the SNR shell and is high enough to exceed the internal column density near the edge of the projection, we will see a centrally filled X-ray projection. Thus, this new model strongly suggests that the thermal X-ray peak inside the radio shell tells us that one part of SNR shock has entered into a denser medium compared with other parts. This makes TXCs promising sites for γ-ray generation via π^o decays.

4 SNR γ-rays from decay of neutral π-mesons

Whereas astrophysical realisations of several emission mechanisms allow us to make conclusions about acceleration of CR electron component on the shock fronts of SNRs, only π^o decay γ-rays [5] give us the possibility to look at the proton component acceleration processes having data in $\varepsilon_\gamma \geq 100$ MeV.

4.1 ESTIMATIONS ON THE π^o DECAY γ-RAY LUMINOSITY

Luminosity of an SNR in π^o decay γ-rays in $\varepsilon \geq \varepsilon_{min}/6$ band is [31, 13]:

$$L_\gamma = \frac{c\overline{\sigma}_{pp}}{6} n_N W_{cr}. \tag{1}$$

where $\varepsilon_{min} \approx 600$ MeV is the minimal proton kinetic energy of the effective pion creation, cross section $\sigma_{pp}(\varepsilon)$ is close to the mean value $\overline{\sigma}_{pp} = 3 \cdot 10^{-26}$ cm^2, n_N is the mean number density of target nuclei and W_{cr} is the total energy of cosmic rays in the SNR with $\varepsilon \geq \varepsilon_{min}$. There are different estimations of the efficiency ν of the flow's kinetic energy transformation into the energy of accelerated particles: $W_{cr} = \nu E_o$. We take acceptable value $\nu = 0.03$. Thus, in the first approach, the theoretical estimation on the π^o decay γ-luminosity of any SNR is

$$L_\gamma^{\geq 100} = 6.3 \cdot 10^{33} \overline{n}_o \nu_3 E_{51} \quad \text{erg/s},$$

where $\nu_3 = \nu/0.03$, E_{51} is the energy of supernova explosion E_o in the units of 10^{51} erg, $\overline{n}_o = \overline{n}_N^o/1.4$ is the average hydrogen number density within SNR which equals to average hydrogen number density of the ambient medium in the region of an SNR location, in cm^{-3}.

The real situation is more complicated. Often only a part of SNR interacts with a denser ISM material. There are factors which increase CR energy density ω_{cr} [13]: 1) The CRs are not uniformly distributed inside an SNR; most of CRs are expected to be in a thin shell near the shock front where most of swept-up mass is concentrated. 2) The reverse shock from interaction with dense cloud also increases the energy density of CRs. These factors enhance ω_{cr} in the region of interaction [13]:

$$\omega_{cr} \approx 1.7 \left(\frac{\gamma}{\gamma-1} \right)^{3/2} \overline{\omega}_{cr}, \tag{2}$$

where CR energy density in the region of interaction is $\omega_{cr} = W_{cr}/V_{int}$ and $\overline{\omega}_{cr} = W_{cr}/V_{snr}$. This gives $\omega_{cr} \approx 6.6\, \overline{\omega}_{cr}$ for $\gamma = 5/3$.

If we put $W_{cr} = \omega_{cr}V_{int}$ into (1), we obtain with (2) that for any SNR

$$L_\gamma^{\geq 100} = 1.7 \cdot 10^{35} \eta n_o \nu_3 E_{51} \quad \text{erg/s} \tag{3}$$

where $\eta = V_{int}/V_{snr}$, n_o is the number density of the ambient medium before the shock wave in the region of interaction, $\gamma = 5/3$. We have to take into account that region of interaction is not extended to the region before the shock, since the energy density of CR should be considerably lower outside the SNR [5, 13]. It is easy to estimate η following the consideration in [13]:

$$\eta = 0.18\mu^2\sqrt{\xi}, \qquad \mu \leq 0.2, \qquad (4)$$

where $\xi = n_o/\overline{n}_o$, $\mu = R_{int}/\overline{R}$, R_{int} is the avarage radius of the surface of interaction, $\overline{R}$ is the avarage radius of SNR.

Thus, considering the hydrodynamic process of SNR-cloud interaction in details, it is possible to increase the expected π^o decay γ-ray flux by $26\eta\xi \simeq 0.2\xi^{3/2}$ times, i. e., up to few orders of magnitude. This is enough e.g. for explanation of TeV γ-rays from SN1006 as a result of decays of neutral pions [33]. Additional factor increasing the effectivity of π^o meson production is instability of contact discontinuity between the SNR and the cloud material which mixes the media with high energy particles and target nuclei [13]. This factor should also be considered in future.

4.2 NUCLEONIC ORIGIN γ-RAYS FROM MSH 11-61A

SNR G290.1-0.8 (MSH 11-61A) is located in the southern hemisphere. The distance to the remnant, 7 kpc, is obtained from the optical observations [24], but not yet confirmed by X-ray observations [30]. X-ray and radio morphologies [27, 35] make the SNR a member of TXC class. In the direction to MSH 11-61A lies the γ-ray source 2EG 1103-6106 (3EG J1102-6103) [32, 23]. Is it possible to consider observed flux of the EGRET source directed toward the MSH 11-61A as π^o decay γ-ray emission?

The γ-flux from the source 2EG 1103-6106 in the EGRET band $\varepsilon_\gamma = 30 \div 2 \cdot 10^4$ MeV is approximated as [18]

$$S_\gamma = (1.1 \pm 0.2) \cdot 10^{-9} \left(\frac{\varepsilon_\gamma}{213 \text{ MeV}}\right)^{-2.3\pm0.2} \frac{\text{photon}}{\text{cm}^2 \cdot \text{s} \cdot \text{MeV}}.$$

Thus, the luminosity of the source in $\varepsilon_\gamma \geq 100$ MeV band is respectively

$$L_{\gamma,\text{obs}}^{\geq100} = 4 \cdot 10^{34} \, d_{\text{kpc}}^2 \quad \text{erg/s}.$$

Most recent study on MSH 11-61A [30] gives the parameters of the object presented in Table 1 (different for different distance assumed). There are also values of ξ in the table which allow to adjust the expected luminosity of MSH 11-61A in π^o decay γ-rays with observed flux from the source 2EG 1103-6106. We see that moderate number density ~ 150 cm^{-3} of cloud located near the one region of the remnant is enough to explain the luminocity of 2EG 1103-6106, by protons accelereted on the shock front of MSH 11-61A. Note, if we take $\nu \simeq 0.1 - 0.2$ [1] instead of the value used here, $\nu = 0.03$, the density of cloud needs to be only $\sim 20 \div 40$ cm^{-3}. It is interesting that the same consideration allows also to adjust the π^o decay γ-luminosity of the source 2EG J0618+2234 (3EG J0617+2238) directed toward IC 443 with the luminosity of this SNR [13].

TABLE 1: Parameters of MSH 11-61A [30], luminosity $L_{\gamma,obs}^{\geq 100}$ of 2EG 1103-6106 and estimations on the proton origin γ-luminosity of the SNR. Presented values of ξ allow us to satisfy condition $L_{\gamma,obs} = L_\gamma$.

$d,$ kpc	Age $t,$ 10^4 yrs	$\overline{R},$ pc	$E_{51},$ 10^{51} erg	$\overline{n}_o,$ cm^{-3}	$L_{\gamma,obs}^{\geq 100},$ 10^{36} erg/s	$L_\gamma^{\geq 100}/\xi^{3/2},$ 10^{32} erg/s	$\xi,$ 10^2
10	1.3	18	1	0.27	4	3.2	5
7	0.9	13	0.4	0.27	2	1.3	6

5 Conclusions

We may expect that nucleonic component of CRs accelerated on SNR shocks have to reveal itself in observations. Which SNRs might emit γ-rays mainly via neutral pion decays? Presented model for TXCs makes the members of this class very promising candidates for the π^o decay γ-ray sources since a central thermal X-ray peaked brightness might be evidence for density gradient and high density in one of the regions before SNR shock. CR distribution inside SNR is not uniform; most of CR should be confined in a relatively thin SNR shell. CRs are accelerated not only by the forward shock, but also by the reverse one. Consideration of these factors causes an enhancemnent of the π^o decay γ-ray flux of at least on an order of magnitude. This allows us to adjust the theoretical flux with observed one in at least few cases. So, we have enigmatic situation. There are the class of prospective sources. Theory can fit the π^o decay γ-ray fluxes. Why we do not have many direct observations of such γ-rays in order to confirm theoretical predictions about CR generation by SNRs? The question remains open.

Acknowledgements. Author is thankful to the organizers for support which allows him to participate in the Course.

References

[1] Aharonian, F., Drury, L., Voelk, H. (1994) GeV/TeV gamma-ray emission from dense molecular clouds overtaken by supernova shells, *A&A* **285**, 645.

[2] Aharonian, F., Atoyan, A. (1999) On the origin of TeV radiation of SN 1006, *A&A* **351**, 330.

[3] Allen, G., Keohane, J. et al. (1997) Evidence of X-Ray Synchrotron Emission from Electrons Accelerated to 40 TeV in the Supernova Remnant Cassiopeia A, *ApJ* **487**, L97.

[4] Ammosov, et al. (1994) Synchrotron emission from type I supernova remnants, *Ast. Let.* **20**, 157

[5] Berezinskii V., Bulanov S., Dogiel V. et al. (1990) *Astrophysics of cosmic rays*, North Holland, Amsterdam.

Gaisser T. K. (1990) *Cosmic rays and particle physics*, Cambridge University Press, Cambridge.

[6] Chevalier R.A. (1996) Compact Objects in Supernova Remnants // *Supernovae and Supernova Remnants* (Proceedings IAU Coloquium 145, May, 24-29, 1993), Cambridge University Press, Cambridge, pp.399-406.

Jones, T., Jun, B., Borkowski, K., et al. (1998) 10^{51} Ergs: The Evolution of Shell Supernova Remnants, *PASP* **110**, 125-151.

[7] Combi, J., Romero, G., Benaglia, P. (1998) The gamma-ray source 2EGS J1703-6302: a new supernova remnant in interaction with an HI cloud? *A&A* **333**, L91.

[8] Drury, L., Aharonian, F., Voelk, H. (1994) The gamma-ray visibility of supernova remnants. A test of cosmic ray origin, *A&A* **287**, 959.

[9] Fichtel, C., Bertsch, D. et al. (1994) The first energetic gamma-ray experiment telescope (EGRET) source catalog, *ApJ Suppl.* **94**, 551.

[10] Green D.A. (2000) 'A Catalogue of Galactic Supernova Remnants (2000 August version)', Mullard Radio Astronomy Observatory, Cavendish Laboratory, Cambridge, United Kingdom (available on the World-Wide-Web at http://www.mrao.cam.ac.uk/surveys/snrs/)

[11] Hartman, R., Bertsch, D. et al. (1999) The Third EGRET Catalog of High-Energy Gamma-Ray Sources, *ApJ Suppl.* **123**, 79.

[12] Hillas (1995) *Proc. 24th ICRC, Inv., par. and highl. papers*, p.701.

[13] Hnatyk, B., & Petruk, O. (1998) Supernova Remnants as Cosmic Ray Accelerators, *Condensed Matter Physics* **1**, 655 [available also as preprint astro-ph/9902158].

[14] Hnatyk B., Petruk O. (1999) Evolution of supernova remnants in the interstellar medium with a large-scale density gradient. 1. General properties of the morphological evolution and X-ray emission, *A&A* **344**, 295.

[15] Keohane, J., Petre, R. et al. (1997) A Possible Site of Cosmic Ray Acceleration in the Supernova Remnant IC 443, *ApJ* **484**, 350.

[16] Koyama, K., Petre, R. et al. (1995) Evidence for Shock Acceleration of High-Energy Electrons in the Supernova Remnant SN:1006, *Nature* **378**, 255.

Willingale, R., West, R. et al. (1996) ROSAT PSPC observations of the remnant of SN 1006, *MNRAS* **278**, 749.

[17] Koyama, K., Kinugasa, K. et al. (1997) Discovery of Non-Thermal X-Rays from the Northwest Shell of the New SNR RX J1713.7-3946: The Second SN 1006? *PASJ* **49**, L7.

[18] Merck M., Bertsch D. L., et al. (1996) Study of the spectral characteristics of unidentified galactic EGRET sources. Are they pulsar-like? *A&A Suppl.* **120**, 465.

[19] Muraishi, H., Tanimori, T. et al. (2000) Evidence for TeV gamma-ray emission from the shell type SNR RX J1713.7-3946, *A&A* **354**, L57.

[20] Petruk, O. (2001) Thermal X-ray Composits as an Effect of Projection, *A&A* accepted [astro-ph/0006161]

[21] Reynolds, S. (1996) Synchrotron Models for X-Rays from the Supernova Remnant SN 1006, *ApJ* **459**, L13.

[22] Rho J., Petre R. (1998) Mixed-Morphology Supernova Remnants, *ApJ* **503**, L167.

[23] Romero, G., Benaglia, P., Torres, D. (1999) Unidentified 3EG gamma-ray sources at low galactic latitudes, *A&A* **348**, 868.

[24] Rosado, M., Ambrocio-Cruz, P. et al. (1996) Kinematics of the galactic supernova remnants RCW 86, MSH 15-56 and MSH 11-61A, *A&A* **315**, 243.

[25] Rowell, G., Naito, T. et al. (2000) Observations of the supernova remnant W28 at TeV energies, *A&A* **359**, 337.

[26] Seward, F. D. (1985) *Comments Astrophys.* XI **1**, 15.

[27] Seward F. (1990) Einstein Observations of Galactic supernova remnants, *ApJ Suppl.* **73**, 781.

[28] Slane, P., Gaensler, B. et al. (1999) Nonthermal X-Ray Emission from the Shell-Type Supernova Remnant G347.3-0.5, *ApJ* **525**, 357.

[29] Slane, P., Huges, J. et al. (2001) RXJ 0852.0-0462: Another Nonthermal Shell-Type SNR (G266.2-1.2), *ApJ* **548**, 814.

[30] Slane, P. et. al., in preparation

[31] Sturner S., Dermer, C. (1995) Association of unidentified, low latitude EGRET sources with supernova remnants, *A&A* **293**, L17.

[32] Sturner S. J., Dermer C. D., Mattox J. R. (1996) Are supernova remnants sources of > 100 MeV γ-rays? *A&A Suppl.* **120**, 445.

[33] Tanimori, T., Hayami, Y. et al. (1998) Discovery of TeV Gamma Rays from SN 1006: Further Evidence for the Supernova Remnant Origin of Cosmic Rays, *ApJ* **497**, L25.

[34] Thompson, D. J., Bertsch, D. et al. (1995) The Second EGRET Catalog of High-Energy Gamma-Ray Sources, *ApJ Suppl.* **101**, 259.

Thompson, D. J., Bertsch, D. et al. (1996) Supplement to the Second EGRET Catalog of High-Energy Gamma-Ray Sources, *ApJ Suppl.* **107**, 227.

Kanbach, G., Bertsch, D. et al. (1996) Characteristics of galactic gamma-ray sources in the second EGRET catalog, *A&A Suppl.* **120**, 461.

[35] Whiteoak J.B., Green A. (1996) The MOST supernova remnant catalogue (MSC), *A&A Suppl.* **118**, 329.

Evolution of Isolated Neutron Stars

Sergei Popov
Sternberg Astronomical Institute

November 10, 2000

Abstract.

In this paper we briefly review our recent results on evolution and properties of isolated neutron stars (INSs) in the Galaxy.

As the first step we discuss stochastic period evolution of INSs. We briefly discuss how an INS's spin period evolves under influence of interaction with turbulized interstellar medium.

To investigate statistical properties of the INS population we calculate a *census* of INSs in our Galaxy. We infer a lower bound for the mean kick velocity of NSs, $< V > \sim (200\text{-}300)$ km s^{-1}. The same conclusion is reached for both a constant magnetic field ($B \sim 10^{12}$ G) and for a magnetic field decaying exponentially with a timescale $\sim 10^9$ yr. These results, moreover, constrain the fraction of low velocity NSs, which could have escaped pulsar statistics, to $\sim$few percents.

Then we show that for exponential field decay the range of minimum value of magnetic moment, μ_b: $\sim 10^{29.5} \geq \mu_b \geq 10^{28}$ G cm^3, and the characteristic decay time, t_d: $\sim 10^8 \geq t_d \geq 10^7$ yrs, can be excluded assuming the standard initial magnetic momentum, $\mu_0 = 10^{30}$ G cm^3, if accreting INSs are observed. For these parameters an INS would never reach the stage of accretion from the interstellar medium even for a low space velocity of the star and high density of the ambient plasma. The range of excluded parameters increases for lower values of μ_0.

It is shown that old accreting INSs become more abundant than young cooling INSs at X-ray fluxes below $\sim 10^{-13}$ erg cm^{-2} s^{-1}. We can predict that about one accreting INS per square degree should be observed at the *Chandra* and *Newton* flux limits of $\sim 10^{-16}$ erg cm^{-2} s^{-1}. The weak *ROSAT* sources, associated with INSs, can be young cooling objects, if the NSs birth rate in the solar vicinity during the last $\sim 10^6$ yr was much higher than inferred from radiopulsar observations.

Keywords: neutron stars, magnetic field, accretion

Introduction

Despite intensive observational campaigns, no irrefutable identification of an isolated accreting neutron star (NS) has been presented so far. Six soft sources have been found in ROSAT fields which are most probably associated to isolated radioquiet NSs. Their present X-ray and optical data however do not allow an unambiguous identification of the physical mechanism responsible for their emission. These sources can be powered either by accretion of the interstellar gas onto old ($\approx 10^{10}$ yr) NSs or by the release of internal energy in relatively young ($\approx 10^6$ yr) cooling NSs (see (Treves et al., 2000) for a recent review). The ROSAT candidates,

M. M. Shapiro et al. (eds.), Astrophysical Sources of High Energy Particles and Radiation, 101–110.

although relatively bright (up to ≈ 1 cts s^{-1}), are intrinsically dim and their inferred luminosity ($L \approx 10^{31}$ erg s^{-1}) is near to that expected from either a close-by cooling NS or from an accreting NS among the most luminous. Their X-ray spectrum is soft and thermal, again as predicted for both accretors and coolers (Treves et al., 2000). Up to now only two optical counterparts have been identified (RXJ 1856, (Walter and Matthews, 1997); RXJ 0720, (Kulkarni and van Kerkwijk, 1998)) and in both cases an optical excess over the low-frequency tail of the black body X-ray spectrum has been reported. While detailed multiwavelength observations with next-generation instruments may indeed be the key for assessing the true nature of these sources, other, indirect, approaches may be used to discriminate in favor of one of the two scenarios proposed so far.

Since early 90^s, when in (Treves and Colpi, 1991) it was suggested to search for IANSs with ROSAT satellite, several investigations on INSs have been done (see for example (Madau and Blaes, 1994), (Manning et al., 1996), and (Treves et al., 2000) for a review). Here we present our recent results in that field.

Stochastic evolution of isolated neutron stars

One can distinguish four main stages of INS evolution: Ejector, Propeller, Accretor and Georotator (see (Lipunov, 1992) for more detailes). Destiny of an INS depends on its spin period, magnetic field, spatial velocity and properties of the interstellar medium (ISM). Schematically typical evolutionary paths can be shown on p-y diagram, where $y = \dot{M}/\mu^2$ – gravimagnetic parameter (Fig. 1).

An INS normally is born as Ejector. Then it spins down, and appear as Propeller. At last it can reach the stage of accretion, if its spatial velocity is low enough, or if the magnetic field is high.

As an INS evolves at the Accretor stage it more and more feels the influence of the turbulized nature of the ISM (Fig. 2). Initially isolated accreting NS spins down due to magnetic breaking, and the influence of accreted angular momentum is small. At last, at the stage of accretion, when the INS spun down enough, its rotational evolution is mainly governed by the turbulence, and the accreting angular moment fluctuates.

Observations of p and $\dot{p}$ of isolated accreting NSs can provide us not only with the information about INSs properties, but also about different properties of the ISM itself.

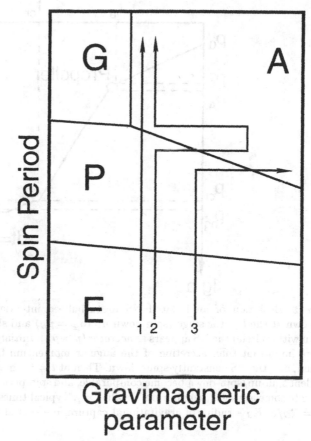

Figure 1. P-y diagram. Three evolutionary tracks are shown: 1 – evolution with constant field in constant medium. The NS spins down passing through different stages (Ejector, Propeller, Accretor); 2 – passing through a giant molecular cloud. The accretion rate is increased for some time, and the NS makes a loop on the diagram; 3 – evolution with field decay. The NS appears as Accretor due to fast field decay.

Neutron Star Census

We have investigated, (Popov et al., 2000a), how the present distribution of NSs in the different stages (Ejector, Propeller, Accretor and Georotator, see (Lipunov, 1992)) depends on the star mean velocity at birth (see Fig. 3). The fraction of Accretors was used to estimate the number of sources within 140 pc from the Sun which should have been detected by ROSAT. Most recent analysis of ROSAT data indicate that no more than ~ 10 non-optically identified sources can be accreting old INSs. This implies that the average velocity of the INSs population at birth has to exceed ~ 200 km s^{-1}, a figure which is consistent with those derived from radio pulsars statistics. We have found that this

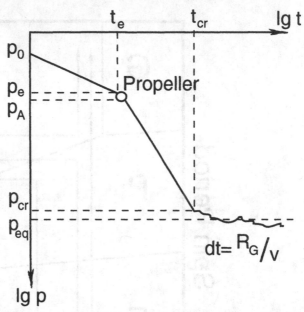

Figure 2. Evolution of an isolated NS in turbulized interstellar medium. After spin-down at the Ejector stage (spin-down up to $p = p_E$) and short Propeller stage (shown with a circle) the NS appears as accretor ($p > p_A$). Initialy magnetic breaking is more important than accretion of the angular momentum from the interstellar medium, and the NS constantly spins down. Then at $t = t_{cr}$ magnetic breaking and turbulent spin-up/spin-down become comparable, and spin period starts to fluctuate coming to some average ("equilibrium") value, p_{eq}. Typical timescale for fluctuations is $dt = R_G/v$, R_G – radius of gravitational capture, v – spatial velocity.

lower limit on the mean kick velocity is substantially the same either for a constant or a decaying B–field, unless the decay timescale is shorter than $\sim 10^9$ yr. Since observable accretion–powered INSs are slow objects, our results exclude also the possibility that the present velocity distribution of NSs is richer in low–velocity objects with respect to a Maxwellian. The paucity of accreting INSs seem to lend further support in favor of NSs as fast objects.

Magnetic Field Decay

Magnetic field decay can operate in INSs. Probably, some of observed ROSAT INS candidates represent such examples ((Konenkov and Popov,

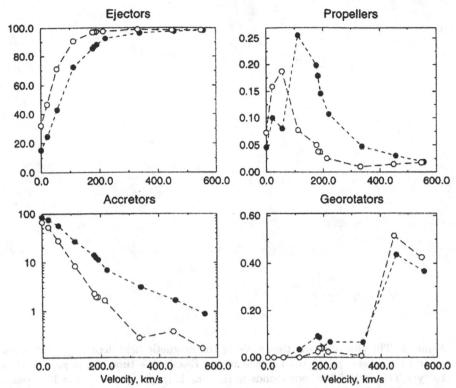

Figure 3. Fractions of NSs (in percents) on different stages vs. the mean kick velocity for two values of the magnetic moment: $\mu_{30} = 0.5$ (open circles) and $\mu_{30} = 1$ (filled circles). Typical statistical uncertainty for ejectors and accretors is $\sim$ 1-2%. Figures are plotted for constant magnetic field.

1997), (Wang, 1997)) We tried to evaluate the region of parameters which is excluded for models of the exponential magnetic field decay in INSs using the possibility that some of ROSAT soft X-ray sources are indeed old AINSs.

In this section we follow the article (Popov and Prokhorov, 2000).

Here the field decay is assumed to have an exponential shape:

$$\mu = \mu_0 \cdot e^{-t/t_d}, \text{ for } \mu > \mu_b \tag{1}$$

where μ_0 is the initial magnetic moment ($\mu = \frac{1}{2}B_p R_{NS}^3$, here B_p is the polar magnetic field, R_{NS} is the NS radius), t_d is the characteristic time scale of the decay, and μ_b is the bottom value of the magnetic momentum which is reached at the time

$$t_{cr} = t_d \cdot \ln\left(\frac{\mu_0}{\mu_b}\right) \tag{2}$$

and does not change after that.

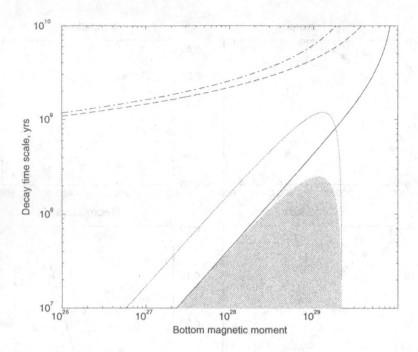

Figure 4. The characteristic time scale of the magnetic field decay, t_d, vs. bottom magnetic moment, μ_b. In the hatched region Ejector life time, t_E, is greater than 10^{10}yrs. The dashed line corresponds to $t_H = t_d \cdot \ln(\mu_0/\mu_b)$, where $t_H = 10^{10}$ years. The solid line corresponds to $p_E(\mu_b) = p(t = t_{cr})$, where $t_{cr} = t_d \cdot \ln(\mu_0/\mu_b)$. Both the lines and hatched region are plotted for $\mu_0 = 10^{30}\,\mathrm{G\,cm^{-3}}$. The dash-dotted line is the same as the dashed one, but for $\mu_0 = 5 \cdot 10^{29}\,\mathrm{G\,cm^3}$. The dotted line shows the border of the "forbidden" region for $\mu_0 = 5 \cdot 10^{29}\,\mathrm{G\,cm^3}$.

The intermediate values of t_d ($\sim 10^7 - 10^8$ yrs) in combination with the intermediate values of μ_b ($\sim 10^{28} - 10^{29.5}\,\mathrm{G\,cm^3}$) for $\mu_0 = 10^{30}\,\mathrm{G\,cm^3}$ can be excluded for progenitors of isolated accreting NSs because NSs with such parameters would always remain on the Ejector stage and never pass to the accretion stage (see Fig. 4). Even if all modern candidates are not accreting objects, the possibility of limitations of magnetic field decay models based on future observations of isolated accreting NSs should be addressed.

For higher μ_0 NSs should reach the stage of Propeller (i.e. $p = p_E$, where p_E- is the Ejector period) even for $t_d < 10^8$ yrs, for weaker fields the "forbidden" region becomes wider. Critical period, p_E, corresponds to transition from the Propeller stage to the stage of Ejector, and is about 10-25 seconds for typical parameters. The results are dependent on the initial magnetic field, μ_0, the ISM density, n, and NSs velocity, V. So here different ideas can be investigated.

In fact the limits obtained above are even stronger than they could be in nature, i.e. "forbidden" regions can be wider, because we did not take into account that NSs can spend some significant time (in the case with field decay) at the propeller stage (the spin-down rate at this stage is very uncertain, see the list of formulae, for example, in (Lipunov and Popov, 1995) or (Lipunov, 1992)). The calculations of this effect for different models of non-exponential field decay were made separately (Popov and Prokhorov, 2001).

Note that there is another reason due to which a very fast decay down to small values of μ_b can also be excluded, because this would lead to a huge amount of accreting isolated NSs in drastic contrast with observations. This situation is similar to the "turn-off" of the magnetic field of an INS (i.e., quenching any magnetospheric effect on the accreting matter). So for any velocity and density distributions we should expect significantly more accreting isolated NSs than we know from ROSAT observations (of course, for high velocities X-ray sources will be very dim, but close NSs can be observed even for velocities ~ 100 km s^{-1}).

Log N – Log S distribution

In this section we briefly present our new results on INSs, (Popov et al., 2000b).

We compute and compare the $\log N - \log S$ distribution of both accreting and cooling NSs, to establish the relative contribution of the two populations to the observed number counts. Previous studies derived the $\log N - \log S$ distribution of accretors ((Treves and Colpi, 1991); (Madau and Blaes, 1994); (Manning et al., 1996)) assuming a NSs velocity distribution rich in slow stars ($v < 100$ km s^{-1}). More recent measurements of pulsar velocities (e.g. (Lyne and Lorimer, 1994)) and upper limits on the observed number of accretors in ROSAT surveys point, however, to a larger NS mean velocity (see (Treves et al., 2000) for a critical discussion). Recently in (Neühauser and Trümper, 1999) the authors compared the number count distribution of the ROSAT isolated NS candidates with those of accretors and coolers. In (Popov et al., 2000b) we address these issues in greater detail, also in the light of the latest contributions to the modeling of the evolution of Galactic NSs.

Our main results for AINSs are presented in Fig. 5 and Fig. 6.

108

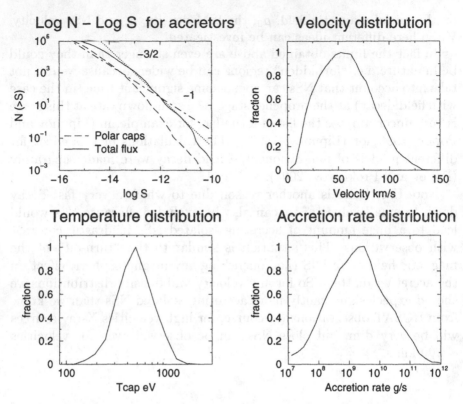

Figure 5. Upper left panel: the $\log N - \log S$ distribution for accretors within 5 kpc from the Sun. The two curves refer to total emission from the entire star surface and to polar cap emission in the range 0.5-2 keV; two straight lines with slopes -1 and -3/2 are also shown for comparison. From top right to bottom right: the velocity, effective temperature and accretion rate distributions of accretors; all distributions are normalized to their maximum value.

In Fig. 5 we show main parameters of accretors in our calculations: Log N – Log S distribution, distributions of velocity, accretion rate and temperature (for polar cap model).

In Fig. 6 we present joint Log N - Log S for accretors in our calculations, observed candidates, naive estimate from (Popov et al., 2000a), and our calculations for coolers in a simple model of local sources (Popov et al., 2000b).

Using "standard" assumptions on the velocity, spin period and magnetic field parameters, the accretion scenario can not explain the observed properties of the six ROSAT candidates.

A key result of our statistical analysis is that accretors should eventually become more abundant than coolers at fluxes below 10^{-13} erg cm^{-2} s^{-1}.

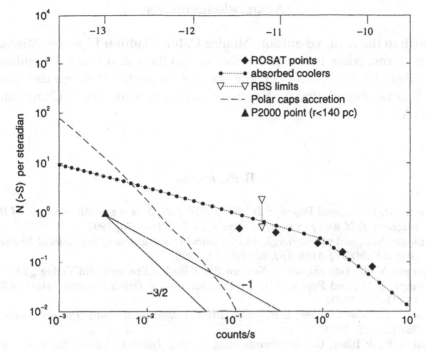

Figure 6. Comparison of the log N – log S distributions for accretors and coolers together with observational points, the naive log$N$ – log$S$ from P2000a and the ROSAT Bright Survey (RBS) limit. The scale on the top horizontal axes gives the flux in erg cm^{-2}s^{-1}.

Conclusions

INSs are now really *hot* objects. In many cases INSs can show different effects in the most "pure" form: without influence of huge accretion rate, for example.

We tried to show how these objects are related with models of magnetic field decay, and with recent and future X-ray observations.

Observed candidates propose "non-standard" properties of NSs. Future observations with XMM (Newton) and Chandra satellites can give more important facts.

INSs without usual radio pulsar activity (SGRs, AXPs, compact X-ray sources in SNRs, dim ROSAT candidates, Geminga, dim sources in globular clusters) together show "non-standard" or better say more complete picture of NS nature. Future investigations are strongly wanted.

Acknowledgements

I wish to thank my co-authors Monica Colpi, Vladimir Lipunov, Mikhail Prokhorov, Aldo Treves and Roberto Turolla. I also thank Organizers for their kind hospitality. The work was supported through the grant of Russian Foundation for Basic Research, Scientific Travel Center and INTAS.

References

Konenkov, D.Yu., and Popov, S.B. *RX J0720.4-3125 as a possible example of the magnetic field decay in neutron stars PAZH*, 23:569-575, 1997

Kulkarni, S.R., and van Kerkwijk, M.H. *Optical Observations of the Isolated Neutron Star RX J0720.4-3125 ApJ*, 507:L49-52, 1998.

Lipunov, V.M. *Astrophysics of Neutron Stars* Berlin: Springer and Verlag , 1992.

Lipunov, V.M., and Popov, S.B. *Period evolution of isolated neutron stars AZh*, 71:711-716, 1995.

Lyne, A.G., and Lorimer, D.R. *High Birth Velocities of Radio Pulsars Nature*, 369:127-129, 1994.

Madau, P., & Blaes, O. *Constraints on Accreting, Isolated Neutron Stars from the ROSAT and EUVE Surveys ApJ*, 423:748-752, 1994.

Manning, R.A., Jeffries, R.D., and Willmore A.P. *Are there any isolated old neutron stars in the ROSAT Wide Field Camera survey? MNRAS*, 278:577-585, 1996.

Neühauser, R., & Trümper, J.E. *On the number of accreting and cooling isolated neutron stars detectable with the ROSAT All-Sky Survey A& A*, 343:151-156, 1999.

Popov, S.B., Colpi, M., Treves, A., Turolla, R., Lipunov, V.M., and Prokhorov, M.E. *The Neutron Star Census ApJ*, 530:896-903, 2000.

Popov, S.B., and Prokhorov, M.E. *ROSAT X-ray sources and exponential field decay in isolated neutron stars A&A*, 357:164-168, 2000

Popov, S.B., and Prokhorov, M.E. *Restrictions on parameters of power-law magnetic field decay for accreting isolated neutron stars Astro. Astroph. Trans.* accepted, 2001 (astro-ph/0001005)

Popov, S.B., Colpi, M., Prokhorov, M.E., Treves, A., and Turolla, R. *Log N - Log S distributions of accreting and cooling isolated neutron stars ApJ*, 544:L53-L56, 2000 (astro-ph/0009225)

Treves, A., and Colpi, M. *The observability of old isolated neutron stars A& A*, 241:107-111, 1991.

Treves, A., Turolla, R., Zane, S., and Colpi, M. *Isolated Neutron Stars: Accretors and Coolers PASP*, 112:297-314, 2000.

Walter, F., & Matthews, L.D. *The optical counterpart of the isolated neutron star RX J185635-3754 Nature*, 389:358-359, 1997.

Wang, J. *Evidence for Magnetic Field Decay in RX J0720.4-3125 ApJ*, 486:L119-L123, 1997.

HIGH ENERGY OUTBURST OF RECURRENT TRANSIENT PULSAR A0535+26

T.N. DOROKHOVA & N.I. DOROKHOV
Astronomical Observatory of Odessa State University, park T.G.Shevchenko, 65014 Odessa, Ukraine

Abstract. The optical outburst of an extremely high energy occurred in X-ray/Be system A0535+26 on 28/29 Oct. 1995. The luminosity of the maximal flash of the outburst amounted to $\sim 1.4 * 10^{38}$ erg/sec, the same order of magnitudes as the luminosity of the 'giant' X-ray outburst of the system. Probably, the flashes were even more powerful and more transient. These superfast and superpower processes could arise in the transient accretion disk of the pulsar with the presence of strong and rapidly rotating magnetic field. It is obvious the nonthermal nature of the phenomenon, although output mechanism of such great energy in the optics during some tenths of second is unclear.

The recurrent X-ray pulsar A0535+26 (HDE 245770, V725 Tau) belongs to the group of hard transient X-ray sources (HTXS) associated with high mass Be/neutron star binaries ([?]). In GCVS[?] V725 Tau is related to the class of XNG (X-ray Novalike Giants).

We discovered the unusual flare of A0535+26 on the night 28/29 Oct. 1995 ([?]) during the observations within the framework of the Russian Ministry of Science Program "Monitoring of Unique Astrophysical Objects". The dual-channel continuous photometry was carried out with the Johnson's B filter and 10 sec integration time at the Mt. Dushak-Erekdag branch of Odessa Astronomical Observatory. The flare consisted of a great number of transitory sharp flashes, the pulses. Every pulse lasted no more than 20 sec, evidently this duration was dictated by the 10 sec time resolution. We called this optical flare "the multi-pulsed outburst" by analogy with X-ray outbursts. The outburst's maximal pulse was $1.^m 74$, that means the star became brighter by the factor of ~ 4.79 during 10 sec. In Fig.1 the pattern of the outburst is shown in energetic units which were converted from the magnitudes. A mean luminosity of the source in the Johnson's B filter

111

M. M. Shapiro et al. (eds.), Astrophysical Sources of High Energy Particles and Radiation, 111–114.
© 2001 *Kluwer Academic Publishers. Printed in the Netherlands.*

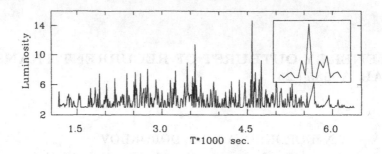

Figure 1. The light curve of the outburst of A0535+26 in energetic units (10^{37} erg/sec) relative to the quiescent state level, $\sim 3 * 10^{37}$ erg/sec. The time is in seconds relative to the start of the observations. In the insert: the maximum at 4600 sec (the scope 150 sec) is stretched in the time.

may be estimated as $\sim 3 * 10^{37}$ erg/sec, taking into account that the star of 9 mag is at a distance of 2.6 kpc ([?]) At the outburst's maximum the luminosity of the star run up to $\sim 1.44 * 10^{38}$ erg/sec ([?]).

The luminosity of the Feb-March 1994 'giant' X-ray outburst of A0535+26 amounted to $L_x \approx 1 - 1.5 * 10^{38}$ erg/sec ([?]). Thus the luminosity of the maximal 'blue' optical flash was the same or even more than the luminosity at the maximum of a 'giant' outburst in X-ray region. It is difficult to compare these events: X-ray outbursts of A0535+26 last usually from some days to some weeks ('giant' outburst of 1994 lasted about 50 days), the total recorded time of this optical outburst was about 2 hours. We could compare only the energetic values.

An information about the X-ray behavior of A0535+26 for that time would be very important, however, it is lacking. Mir-Kvant ([?]), CGRO ([?]) and ASCA ([?]) have not detected the source on October - November 1995. Other optical observations are also not known on the outburst's interval TJD 50019.45 - 50019.55.

Were the references on analogous optical outbursts in the previous history of the pulsar?

In spite of more than 20 years history of systematic photoelectric observations of the star we had not found mentions about similar events in the literature except a very brief report by Urasin and Shaimukhametov [?]. They observed the object with the time resolution 0.1 sec. and 0.05 sec (by factor of 100 and 200 higher than our resolution) and recorded the short-time optical flashes with amplitudes 2-3^m and duration 0.2 - 0.3 sec. They did not obtain a total picture of the outburst with their recorder, but only separate flashes which were very similar to the individual pulses of our outburst. Apparently, they and we observed the same processes in the system, but we detected the merged and smoothed pulses. The luminosity of the flashes of Urasin and Shaimukhametov should be $\sim 10^{39}$ erg/sec in 0.1 - 0.2 sec. These extremely powerful and fast processes should be located in rather small area of 30000 km. It is visible the nonthermal nature of the both outbursts because the bottom level between the pulses was stable and equal to the brightness of the object in a quiescent state.

Generally, the system is characterized by the sudden and irregular brightening, so called transitory optical flares, with amplitudes $0.^m1 - 0.^m3$ in the time scale from some hours to some days ([?],[?]). In the review by Giovannelli and Graziati [?] the information about the optical flares preceding 'giant' X-ray outbursts is quoted. It may be interpreted as a result of the expulsion of a dense Be-star's shell and subsequent accretion of this shell material to the neutron star ([?]). We did reveal the similar regularity: the optical flare of 0.1 mag V-band intensity was detected ([?]) at the start of October 1995, two weeks before our outburst.

This optical outburst did not occur in the periastron passage of the pulsar if to apply the last elements of the orbit ([?]). Usually the 'normal' X-ray outbursts occur near the periastron passage of the compact object, but the 'giant' outbursts appear with a delay in a phase. From this fact and X-ray light curve modeling Motch et al. ([?]) made a conclusion about a transient accretion disk formation. BATSE detection of the quasi-periodic oscillations during 'giant' outburst 1994 ([?]) and the following model calculations by Li([?]) confirmed these conclusions. Probably, the transient accretion disk is formed around the neutron star as a consequence of an expulsion of a Be- star's shell. 'Giant' outbursts occur when the subsistence of the disk-fed ([?]) drop out on the surface of the neutron star.

So the 'usual' optical flares are located near the surface of Be-giant. The X- ray outbursts arise at the neutron star surface. Optical outbursts could not occur at the surface of the neutron star because only a short-wave radiation can take off such great energy from the compact object. However X-ray radiation can be transformed to longer waves in the disk-fed around the neutron star. Here the hypothesis by Gurzadyan about re-emission of the radiation energy on fast and relativistic electrons with the inverse Compton-effect ([?]) may be suitable. The author offered the explaination of the dMe-stars' flares, where the energy of the flare amounted to 10^{36} erg during some seconds. For the case of the X-ray pulsar the direct Compton-effect seems more probable: the re-emission from the X-ray region to the optical one in the flux of fast particles.

May be another interpretation. It should recall that the pulsar A0535+26 has the largest magnetic field for this class of objects, $2*10^{13}$ Gs ([?]). Possibly the rapidly rotating (spin period 103 sec., [?]) magnetic field prevents an accretion of the plasma to the neutron star. The arising synchrotron emission could transfer the energy into the optical region of spectrum ([?]).

The observations in X-ray region testify to the different regimes of accretion in this system and the changing of an accretion rate more than by factor of 20 ([?]).

In this course we have heard about the most high energy processes, particles and rays which exist in the Universe. However, it appears, no less high energetic, no less intriguing and still unclear phenomena one can observe in the ordinary optics.

References

1. Borkus V.V., Kaniovsky A.S., Sunyaev R.A., Efremov V.V., Kretchmar P., Staubert R., Englhauser J., Pietsch W., 1998, *Astronomy Letters*, **24**, N6, 415
2. Clark J.S., Lyuty V.M., Zaiseva G.V., Larionov V.M., Larionova L.V., Finger M., Tarasov A.E., Roche P., Coe M.J., 1999, *Monthly Notices of the Royal Astronomical Society*, **302**, 167
3. Dorokhov N.I., Dorokhova T.N., 1996, *IBVS*, N4357
4. Dorokhova T.N., Dorokhov N.I., 2000, *Gravitation & Cosmology Supplement*, **6**, 234
5. Finger M.H., Wilson R.B., Harmon B.A., 1996, *Astrophysical Journal*, **459**, 288
6. Giovannelli F., Graziati L.S., 1992, *Space Sci. Reviews*, **59**, 1
7. Gurzadyan G.A., 1985, *Stars' flares*, Nauka, Moscow, p.511
8. Janot-Pacheco E., Motch C., Mouchet M., 1987, *Astrophysical Journal*, **177**, 91
9. Kholopov P. N., Samus' N. N., Goranskij V. P. et al., 1985, *General Catalog of Variable Stars*, **v.3**, Nauka, Publishing House, Moscow, p. 260
10. Li X.-D., 1997, *Astrophysical Journal*, **476**, 278
11. Lipunov V.M., Popov S.B., 2000, *Astron. and Astrophys. Transactions*, in press
12. Lyuty V.M., Zajtseva G.V., Latysheva I.D., 1989, *Astronomy Letters*(rus.), **15**, 421
13. Lyuty V.M., Zajtseva G.V., 2000, *Astronomy Letters*, **26**, 9
14. Motch C., Stella L., Janot-Pacheco E., Mouchet M., 1991, *Astrophysical Journal*, **369**, 490
15. Nagase F., 1999, private communication
16. Negueruela I., 1998, *Astronomy and Astrophysics*, **338**, 505
17. Shvartsman V.F., 1970, *Astronomy Reports* (rus.), **47**, 824
18. Urasin L.A., Shaimukhametov R.R., 1987, *Astron. Circ.* (rus.), N1492
19. Wilson R.B., 2000, private communication

HIGH ENERGY PARTICLES IN ACTIVE GALACTIC NUCLEI

PETER L. BIERMANN[1,2]

[1] *Max-Planck Institute for Radioastronomy, and*
[2] *Department for Physics and Astronomy, University of Bonn*
Bonn, Germany
plbiermann@mpifr-bonn.mpg.de
http://www.mpifr-bonn.mpg.de/div/theory

Abstract. The detection of particles with energies above 10^{20} eV in giant airshowers continues to defy our attempts to understand them in physical terms. Of the many attempts to explain them I focus here on the by now classical suggestion, that they come from powerful radio galaxies. In this review I will embed the discussion into a general framework of hadronic particle acceleration in Active galactic Nuclei.

1. Introduction

High energy events have been recorded in airshowers since the beginning of the 1960ies, and since 1993 the evidence is rapidly increasing, that there are events, which imply an original particle energy in excess of 10^{20} eV, e.g. (Ave *et al.*, 1999; Ave *et al.*, 2000; Takeda *et al.*, 1998; Takeda *et al.*, 1999; Hayashida *et al.*, 1999b; Antonov *et al.*, 1999; Ivanov *et al.*, 1999). These events are quite rare, with rates of about one particle per km^2 per century. The entire tally by now is about two dozen, some of which have not been fully published. These events are usually assumed to be incoming protons, and there are various lines of argument for such an interpretation, mainly based on the airshower profile.

If one assumes that these events originate in the nearby universe, then one might expect that their arrival direction should correlate with that of the nearby astronomical objects, mainly galaxies (Stanev, 1997; Medina-Tanco, 2000). Galaxies are clustered inhomogeneously, and are aggregated mostly in the supergalactic sheets, the local manifestation of the soap-bubble-like cosmological galaxy distribution; these bubbles have walls or

M. M. Shapiro et al. (eds.), Astrophysical Sources of High Energy Particles and Radiation, 115–133.

bounding sheets, which constitute the main locus for galaxies. Filaments occur when two sheets meet, and clusters and superclusters of galaxies are at the junction of two or more filaments. There is not a really strong and convincing correlation in arrival direction of these high energy events with the local supergalactic plane, or the more general distribution of galaxies and clusters (Stanev et al., 1995).

The main difficulty in traversing the universe for protons is the microwave background, the residual radiation from the big bang, predicted in the 1940ies, and then detected in the 1960ies; this microwave background has been well measured by the COBE satellite, and various special ground and balloon systems. A proton encounters one of these photons, and loses on average 20 % of its energy, in creating a pion. Accounting for such an energy loss, protons have to start with an extremely high initial energy, if they originate from more than about 50 to 100 Mpc; if they were to start at a distance of 200 Mpc, their initial energy needs to be of order 10^{24} eV; for an initial travel distance of 20 Mpc the initial energy can be as low as a few times 10^{20} eV.

Another way to consider this is to assume that the sources are distributed homogeneously in the cosmos, and ask what the resulting spectrum of arriving protons would be, and compare this with data. This has been done many times since the early 1990ies, and the result is invariably, that there is a strong excess of observed events over what is expected. The expected spectrum shows a sharp cutoff at about $5\,10^{19}$ eV, usually referred to as the Greisen-Zatsepin-Kuzmin (or GZK) cutoff (Greisen, 1966; Zatsepin & Kuzmin, 1966). Even allowing for an extra factor of 2 in energy leaves many events beyond 10^{20} eV. This discrepancy is alleviated, when one assumes, that the sources are inhomogeneously distributed, such as the real galaxies. Since the local distribution is two-dimensional, the expected cutoff in the observed spectrum is not quite as sharp. There are just not as many moderately distant sources compared to nearby sources in a 2D distribution as in a distribution; thus the spectrum does not cut off quite as abruptly (Blanton et al., 2000). Also, if the particles were to experience some measure of confinement to the supergalactic sheet, then their flux may decrease with distance d only as $1/d$, and not as $1/d^2$; this would increase their arriving flux by another moderate factor.

The discovery of magnetic fields in the wider cosmos (Vallée, 1990; Clarke et al., 1999; Clarke et al., 2001; Enßlin et al., 1999; Kronberg, 1984) implies that we need to consider the strength of magnetic fields, and their possible origin as well; these detections strongly suggest that magnetic fields are ubiquitous in the cosmos. The strength of magnetic fields in clusters has now been determined from Rotation Measure data, using lines of sight to distant radio sources that cut right through a cluster (Clarke *et al.*, 2001). This has established a lower limit of magnetic field strength in clusters of 5 microGauss, and has also shown that these magnetic fields are coherent over fairly large scales in clusters. The sample studied excludes any peculiar clusters such as cooling flow clusters and the like. The renewed detection of vastly extended radio emission at low frequency in the region outside the Coma cluster (Kim *et al.*, 1989; Enßlin *et al.*, 1999) suggests equipartition magnetic fields of order 0.1 microGauss. Cosmic simulations show that for any arbitrary source of cosmic magnetic fields the accretion flow towards the large sheets of the galaxy distribution carries the magnetic fields along, and so aggregates them in the sheets, and then carries them along the sheets in a very large scale shear flow towards the filaments, and finally into the clusters and superclusters. Therefore, field strengths of order 0.1 microGauss are not seriously surprising; however, the uncertainty in this number is a factor of ten. One cogent consequence is unfortunately, that protons near 10^{21} eV no longer travel in straight line paths, and so cannot be expected to point back to their site of origin; classical astronomy with very high energy charged particles seems unlikely to be successful.

Recent reviews in this field are (Nagano & Watson, 2000; Bhattacharjee & Sigl, 2000; Olinto, 2000; Biermann *et al.*, 2000b). They give many further references. A general recent review on cosmic rays is (Wiebel-Sooth & Biermann, 1999). Classical books in the field are, e.g., (Berezinskii *et al.*, 1990; Gaisser, 1990).

In this review I will focus in on the concept that radio galaxies, with their jets and hot spots can accelerate protons to very high energy. I will argue that such an acceleration is a necessary implication of optical data. I will show that the nearby radio galaxy M87 is a good candidate to explain all those events beyond 10^{20} eV. I will embed this in a simple synopsis of a) where cosmic rays might come from, and speculate on simple consequences, and b) the possible role of cosmic ray particles in Active Galactic Nuclei.

Here we take the point of view that the highest energy particles should give a distinctive clue whether they play a role in Active Galactic Nuclei, and their physical phenomena.

1.1. QUESTIONS

The difficulties presented then by the data can be summarized in three short questions:

- How do the particles get their energy?
- How do they survive the interaction with the microwave background?
- How do they achieve the fairly high degree of isotropy in arrival directions observed?

- Can we understand the excess of pairs and triplets in the arrival directions of events with very different energy?
- Are these very energetic particles signatures of hadronic processes in Active galactic Nuclei?

2. Active Galactic Nuclei - basic concepts

- Jets and disks:
  Active Galactic Nuclei are now generally accepted to have a massive central black hole, and accretion disk, and a jet. The jet is initiated very close to the central black hole, possibly from the inner edge of the accretion disk. Considering then the accretion, the disk and the jet as a system, one can derive the "jet-disk symbiosis" picture, as developed by Falcke and collaborators, see e.g. (Falcke & Biermann, 1995; Falcke et al., 1995; Falcke & Biermann, 1999).
- Broad line emission:
  Active Galactic nuclei are often characterized by very broad emission lines, which arise from somewhere rather close to the central region, near to the black hole. One possible model for these emission clouds is stellar winds.
- Torus:
  X-ray absorption, optical line polarization, and statistics of detecting the broad line region all suggest that a torus of dusty material is covering a large solid angle of the central region, the broad line region. When seen from the side, the broad line region is only detectable in polarized line emission, and the central X-ray emission is heavily absorbed.
- Triggering activity:
  The accretion events, which lead to observable central activity may be initiated by a galaxy-galaxy merger, or by enhanced accretion towards the central region, caused by some disk instability. If mergers are at the root, cosmologically starbursts and nuclear activity should correlate.
- Relativistic jets:
  The jets have been recognized as usually relativistic, which leads to considerable boosting of the emission in the observer frame, and also to very strong selection effects, looking at a complete sample selected at a frequency of emission, which arises from jet radiation.
- TeV emission:
  Gamma ray emission in the TeV photon energy range has been detected in three radio-loud active galactic nuclei sofar. This gamma ray emission is coming from the relativistic jet; whether it comes from a part of the jet very close to the central engine, or from a region far out, is not clear. Gamma-gamma absorption arguments of the TeV photons

in the radiation field of the torus (Protheroe & Biermann, 1997) would suggest that the emission comes from a region in the jet outside the torus; such an argument is corrobated by the high variability in the radio emission, which certainly arises very far out in the jet. However, if in some such sources there were no torus, or only a very weak mone, then such an argument could be circumvented.

– Hadronic versus leptonic:

It is an open question at this time, whether the emission observed from the jet in any wavelength is initiated by hadronic cascades or arises from a purely leptonic process. It is likely that a very careful study o=f the variability episodes of these sources will provide an answer, both based on observations, and a detailed prediction based on these two approaches.

3. Radio Galaxies

Radio galaxies with their hot spots are perfect accelerators , e.g. see (Bednarz & Ostrowski, 1998), since they have a nearly relativistic shock wave, extended over a scale of kpc, and, furthermore, imply from optical observations that these extreme energies of protons are required. One difficulty is that we do not have many such radio galaxies in our cosmic neighborhood. We will discuss this option below.

Radio galaxy hotspots have been suggested already by Ginzburg (Ginzburg & Syrovatskii, 1964) and Hillas (Hillas, 1984), and were worked out in detail using new observations of radio quasars and their jets by Biermann & Strittmatter (Biermann & Strittmatter, 1987). Ginzburg specifically already focussed on the nearby radio galaxy M87.

Here we will go through this argument (Biermann & Strittmatter, 1987) in some critical detail and develop it further:

The mean free path λ for resonant scattering of a charged particl=e is given by

$$\lambda \ r_g \frac{B^2/8\pi}{I(k)k} \tag{1}$$

where we have used r_g for the Larmor radius of the particle, $I(k)$ for the energy density of the turbulence, per wavenumber $k = 2\pi/r_g$, and B for the magnetic field. This expression assumes isotropic turbulence, and is in the small angle approximation.

Turbulence can be described usually as a powerlaw

$$I(k) = I_0(k/k_0)^{-\beta} \tag{2}$$

The exponent β can have a variety of values, with $\beta = 5/3$ for Kolmogorov turbulence, what is found almost everywhere. $\beta = 1$ fo=r saturated turbulence, or an inverse cascade, $\beta = 3D3/2$ for Kraichnan turbulence (found for turbulence in a dominating "stiff" magnetic field), and $\beta = 3D2$ for sawtooth turbulence, such as produced by a multitude of shockwaves. Interesting and useful points of reference are $\beta = 3D\ 1,\ 5/3$, and 2. The lowest wavenumber k_0 corresponds to the largest wavelength, $r_{g,max}$, while the maximum wavenumber is k_{max}, the dissipation scale. We define then

$$\Lambda = \ln\frac{k_{max}}{k_0} \qquad (3)$$

The energy contained in the turbulence can be parametrized as

$$b = \int_{k_0}^{k_{max}} I(k)dk/(B^2/8\pi) = \frac{8\pi I_0 k_0}{(\beta - 1)B^2} \qquad (4)$$

in the case β different from unity, and k_{max} very much larger than k_0; while for $\beta = 1$ we have

$$b = \frac{8\pi I_0 k_0}{B^2} \ln\frac{k_{max}}{k_0} \qquad (5)$$

The mean free path can then be rewritten as

$$\lambda = \frac{r_g}{b(\beta - 1)} \left(\frac{r_{g,max}}{r_g}\right)^{\beta - 1} \qquad (6)$$

for β not equal to unity, and correspondingly slightly different for $\beta = 1$

$$\lambda = \frac{r_g}{b}\Lambda \qquad (7)$$

For relativistic particles of mass m and Lorentz factor γ the Larmor radius is $r_g = \gamma mc^2/(eB)$. The mean free path becomes strictly independent of energy for $\beta = 2$, the sawtooth turbulence.

The diffusion coefficient is then given for parallel shocks (i.e. where the shock normal is parallel to the flow on both sides) by (β different from unity)

$$\kappa \simeq \frac{1}{3}\lambda c \simeq \frac{c}{3}\frac{r_g}{b(\beta - 1)} \left(\frac{r_{g,max}}{r_g}\right)^{\beta - 1} \qquad (8)$$

and for $\beta = 1$ by

$$\kappa \simeq \frac{c}{3}\frac{r_g}{b}\Lambda \qquad (9)$$

Jokipii (Jokipii, 1987) has shown, that for highly oblique shocks a limiting situation can occur where

$$\kappa \simeq r_g U_{sh} \tag{10}$$

where U_{sh} is the shock speed (i.e. the flow speed of the upstream flow in the shock frame); this is a factor of $3bU_{sh}/(c\Lambda)$ smaller than the limit of the earlier expression, allowing much shorter acceleration times, and consequently, much larger maximum energies of particles, accelerated in diffuse shock acceleration.

The acceleration time is given by (e.g., (Jokipii, 1987))

$$\tau_{acc} = \frac{(pc)}{d(pc)/dt} = \frac{3r(\kappa_1 + r\kappa_2)}{U_{sh}^2(r-1)} \tag{11}$$

where $\kappa_{1,2}$ are the diffusion coefficients on the two sides of the shock, index 1 referring to upstream, and r is the density jump at the shock. We note that the density jump can have several obvious limits: For a strong shock in a gas with adiabatic gas constant 5/3, the density jump is 4; for a gas with a relativistic equation of state, which means an adiabatic gas constant of 4/3, the density jump in a strong shock is 7; for a strong shock in a cooling limit, the density jump can approach infinity, as it can for a relativistic shock, where the density jump can be approximated by $4\gamma_{sh}$; γ_{sh} is the Lorentz factor of the velocity difference at the shock. Of these various choices we adopt here a density jump of 4. In standard diffusive shock acceleration these various density jumps translate into a spectrum of accelerated particles of p^{-2} for a density jump of 4, to $p^{-3/2}$ for 7, and to p^{-1} for a density jump approaching infinity. In optically thin synchrotron emission this in turn gives a frequency spectrum, respectively, of $\nu^{-1/2}$, $\nu^{-1/4}$, and ν^0.

We then find ($\kappa_1 = \kappa_2$, since this is a parallel shock) an aceleration time of

$$\tau_{acc} = \frac{20}{3} \frac{c}{U_{sh}^2} \frac{r_g}{b(\beta-1)} \left(\frac{r_{g,max}}{r_g}\right)^{\beta-1} \tag{12}$$

The corresponding expression for $\beta = 1$ is:

$$\tau_{acc} = \frac{20}{3} \frac{c}{U_{sh}^2} \frac{r_g}{b} \Lambda \tag{13}$$

In the Jokipii-case ($\kappa_1 = r\kappa_2$, since this is a perpendicular shock) this is

$$\tau_{acc} = \frac{8}{3} \frac{r_g}{U_{sh}} \tag{14}$$

Next we determine the losses, and fix our attention on synchrotron losses, both for electrons and protons; a generalization to include photon interaction losses is straightforward (Biermann & Strittmatter, 1987; Mannheim et al., 1991; Mannheim, 1993a; Mannheim, 1993b).

Synchrotron losses are given by

$$\tau_{p,syn} = \frac{6\pi m_p^3 c}{\sigma_T m_e^2 \gamma_p B^2} \tag{15}$$

for protons, and by

$$\tau_{e,syn} = \frac{6\pi m_e c}{\sigma_T \gamma_e B^2} \tag{16}$$

for electrons.

Putting the rate of energy gain and loss equal then yields for protons

$$\gamma_{p,max} = (\frac{27}{80} b(\beta - 1))^{1/2} (\frac{e}{r_0^2 B})^{1/2} (\frac{U_{sh}}{c}) (\frac{m_p}{m_e}) \tag{17}$$

where r_0 is the classical electron radius. The corresponding cases for $\beta = 1$ and the Jokipii-case are

$$\gamma_{p,max} = (\frac{27}{80} \frac{b}{\Lambda})^{1/2} (\frac{e}{r_0^2 B})^{1/2} (\frac{U_{sh}}{c})(\frac{m_p}{m_e}) \tag{18}$$

and

$$\gamma_{p,max} = (\frac{27}{32} \frac{e}{r_0^2 B})^{1/2} (\frac{U_{sh}}{c})^{1/2} (\frac{m_p}{m_e}) \tag{19}$$

These protons at maximum energy are assumed to have enough space for the gyromotion, and then initiate a cascade in plasma waves, which establishes the wave field for the electrons to scatter in. We will return to this restrictive condition of the available space below. The corresponding maximum energy for electrons is then given by

$$\gamma_{e,max} = (\frac{27}{80} b(\beta - 1))^{1/2} (\frac{e}{r_0^2 B})^{1/2} (\frac{U_{sh}}{c}) (\frac{m_p}{m_e})^{2(\beta-1)/(3-\beta)} \tag{20}$$

For the case $\beta = 1$ this is

$$\gamma_{e,max} = (\frac{27}{80} \frac{b}{\Lambda})^{1/2} (\frac{e}{r_0^2 B})^{1/2} (\frac{U_{sh}}{c}) \tag{21}$$

and for the Jokipii case

$$\gamma_{e,max} = (\frac{27}{32})^{1/2} (\frac{e}{r_0^2 B})^{1/2} (\frac{U_{sh}}{c})^{1/2} \tag{22}$$

Using then the Jokipii-limit then is basically weakening the dependence on the shock speed.

The maximum frequency for synchrotron emission is given by

$$\nu^\star \simeq \frac{3}{16} \frac{e}{mc} \gamma_{max}^2 B \tag{23}$$

for either electrons or protons. For electrons this yields

$$\nu_e^\star \simeq \left(\frac{81}{1280} b(\beta - 1)\right) \left(\frac{c}{r_0}\right) \left(\frac{U_{sh}}{c}\right)^2 \left(\frac{m_e}{m_p}\right)^{4(\beta-1)/(3-\beta)} \tag{24}$$

Adopting a Kolmogorov spectrum for the turbulence we then obtain

$$\nu_e^\star \simeq 3\,10^{14} \, (3bU_{sh}^2/c^2)\,\mathrm{Hz} \tag{25}$$

For the highest shock speed compatible with a non-relativistic approximation and maximal turbulence this yields

$$\nu_e^\star \simeq 3\,10^{14}\,\mathrm{Hz} \tag{26}$$

quite independent of the magnetic field; this independence is a strong reason, why it is possible to observe a similar limiting cutoff frequency in so very different source environments, such as compact quasars, jets and hot spots. This maximum frequency just depends on natural constants

$$\nu_e^\star \sim 0.01 \frac{c}{r_0} \left(\frac{m_e}{m_p}\right)^2 \tag{27}$$

demonstrating its fundamental nature.

It is worth then also to investigate the limit for the maximum energy of electrons, when the turbulence is not derived from a plasma cascade initiated by protons, but is saturated, i.e. $\beta = 1$. This then produces a maximum emission frequency approximately $(m_p/m_e)^2$ larger than above, in the 100 keV to MeV range of photon energies at maximum, but we also need to replace the factor $\beta - 1$ by $1/\Lambda$, which then yields

$$\nu_e^\star \simeq \left(\frac{81}{1280} \frac{b}{\Lambda}\right) \left(\frac{c}{r_0}\right) \left(\frac{U_{sh}}{c}\right)^2 \tag{28}$$

which then gives (using $\Lambda = 10$) a limiting frequency of

$$\nu_e^\star \simeq 3\,10^{20} \, (3bU_{sh}^2/c^2)\,\mathrm{Hz} \tag{29}$$

corresponding to about 1 MeV in the high limit of U_{sh}/c. This limit has been invoked to explain the so-called "canonical" 100 keV cutoff seen in the X-ray spectrum of compact X-ray binaries by Markoff et al.(Markoff *et al.*, 2001). Using then the Jokipii limit again for the diffusion coefficient translates this limit to

$$\nu_e^* \simeq 2\,10^{22}\,(U_{sh}/c)\,\text{Hz} \tag{30}$$

corresponding to about a hundred MeV photon energy.

For protons the equivalent maximum frequency for synchroton emission is

$$\nu_p^* \simeq \left(\frac{81}{1280}b(\beta-1)\right)\left(\frac{c}{r_0}\right)\left(\frac{U_{sh}}{c}\right)^2\left(\frac{m_p}{m_e}\right) \tag{31}$$

which is $(m_p/m_e)^3$ larger than the electron maximum emission frequency. Thus we find another fundamental emission frequency here

$$\nu_p^* \simeq 5\,10^{24}\,(3bU_{sh}^2/c^2)\,\text{Hz} \tag{32}$$

basically in the 10 GeV range of photon energy. In the case $\beta = 1$ the frequency is very similar, and in the Jokipii case the frequency is somewhat higher still, by a factor up to about 10 in the limit of the shock velocity U_{sh}/c approaching near relativistic speeds. Therefore, in the GeV photon energy range the spectrum may well have some contribution from proton synchrotron emission.

We have used here the entire time the assumption, that the protons have enough space for their gyromotion. For the M87 jet this has been checked, and is possible (Biermann & Strittmatter, 1987). However, this may not be true in general. As noted earlier, the maximal energy permitted by spatial constraints can be translated to (Biermann, 1998)

$$\gamma_{p,max}m_pc^2 \simeq 5\,10^{20}\,L_{disk,46}^{1/2}\,\text{eV} \tag{33}$$

Lovelace had derived this scaling on rather general grounds already in 1976 (Lovelace, 1976). Here this has been established from work by (Falcke & Biermann, 1995; Falcke *et al.*, 1995) investigating the connections between accretion power, jet power, and black hole mass, in the concept referred to as "jet-disk symbiosis" and mentioned earlier. Inserting then also the scaling of the magnetic field with luminosity gives another limit for the maximum energy of protons, the loss limit as connected to the luminosity of the radio quasar or radio galaxy

$$\gamma_{p,max}m_pc^2 \simeq 10^{21}\,L_{disk,46}^{-1/4}\,\text{eV} \tag{34}$$

This shows, that near and somewhat above 10^{46} erg/s there is a window of opportunity to obtain particles near 10^{21} eV. These relations here obviously are statistical statements, and may not hold very accurately for any specific source; as the example of M87 shows, the maximum energy may deviate from the average by somewhat less than order of magnitude. Conversely, considering any other source, a discrepancy should be larger than an order of magnitude to be convincing.

This means, since the magnetic field in a jet scales as $L^{1/2}$, that at lower luminosities than those of powerful radio galaxies or radio quasars, the maximum energy of protons is no longer given by the loss limit, but by the spatial constraint limit. Hence the observation of such a cutoff in the nonthermal emission near $3\,10^{14}$ Hz itself suggests that the source can be expected to have a power in the range near to or above 10^{46} eg/s. Using the spatial constraint then gives a dependence on magnetic field instead of simple independence.

Both sources at very much higher luminosity and also at much lower luminosity cannot accelerate protons to energies well beyond about 10^{21} eV. At lower luminosities the space limit comes in, and at much higher luminosities the loss limit. Obviously, any relativistic boosting of the maxium energy in the case the shock region itself moves relativistically, does push the energy somewhat higher.

The ensuing particle interaction cascades have been thoroughly reviewed by (Rachen, 1999; Learned & Mannheim, 2000).

The key argument is that the nonthermal spectra of radio galaxy hotspots, and also jets and compact flat spectrum radio sources sometimes show a cutoff at or below $3\,10^{14}$ Hz. The generality of this cutoff requires a mechanism which is free of any parameters, and the argument given above fulfills this condition, and leads to the requirement, we repeat, of protons of 10^{21} ev in the sourec.

This is the only site, radio galaxy hotspots and knots in jets, of all proposed, which thus has an argument that *protons of such energy are required in the source* by other independent observations. If radio galaxies do accelerate protons to near 10^{21} eV, they surely also accelerate protons at all lower energies - ensuring that proton interactions play an important role in all radio-loud Active Galactic Nuclei. This is the reason we went through this argument in so much detail here.

One can then proceed to estimate analogously the maximum energy in other powerful radio galaxy jets and hot spots, disregarding the distance to us: This then leads to a scaling of maximum energy with jet- and so disk power (Falcke & Biermann, 1995), which is analogous to that derived by Lovelace (Lovelace, 1976), but here with a much harder limit:

$$E_{max} = 10^{22}\text{eV} \tag{35}$$

using the most powerful radio galaxy in the sample considered b= y (Falcke *et al.*, 1995), and allowing for some boosting in a weakly relativistic boundary shock. This is a serious upper limit; the more plausible maximum energy for protons should be well below this number, as discussed above; the loss limit comes in at such high powers, since there the magnetic field is higher.

However, Helium nuclei have a smaller Larmor radius at the same energy per particle, and so, given the same space to circle, could reach higher energies by a factor of 2; conversely, by such a selection, the top energy particles could well be all Helium nuclei rather than Hydrogen, i.e. protons. Also, for Helium, the synchrotron losses are yet weaker than for protons.

The general flux at about 10^{19} eV can easily be accounted for with very modest assumptions about radio galaxies, and the spectrum up to about $5\,10^{19}$ eV can be completely understood in flux and shape (Berezinsky & Grigor=92eva, 1988; Rachen et al., 1993).

The main difficulty is, again as noted, that there are very few radio galaxies in our cosmic neighborhood. M87 is the only one with sufficient power (Ginzburg & Syrovatskii, 1964; Cunningham et al., 1980; Watson, 1981), Cen A is too weak, and NGC315 is too far. Cen A is unlikely to be a really powerful accelerator; however, at about 5 Mpc in the southern sky, it is considerably closer than M87 in the northern sky, which is at about 15 - 20 Mpc. M87 is therefore just about close enough and also powerful enough, and as such the only serious candidate radio galaxy, and the main problem with M87 is the isotropy of the arriving events.

4. A Galactic wind model

Here we wish to outline a possible path to surmount these problems by using the concept that our Galaxy has a wind, akin to the Solar wind (L. Biermann, 1951; Parker, 1958; Weber & Davis, 1967); the notion of galactic winds is quite old by now (Burke, 1968; Johnson & Axford, 1971; Mathews & Baker, 1971). In such a wind the dominant magnetic field is an Archimedian spiral, with

$$B_\phi \sim \sin\theta/r \qquad (36)$$

in polar coordinates, while $B_r \sim 1/r^2$, and $B_\theta = 0$. What drives such a wind, and what are its properties? The wind clearly needs driving, and the most powerful source available is the energy of normal cosmic rays, about $\lesssim 3\,10^{41}$ erg/s. Then the wind has to escape the gravitational potential of the Galaxy, and so has to be faster than about 500 km/s. Furthermore, the line of sight integral of electron density multiplied with the line-of-sight component of the magnetic field through this wind has to be less than what the data show, about 30 rad/m^2 (Simard-Normandin & Kronberg, 1979; Kronberg, 1984). And fourth, the wind driving needs coupling from the relativistic fluid, the cosmic rays, to the gas, and this can be achieved best via magnetic fields, as proposed for WR stars (Seemann & Biermann, 1997), and this suggests that the wind may be only slightly super-Alfvénic. And, finally, the mass loss in the wind should be less than the total gaseous mass turnover in the Galactic disk, and so less than about 10 $M_\odot$/yr. To make matters simple at first, we assume the wind to be fairly close to its asymptotic state early on, at small radii. These five conditions suggest a magnetic field of 5 - 7 microGauss at the wind-base consistent with the numbers derived for the Solar neighborhood (Beck $et\ al.$, 1996), and a density of $5\,10^{-4}$ cm^{-3}, as reference values at the wind-base. This rather low density is consistent with the high ASCA temperature (Kaneda $et\ al.$, 1997; Valinia & Marshall, 1998), and taking pressure equilibrium with the ROSAT hot gas (Snowden $et\ al.$, 1997; Pietz $et\ al.$, 1998). The wind size can thus be estimated to be of order 1 Mpc, as long as no other galactic wind is encountered. The next galaxy is M31, but its cosmic ray output can be estimated through the far-infrared emission, since it comes from the same stellar population, and that is very much weaker than that of our Galaxy (Mezger $et\ al.$, 1996; Niklas, 1997); the next galaxy which is equivalent to ours in expected wind power is M81, at about 3 Mpc. Comparing such a wind model with those by Breitschwerdt et al. (Breitschwerdt $et\ al.$, 1991; Breitschwerdt $et\ al.$, 1993; Zirakashvili $et\ al.$, 1996; Ptuskin $et\ al.$, 1997; Breitschwerdt & Komossa, 1999), we note that our wind is consistent with their approach, but just has a much higher sustained magnetic field, since we use the asymptotic regime from small radii on, following (Seemann & Biermann, 1997).

Such a wind as the agent bends orbits of energetic particles fairly close by, and so does not add substantially to the travel distance for the particle. The Archimedian spiral also has the advantage that the bending, which is really an integral over the Lorentz-force, gives a logarithmic divergence, and so the bending is considerably more - by the logarithm of the ratio of the outer and inner radius of the wind - than trivially estimated.

This wind as the key bending agent has the disadvantage, that it is a priori not clear how to avoid extreme anisotropies (Billoir & Letessier-Selvon, 2000), and how to let lower energy particles through from the outside, down to the disk. The analogy with Solar wind modulation of cosmic rays below a few hundred MeV would suggest, that below a critical energy no particles get through down to the bottom of the wind anymore. And this critical energy cannot be very different from that where bending becomes important. For the Solar wind a small amount of "modulation" is relevant at 10 GeV, and the total cutoff is at around 500 MeV original energy, outside the Solar wind. This suggests a factor of 20 in that case. Extrapolating naively to the Galactic wind case, this would suggest, that no particles whatsoever should come through below about $2\,10^{19}$ eV, while the present data are quite consistent with no serious modulation between about $2\,10^{18}$ eV, and 10^{20} eV.

However, the Solar wind also provides tentative answers to these problems: The irregularities are strong in the magnetic field of the Solar wind, and do not decay towards the poles with the azimuthal component, therefore in the pole regions the irregularities do dominate over the regular systematic component. This means that the transport downwards of lower energy particles is more diffusive, and not on straight line orbits, and what we need to require is that the diffusive time scale is actually shorter than the convective time scale on which particles would be carried outwards again by the wind. This condition should hold over the energy range from $3\,10^{18}$ eV through about 10^{20} eV, where diffusive transport surely has changed over to a direct bending of the path, if not earlier. This requires just the spectrum of irregularities of $I(k) \sim k^{-2}$, a sawtooth turbulence, expected from the many shockwaves running up into the halo wind from all those supernovae driving it. For such a spectrum the scattering becomes independent of energy, allowing even particles at energies much below 300 EeV to come into the Galaxy with a reasonable flux. This is a condition on the irregularities, the dependence on zenith distance angle, and their spectrum. Such a spectrum may be quite natural, since supernova shock waves carry the energy and cosmic rays out into the wind.

There is another important point as regards the sign of the magnetic field: In the Galactic disk the azimuthal component changes sign every now and then (Han & Qiao, 1994; Han et al., 1997; Han & Manchester, 1999), and so one might expect that this sign change carries over into the wind; these sign changes would then nullify to first order all systematic bending. However, the observations of (Krause & Beck, 1998) of disk galaxies show, that the sign of the magnetic field and its pattern have a very clear symmetry: the magnetic field is spiral-like, and the dominant component always points along the spiral pattern inwards. This means that there is a dominant sign, and it is consistent actually with the local sign of the magnetic field measured near the Sun in our Galaxy. This also means that different from the Solar wind, the azimuthal component does not change sign in mid-plane. In order to satisfy $\mathrm{div}B = 0$ and at the same time follow a consistent spiral pattern, the radial component must change sign at higher Galactic latitude. This pattern entails that all very high energy events observed at Earth, if interpreted as protons, point ultimately upwards, when traced backwards through the halo wind.

This model has been described in (Ahn et al., 1999) and (Biermann et al., 2000a). Related calculations have been done by (Billoir & Letessier-Selvon, 2000; Harari et al., 2000).

One critical test is the possibility to explain the fair abundance of pair and triplet events with very different particle energies. Also, the highest energy particles should show some preference for the northern hemisphere of the sky, while the southern sky would be expected to be depleted at the highest energies, at least in the most simple version of this model. Therefore the Southern sky promises to be the most fruitful for finding events which require less mundane physics.

5. Possible synthesis

What are the sources of cosmic rays, and what is their role in the activity of Active galactic Nuclei? The following synthesis emerges (Biermann, 1997a), with some options still open:

- One part of the population of cosmic rays up to about 10^{14} eV arises from supernova explosions into the interstellar medium, ISM-supernovae. This sub-population is mostly Hydrogen, with only the ISM abundances in heavier elements and Helium.
- Another part arises from supernova explosions into the predecessor stellar wind; such winds have quite high magnetic fields, and so the maximum energies can be up to 3 10^{18} eV. In such a picture the knee arises from a loss of drift acceleration efficiency at a specific energy/charge ratio (Biermann, 1993; Biermann, 1994), given by the space available in a shock racing through a stellar wind with $B_\phi \sim \sin \theta / r$. The chemical abundances of these cosmic rays reflect the highly enriched winds of very massive stars, and so readily explain the enrichment in Helium, and heavier elements observed for cosmic rays.
- There is a competing argument, e.g. (Meyer et al., 1997), that suggests that dust particles are accelerated first in ISM-SN, and then all those selections of chemical elements that enter into dust are reflected in the abundances of cosmic rays. In such a scenario, there is no consistent and quantitative theory yet to account for the knee and the maximum energy. Also, in this concept there is no easy explanation of the gamma-spectrum of the Galaxy; in such a case it is close to impossibl= e to account for it via pion decay.
- The cosmic rays from about 3 10^{18} eV arise from outside our Galaxy; radio galaxy jets and hot spots are one plausible source for this component. Radio galaxies may reach the top energy observed, and M87 may be the last dominant source at the maximum energy.
- However, it cannot be excluded that near the top energy, beyond the GZK-cutoff at 5 10^{19} eV, a new population comes in, either protons from neutrino-interaction or decay of big bang relics, Fe nuclei, or a

new type of particle, which may survive from high redshift yet makes normal airshowers.

- Should the high energy cosmic rays come from radio galaxies, a special class of Active Galactic Nuclei, the proof of principle would have been made: Active Galactic Nuclei do accelerate particles to very high energy, and do so in abundance.
- Since hadrons dominate over leptons in normal cosmic rays, it appears plausible that they also dominate in Active galactic Nuclei. Hadrons and their interactions and cascades may produce many of the leptons which we observe through their emission. Hadrons produce their own synchrotron emission, and some gamma emission (e.g. from pion decay) directly, without even involving leptons. This remains a prime area for study.

6. Conclusions - Future

We are beginning to be able ask much more specific questions now, about cosmic rays at all energies, and may expect to obtain answers in the next 10 - 20 years. The detailed study of the gamma emission from Active Galactic Nuclei, and their possibly connected neutrino emission will hopefully allow soon a judgement on the relevance of cosmic ray hadrons.

Here the next generation Cherenkov telescopes on the various continents will allow much progress to be made, be it STACEE, CELESTE, CAT, CLUE, PACHMARI, MAGIC, HESS, VERITAS, CANGAROO or another one being built or planned.

The effects of the magnetic fields, with their pronounced asymmetries, should clearly be visible for the transport of very high energy hadrons through the universe, if these particles are normal protons.

The scenario, that M87 is the source for all the most energetic particles, suggests that any extremely energetic particle population detected in the deep southern hemisphere must be a different type of population, and therefore AUGER will show us the new physics, beyond the very simple concept discussed here.

For the very high energy particles the existing data base, such as from HAVERAH PARK, and all the arrays, YAKUTSK, AGASA, HIRES, AUGER, and in th= e future, EUSO, will clearly allow us to test all the models proposed; they may be all wrong. By Occams razor, let us eliminate the simple models first; the deep South, AUGER and EUSO, will give the strongest tests yet for new physics.

131

7. Acknowledgements

First of all I would like to thank my partners in all these discussions over the last few years, E.-J. Ahn, P.P. Kronberg, G. Medina-Tanco, and T. Stanev; I have learnt a lot from talking extensively with P. Blasi, J. Cronin, H. Falcke, G. Farrar, T.K. Gaisser, G. Gelmini, H. Kang, A. Kusenko, S. Markoff, W. Matthaeus, U. Mebold, H. Meyer, R.J. Protheroe, D. Ryu, G. Sigl, A.A. Watson, T. Weiler and many others. Finally I would like to express my appreciation for the generous hospitality at Erice in November 2000. Work on these topics with PLB has been supported by grant= s from NATO, the BMBF-DESY, the EU, the DFG and other sources.

References

Ahn, E.-J., Medina-Tanco, G., Biermann, P.L., Stanev, T., (1999), astro-ph/9911123.
Antonov, E.E. *et al. JETP Letters* **69**, 650, (1999).
Ave, M., *et al., Proc. 26th ICRC,* Eds. D. Kieda *et al.*, vol. 1, p. 365 - 368, (1999).
Ave, M., Hinton, J.A., Vazquez, R.A., Watson, A.A., Zas, E., *Phys.Rev.Lett.* **85**, 2244-2247 (2000), astro-ph/0007386.
Beck, R., Brandenburg, A., Moss, D., Shukurov, A., & Sokoloff, D., *Ann. Rev. Astron. & Astrophys.* **34**, 155 - 206 (1996).
Bednarz, J., Ostrowski, M., *Phys. Rev. Letters* **80**, 3911 - 3914 (1998).
Berezinskii, V.S., *et al.*, "Astrophysics of Cosmic Rays", North-Holland, Amsterdam (especially chapter IV) (1990).
Berezinsky V.S., Grigor=92eva S.I., *Astron. & Astroph.* **199**, 1 - 12 (1988).
Bhattacharjee, P. & Sigl, G., *Physics Reports*, **327**, 109 - 247 (2000), astro-ph/9811011.
Biermann, L., *Zeitschr. für Astrophys.* **29**, 274 (1951).
Biermann, P.L. & Strittmatter, P.A., *Astrophys. J.*, **322**, 643, (1987).
Biermann, P.L., *Astron. & Astroph.* **271**, 649 (1993), astro-ph/9301008.
Biermann, P.L., at 23rd International Conference on Cosmic Rays, in Proc. *Invited, Rapporteur and Highlight papers*; Eds. D. A. Leahy et al., World Scientific, Singapore (1994), p. 45.
Biermann, P.L., in *Cosmic winds and the Heliosphere*, Eds. J. R. Jokipii et al., Univ. of Arizona press, p. 887 - 957 (1997a), astro-ph/9501030.
Biermann, P.L., 1998, in "Wprkshop on observing giant cosmic ray air showers from $> 10^{20}$ eV particles from space", Eds. J.F. Krizmanic *et al.*, AIP conf. proc. No. 433, p. 22 - 36
Biermann, P.L., Ahn, E.-J., Medina-Tanco, G., Stanev, T., at TAUP99, the *6th international workshop on topics in Astroparticle Physics and Underground Physics*, College de France, Eds. J. Dumarchez, M. Froissart, D. Vignaud, (Sep 1999), Nucl. Phys. B (Proc. Suppl.), vol. 87, p. 417 - 419, (June 2000).
Biermann, P.L., Ahn, E.-J., Medina-Tanco, G., Stanev, T., *Erice school (Dec 1999)*, Ed. N. Sanchez, (in press) (2000).
Billoir, P. & Letessier-Selvon, A., (2000), astro-ph/0001427.
Blanton, M., Blasi, P., Olinto, A.V., (2000) astro-ph/0009466.
Breitschwerdt, D. McKenzie, J.F., Völk, H.J., *Astron. & Astroph.* **245**, 79 - 98 (1991).
Breitschwerdt, D. McKenzie, J.F., Völk, H.J., *Astron. & Astroph.* **269**, 54 - 66 (1993).
Breitschwerdt, D. & Komossa, S., in *Astrophys. & Sp. Sc.*, 1999, Proc. "Astrophys. Dynamics", Eds. D. Berry et al. (in press).
Burke, J.A., *Month. Not. Roy. Astr. Soc.* **140**, 241 (1968). general discussion of galactic winds, using those words s

132

Clarke, T., Kronberg, P.P., Böhringer, H., in *Clusters of Galaxies*, Ringberg Conference, Ed. H. Böhringer, in press (1999).

Clarke, T., Kronberg, P.P., Böhringer, H., *Astrophys. J. Letters* **547**= (in press) (2001).

Cunningham, G. *et al. Astrophys. J.* **236**, L71 (1980).

Enßlin, T. *et al.*, in *Clusters of Galaxies*, Ringberg Conference, Ed. H. Böhringer, in press (1999).

Falcke, H. & Biermann, P.L., 1995, *Astron. & Astroph.* **293**, 665; astro-ph/9411096

Falcke, H., Malkan, M.A., Biermann, P.L., *Astron. & Astroph.* **298**, 375 (1995), astro-ph/9411100.

Falcke, H., Biermann, P.L., *Astron. & Astroph.* **342**, 49 - 56, 1999, astro-ph/9810226

Gaisser, T.K., *Cosmic Rays and Particle Physics*, Cambridge Univ. Press (1990)

Ginzburg, V.L. & Syrovatskii, S.I., *The origin of cosmic rays*, Pergamon Press, Oxford (1964), Russian edition (1963).

Greisen, K., *Phys. Rev. Letters* , **16**, 748 (1966).

Han, J.L., Qiao, G.J., *Astron. & Astroph.* **288**, 759 (1994). field in the disk of our Galaxy

Han, J.L., Manchester, R.N., Berkhuijsen, E.M., Beck, R., *Astron. & Astroph.* **322**, 98 (1997).

Han, J.L., Manchester, R.N., Qiao, G.J., *Month. Not. Roy. Astr. Soc.* **306**, 371 (1999).

Harari, D., Mollerach, S., Roulet, E., Proceedings of the *International Workshop on Observing Ultra High Energy Cosmic Rays from Space and Earth*, Metepec, Puebla, Mexico, August 9-12, (in press), (2000) astro-ph/0010068.

Hayashida, N. *et al.*, *Astropart. Phys.* **10**, 303 (1999a), astro-ph/9807045.

Hayashida, N. *et al.*, *Astrophys. J.* **522**, 225 (1999b).

Hillas, A. M., *Annual Rev. of Astron. & Astrophys.* **22**, 425 (1984).

Ivanov, A.A. *et al.*, Proc. 26th ICRC, Salt Lake City, eds. D. Kieda *et al.* **1**, 403 (1999).

Johnson, H.E. & Axford, W.I., *Astrophys. J.* **165**, 381 (1971).

Jokipii, J.R.: 1987 , *Astrophys. J.* **313**, 842 - 846

Kaneda, H., Makishima, K., Yamauchi, S., Koyama, K., Matsuzaki, K., Yamasaki, N. Y., *Astrophys. J.* **491**, 638 (1997).

Kim, K.T., *et al.*, *Nature* **341**, 720 - 723 (1989).

Krause, F. & Beck, R., *Astron. & Astroph.* , **335**, 789 (1998).

Kronberg, P.P., *Rep. Prog. Phys.*, **57**, 325 - 382 (1994).

Learned, J.G. & Mannheim, K., *Ann. Rev. Nucl. & Part. Sci.* (in press) (2000)

Lovelace, R.V.E., *Nature* **262**, 649 - 652 (1976).

Mannheim, K., Biermann, P. L. & Kruells, W. M., 1991, *Astron. & Astroph.* **251**, 723–731

Mannheim, K., 1993, *Phys. Rev.* D **48**, 2408; astro-ph/9306005

Mannheim, K., 1993, *Astron. & Astroph.* **269**, 67; astro-ph/9302006

Markoff, S., Falcke, H., & Fender, R. 2001, *Astron. & Astroph. Letters* , (submitted).

Mathews, W.G., Baker, J.C., *Astrophys. J.* **170**, 241 (1971).

Medina-Tanco, G., *Astrophys. J.* (in press) (2000), astro-ph/0009336.

Meyer, J.-P., Drury, L. O'C., Ellison, D. C.: 1997 *ApJ* **487**, 182

Mezger, P. G., Duschl, W. J., Zylka, R., *Astron. & Astroph. Rev.* **7**, 289 - 388 (1996).

Nagano, M., Watson, A.A., *Rev. Mod. Phys.*, **72**, 689 - 732 (2000).

Niklas, S., *Astron. & Astroph.* **322**, 29 (1997).

Olinto, A. V., in *David Schramm Memorial Volume*, *Phys.Rept.* **333-334**, 329 - 348 (2000), astro-ph/0002006.

Parker, E.N., *Astrophys. J.* **128**, 664 (1958).

Pietz, J., Kerp, J., Kalberla, P. M. W., Burton, W. B., Hartmann, D., Mebold, U., *Astron. & Astroph.* **332**, 55 - 70 (1998).

Protheroe, R.J., Biermann, P.L., *Astropart. Phys.* **6**, 293 - 300, 1997; astro-ph/9608052

Ptuskin, V.S., Völk, H.J., Breitschwerdt, D., Zirakashvili, V.N., *Astron. & Astroph.* **321**, 434 (1997).

Rachen, J.P., Stanev, T., & Biermann, P.L., *Astron. & Astrophys.* **273**, 377 (1993).

Rachen. J.P., "GeV to TeV gamma ray astrophysics workshop" Eds. B.L. Dingus *et al.*, AIP conference No. 515, p. 41 - 52, 1999

Seemann, H. & Biermann, P.L., *Astron. & Astroph.* **327**, 273 (1997), astro-ph/9706117.

Simard-Normandin, M., & Kronberg, P.P., *Nature* **279**, 115 (1979).

Snowden, S.L., *et al.*, *Astrophys. J.* **485**, 125 (1997).

Stanev, T., *et al.*, *Phys. Rev. Letters* **75**, 3056 (1995).

Stanev, T., *Astrophys. J.* **479**, 290 (1997).

Takeda, M. et al. *Phys. Rev. Letters* **81**, 1163, (1998).

Takeda, M. *et al. Astrophys. J.* **522**, 225 (1999), astro-ph/9902239.

Valinia, A., Marshall, F. E., *Astrophys. J.* **505**, 134 - 147 (1998).

Vallée, J.P., *Astron. J.* **99**, 459 (1990).

Watson, A.A., 1981, *Cosmology and Particles*, 16th Rencontres de Moriond, Ed. J. Adouze et al., p. 49 - 67, Editions Frontieres.

Weber, E.J., Davis, L., Jr., *Astrophys. J.* **148**, 217 (1967).

Wiebel-Sooth, B., Biermann, P.L.,in Landolt-Börnstein, vol. VI/3c, Springer Publ. Comp. (1999), p. 37 - 90.

Zatsepin, G. T., & Kuzmin, V. A., *Sov. Phys.-JETP Lett.*, **4**, 78, (1966).

Zirakashvili, V.N., Breitschwerdt, D., Ptuskin, V.S. Völk, H.J., *Astron. & Astroph.* **311**, 113 (1996).

GALACTIC CLUES ABOUT THE HIGH-ENERGY PROCESSES IN BL LAC (AGN) JETS

S. Markoff, H. Falcke, P.L. Biermann and R.P. Fender*

Introduction

Active galactic nuclei (AGN) jets are the source of many outstanding puzzles in astrophysics. Not least of these is the question of how these jets are formed, and how they are related to the central massive black hole (BH) and its accompanying accretion flow. Understanding this coupling, as well as the overall system energetics, is not possible without knowing the jet content (i.e., pure leptonic or hadronic). Unfortunately, this is still a matter of significant debate in the community.

While we wait for a 25 TeV photon from Mrk 501 (difficult for leptonic models), we can consider if there are any other sources which could provide crucial information for our modeling of jet processes. It turns out that our own Galaxy contains many accreting BH systems which are either known to contain jets, or which we have good reason to believe do. These sources may be useful analogues especially for BL Lacs because, in certain states, their spectra are similarly nonthermally dominated. Their sheer number also makes them valuable for statistical comparisons and, because they show limited disk activity, we can see the jet emission reasonably unobstructed, perhaps even near the base.

Sgr A*

Sgr A* is a compact radio source lying at the dynamical center of the Galaxy (Reid et al. 1999; Backer & Sramek 1999). Measurements of nearby stellar proper motions have determined the mass to be $\sim 3 \cdot 10^6 M_\odot$ (Ghez et al. 2000; Genzel et al. 2000). Up until recently Sgr A* was only conclusively identified in the radio, which was rather odd even compared to other nearby low-luminosity AGN (LLAGN; Ho

*SM, HF, PB: MPIfR, Bonn, Germany; RF: Astronomical Institute "Anton Pannekoek", University of Amsterdam, The Netherlands; SM is supported by a Humboldt Research Fellowship.

M. M. Shapiro et al. (eds.), Astrophysical Sources of High Energy Particles and Radiation, 135–141.
© 2001 *Kluwer Academic Publishers. Printed in the Netherlands.*

1999), and it appears to be sub-Eddington by roughly nine orders of magnitude. It is close enough to estimate the accretion rate from stellar winds, which gives a luminosity much higher than what we are seeing. This has resulted in several competing models for the radio luminosity, all with inventive ways to "hide" the emission, and none of which could be eliminated because of the lack of high-energy constraints (for a review, see Melia & Falcke 2001).

This picture has changed drastically with the new detection of Sgr A* by *Chandra* in the X-ray (Baganoff et al. 2000). They find a very soft power-law (photon index $\sim$ 2.75), with a total luminosity of $0.5L_\odot$ in the $2 - 10$ keV range, providing an extremely tight modeling constraint.

Based on the spectrum (see Fig. 1a), a natural explanation would be the inverse Compton (IC) upscattering in energy of the submm "bump" by about 10^4, suggesting an electron population with characteristic Lorenz factor $\gamma_e \sim 100$ (see Falcke 1996; Beckert & Duschl 1997). Of the several models proposed, a radio jet can easily provide this scenario (Falcke 1996 and refs. therein), and although one has not yet been conclusively imaged, the circumstantial evidence is quite strong. Firstly, the flat-to-inverted radio spectrum seen from Sgr A* is actually the signature emission of a radio jet (Blandford & Königl 1979). In addition, the compact core shows slight elongation (Lo et al. 1998), similar to what was seen in M81 when the feeble jet it contains was finally imaged after several years of VLBI (Bietenholz et al. 2000). Our Galaxy is very similar to M81, and they seem to be the only two galaxies of the nearby group showing circular but no linear polarization from the core (Mellon et al. 2000; Brunthaler et al., in prep.). If the jet in Sgr A* is of a similar type, it would be even more difficult to detect, since we are looking through the dense Galactic plane.

With these issues in mind, we revisit the jet model to see if its synchrotron self-Compton (SSC) could account for the new X-ray addition to the broadband spectrum. For details of the jet model in general, see Falcke & Markoff (2000), Falcke & Biermann (1995) and references therein.

Leptonic Attempt

As a first attempt, we consider the equivalent of a "leptonic" model, in which we inject an electron distribution into the jet, without addressing its origin, in order to see if we could reproduce the spectrum with reasonable physical parameters. The synchrotron spectrum seems to cut off somewhere between the submm bump and the possible infrared (IR) detection, which we treat as an upper limit. But the lack of the usual (for AGN and even LLAGN) optically-thin power-law suggests that the electron distribution is truncated at relatively low energy. We test two commonly-invoked distributions, a thermal Maxwellian and a power-law, both with characteristic (peak or minimum energy) near $\gamma_e \sim 100$. Both cases give a good fit, as shown in Fig. 1a (see Falcke & Markoff 2000 for more details).

Hadronic Attempt

While it is encouraging how naturally everything comes out of the leptonic model, the treatment of the electron distribution itself is rather ad hoc. Ideally, we would like to tie this self-consistently into the overall accretion picture, which we have significant information about due to radio and IR observations. Along these lines, it did not escape our attention that the γ_e required to fit the submm bump and the X-rays with synchrotron and SSC is suggestively close to the peak energy of the secondary lepton distribution resulting from a typical hadronic collision and subsequent decays:

$$p+p(\gamma) \to \pi^{\pm,0}+X\,, \quad \pi^0 \to \gamma\gamma \quad \text{or} \quad \pi^\pm \to \mu^\pm+\nu_\mu\,, \quad \mu_e^\pm \to e^\pm+\nu_\mu\nu_e.$$

The lepton distribution will peak at ~ 30 MeV, i.e. $\gamma_{e\pm} \sim 60$.

A very popular model for the accretion onto low-luminosity BHs is the advection-dominated accretion flow (ADAF), in which the luminosity is suppressed by a lack of coupling between electrons and protons in the plasma, resulting in electrons with temperatures up to orders of magnitude cooler than the protons. Recently, Manmoto (2000) (and see references therein) has done the fully relativistic calculations around a rotating BH and found that the temperatures of the protons at the inner edge of the accretion disk in the case of a maximally spinning black hole can be $\gtrsim 10^{12}$ K. This is mildly relativistic, and inelastic collisions between the hot protons will be unavoidable. The question is then, could there be enough collisions to power the jet via secondaries?

We find that there are. Qualitatively, because we have already shown above that a peaked distribution of roughly this γ_e can fit the broadband data, it is not surprising that we get a good fit to the data (Fig. 1b, and see our upcoming paper Markoff, Falcke & Biermann, in prep., for full details). What is surprising, however, is the successful self-consistency with the observed and modeled accretion scenario. The hadronic inelastic collisions provide the link between the inflow and observed outflow, via secondary particles.

These results are extremely sensitive to the temperature at the inner edge of the ADAF. To illustrate, we include in Fig. 1b the spectrum from $T_p = 2 \cdot 10^{11}$ K protons. An order of magnitude difference in the

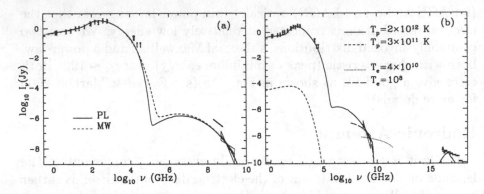

Figure 1. (a) Broad-band spectrum of Sgr A*. See Falcke & Markoff 2000 for full details. The *Chandra* X-ray spectrum is indicated, with spectral-index errors, and above it we include for reference the earlier ROSAT spectrum. We show our model spectra for a power-law distribution of electrons (PL) and a relativistic Maxwellian distribution (MW). (b) Radio through γ-ray spectrum for Sgr A* resulting from proton-induced $e^{\pm}$'s in a hot ($T_p = 2 \cdot 10^{12}$ K) accretion flow fed into a plasma jet. The γ-ray data (Mayer-Hasselwander et al. 1998) are considered upper limits because of the large beam of the observations. The dashed line shows the spectrum for the same parameters but with $T_p = 3 \cdot 10^{11}$ K. The two lines intersecting the *Chandra* data represent free-free emission from the accreting plasma, which comprise the most critical constraint on the modeling.

temperature results in around five orders of magnitude difference in radio power! This is a consequence of the threshold for π production combined with the peaked quasi-thermal proton distribution. As the temperature at the inner edge of the disk goes down with decreasing BH spin, only protons in the tail will have the energy capable of producing secondaries, a number which drops drastically. This effect acts as a switch: if T_p is not close to 10^{12} K, there will not be enough secondaries to explain the radio emission. As a consequence, if this model is correct, it would imply that Sgr A* is a rapidly rotating BH. This also suggests that perhaps a "proton switch" is behind the radio-loud/radio-quiet (RL/RQ) dichotomy in AGN (Markoff et al., in prep.).

The jet model for Sgr A*, leptonic or hadronic, predicts correlated variability between the submm bump and the X-rays, as well as the centroid of emission shifting with frequency. For the hadronic model only, with much better resolution, we would also hope to see signs of the pion bump in the Galactic center γ-ray spectrum, but this may be a long ways away. If the picture of hot protons in the accretion flow is confirmed, it seems likely that the plasma at least in RL jets is actually a mixture of protons and leptons, which will significantly change the arguments about the matter content of jets.

X-ray Binaries

If one believes in a disk/jet symbiosis, these systems should scale with BH mass, taking into account variations in luminosity efficiency. In this case, stellar BH systems can also be important analogues for AGN studies (already noted by Mirabel & Rodriguéz 1994, who coined the term "microquasar" for Galactic source 1E1740.7-2942). BH candidate X-ray binaries (BHC XRBs) in the Low/Hard state are known to contain jets in some (and likely all) systems (Fender 2000). The Low/Hard state is a something of a stellar analogue to BL Lacs/Blazars, in that it appears to be nonthermally dominated even at high-energy. In the past, any possible contribution by jets to the X-rays has been ignored. In contrast, we consider the case that the jets may in some cases even dominate the radio and X-ray emission, as in BL Lacs. Similar to Sgr A*, the disk contribution is small in the Low/Hard state so that if it is jets we are seeing, we are seeing them with little obstruction. Under this assumption, XRBs can provide a fertile testing ground for our ideas about jet emission.

We started with the recently discovered XTE J1118+480 (Remillard et al. 2000), whose broadband spectrum is exceptionally well-covered by observations. Although jets were not directly resolved with MERLIN to a limit of $< 65(d/\text{kpc})$ AU (at 5 GHz; Fender et al. 2000), its radio spectrum shows the signature inversion, flattening at higher frequencies, and so it is a promising source to test the jet hypothesis. One big difference between this system and Sgr A* is the presence of weak thermal disk emission in the optical range. There is also a prominent power-law extending to high frequencies, which is typical of the Low/Hard state and would require acceleration of the electrons in the jet.

We explore a similar model as for Sgr A*, with the addition of a shock, and assuming a combined standard thin disk/ADAF paradigm as described in Esin et al. (1997). The results are shown in Fig. 2 for the parameters in the caption (see Markoff et al. 2000 for details of the modeling). The jet can account for the almost the entire spectrum from radio to X-rays via synchrotron, using only $\lesssim 1\%$ of the total accretion power $\dot{M}c^2$. The current limit on γ_e comes from synchrotron losses, as compared to external Compton such as in BL Lacs. For a much stronger external photon density, e.g., if the optically thick disk extends much closer in or is much more luminous, this maximum energy could be further reduced to IC losses (and the IC component would be higher). In the case of the low disk contribution here, the cutoff could venture into the realm of future high-energy missions like INTEGRAL.

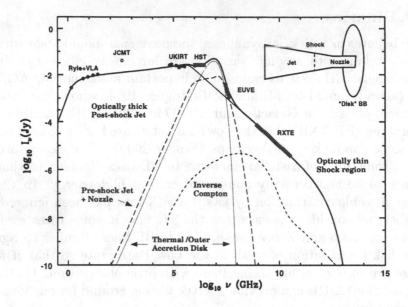

Figure 2. XTE J1118+480 model fit to data referenced in Markoff et al. (2000). The EUVE data shown are for the highest absorption case, and the JCMT point is non-simultaneous. Included is a schematic indicating the different emission components.

It is interesting to note that the lack of a 100 keV cutoff in the X-ray spectrum makes this an unusual source (McClintock et al., in prep.).

A hot hadronic component results in too much synchrotron from the additional pairs to satisfy the tight EUV limits, and so we conclude that this is not a hadronically-induced RL source. This is not surprising, considering that only $\sim$ 10% of all detected quasars seem to be RL. Requiring charge balance between the protons and electrons in the jet plasma, and a low T_e in order to agree with the ADAF scenario, results in the particles being in sub-equipartition with the magnetic field. For a pure pair jet or higher T_e (non-ADAF), assuming equipartition, the IC component would go up.

Summary

Our initial results, which we hope are just the tip of the iceberg, suggest that a scaling jet/disk model can successfully explain data ranging from our very weak Galactic nucleus down to stellar binary systems. This implies that these Galactic sources can provide valuable information about the behavior of jets, and possibly their formation and energetics, for us to later apply to the more complex and distant AGN systems we seek to understand. The amount and quality of the data can also help us more tightly constrain the model. And at least as far as Sgr A* and XRBs (in the Low/Hard state) are concerned, the meager disk contribution allows us to see the jet processes to a much better extent, similar to what we assume is happening with BL Lacs/Blazars.

Future plans involve applying this type of model to many, many more XRBs, and eventually also to LLAGN. If we do not see somewhere around 10% of the sources showing a hadronic/RL component, then this trigger idea is likely incorrect. If protons are really fed from the disk into the jet, and if ADAFs achieve the high proton temperatures predicted, we would expect to see the effects of the additional pair production. Only by studying many more sources can we make a conclusive statement about the particle content and efficiency of jets in accreting BH systems.

References

[1] Backer, D.C. & Sramek, R.A. 1999, ApJ, 524, 805

[2] Baganoff, F. K., Maeda, Y., Morris, M., et al. 2000, ApJ, submitted

[3] Beckert & Duschl 1997, A&A, 328, 95

[4] Bietenholz, M.F., Bartel, N. & Rupen, M.P. 2000, ApJ, 532, 895

[5] Blandford, R.D. & Königl, A. 1979, ApJ, 232, 34

[6] Esin, A.A., McClintock, J.E., Narayan, R., 1997, ApJ, 489, 865

[7] Falcke, H. 1996, IAU Symp. 169: Unsolved Problems of the Milky Way, p. 169

[8] Falcke, H. & Biermann, P.L. 1995, A&A, 293, 665

[9] Falcke, H. & Markoff, S., 2000, A&A, 362, 113

[10] Fender, R.P. 2000, MNRAS, in press (**astro-ph/0008447**)

[11] Fender, R.P., Hjellming, R.M., Tilanus, R.J., et al. 2000, MNRAS, submitted

[12] Genzel, R., Pichon, C., Eckart, A., et al. 2000, MNRAS, 317, 348

[13] Ghez, A.M., Morris, M., Becklin, E.E., et al. Nature, 407, 349

[14] Ho, L.C 1999, ApJ, 516, 672

[15] Lo, K.Y, Shen, Z.-Q., Zhao, J.-H. & Ho, P.T.P. 1998, ApJ, 508, L61

[16] Manmoto, T. 2000, ApJ, 534, 734

[17] Markoff, S., Falcke, H. & Fender, R., 2000, A&AL, submitted

[18] Mayer-Hasselwander, H., Bertsch, D., Dingus, B., et al. 1998, A&A, 335, 161

[19] Melia, F. & Falcke, H. 2001, ARA&A, 39, 2001

[20] Mellon, R.R., Bower, G.C., Brunthaler, A. & Falcke, H., 2000, AAS 197, #39.02

[21] Mirabel, I.F. & Rodriguéz, L.F. 1994, Nature, 371, 46

[22] Reid, M.J., Readhead, A.C.S., Vermeulen, R.C. & Treuhaft, R.N. 1999, ApJ, 524, 816

[23] Remillard, R., Morgen, E., Smith, D., Smith, E., 2000, IAU Circ. 7389

NONTHERMAL PHENOMENA IN CLUSTERS OF GALAXIES

YOEL REPHAELI

School of Physics & Astronomy
Tel Aviv University, Tel Aviv, Israel

Abstract.

Recent observations of high energy (> 20 keV) X-ray emission in a few clusters extend and broaden our knowledge of physical phenomena in the intracluster space. This emission is likely to be nonthermal, probably resulting from Compton scattering of relativistic electrons by the cosmic microwave background radiation. Direct evidence for the presence of relativistic electrons in some ~ 30 clusters comes from measurements of extended radio emission in their central regions. I first review the results from RXTE and BeppoSAX measurements of a small sample of clusters, and then discuss their implications on the mean values of intracluster magnetic fields and relativistic electron energy densities. Implications on the origin of the fields and electrons are briefly considered.

1. Introduction

Extensive X-ray measurements of thermal bremsstrahlung emission from the hot, relatively dense intracluster (IC) gas, have significantly widened our knowledge of clusters of galaxies. Clusters are the largest bound systems in the universe, and as such constitute important cosmological probes. Their detailed dynamical and hydrodynamical properties are therefore of much interest. An improved understanding of the astrophysics of clusters, in particular, a more precise physical description of the IC environment, necessitates also adequate knowledge of the role of nonthermal phenomena. The thermal state of IC gas may be appreciably affected by the presence of magnetic fields, and relativistic electrons and protons. Important processes involving the fields and particles are radio synchrotron emission, X-and-γ-ray emission from Compton scattering of electrons by the cosmic microwave background (CMB) radiation, nonthermal bremsstrahlung, de-

143

M. M. Shapiro et al. (eds.), Astrophysical Sources of High Energy Particles and Radiation, 143–150.
© 2001 *Kluwer Academic Publishers. Printed in the Netherlands.*

cay of charged and neutral pions from proton-proton collisions, and gas heating by energetic protons. Detection of nonthermal X-ray emission from the electrons, when combined with radio measurements, yields direct information on the particles and fields, and establishes the basis for the study of nonthermal phenomena in clusters.

Clusters directly link phenomena on galactic and cosmological scales. As such, knowledge gained on magnetic fields and cosmic ray energy densities will form a tangible basis for the study of the origin of fields and (relativistic) particles, and their distributions in the intergalactic space. Quantitative information on IC magnetic fields and relativistic electrons is also very important for a realistic characterization of the processes governing their propagation in galactic halos and ejection to the IC space.

Nonthermal X-ray emission has recently been measured in a few clusters by the RXTE and BeppoSAX satellites. While we do not yet have detailed spectral and no spatial information on this emission, we are now able to determine more directly the basic properties of the emitting electrons and magnetic fields. Attempts to measure this emission will continue in the near future; spatial information could be obtained for the first time by observations with IBIS imager aboard the INTEGRAL satellite. In this short review, I describe the current status of the measurement of nonthermal X-ray emission by the RXTE and BeppoSAX satellites, and briefly discuss some of the implications on the properties of relativistic electrons and magnetic fields in clusters.

2. Measurements

2.1. RADIO EMISSION

At present the main evidence for relativistic electrons and magnetic fields in the IC space of clusters is provided by observations of extended radio emission which does not originate in the cluster galaxies. In a recent VLA survey of 205 nearby clusters in the ACO catalog extended emission was measured in 32 clusters (Giovannini et al. 1999, 2000). Only about a dozen of these were previously known to have regions of extended radio emission. In many of the clusters the emitting region is central, with a typical size of $\sim 1-3$ Mpc. The emission was measured in the frequency range $\sim 0.04-1.4$ GHz, with spectral indices and luminosities in the range $\sim 1-2$, and $10^{40.5} - 10^{42}$ erg/s ($H_0 = 50\ km\ s^{-1}\ Mpc^{-1}$).

Radio measurements yield a mean, volume-averaged field value of a few μG, under the assumption of global energy equipartition. The field can also be determined from measurements of Faraday rotation of the plane of polarization of radiation from cluster or background radio galaxies (e.g. Kim et al. 1991). This has been accomplished statistically, by measuring the distribution of rotation measures (RM) of sources seen through a sample of clusters. In the recent study of Clarke et al. (2001) the width of the RM distribution in a sample of 16 nearby clusters was found to be about eight times larger than that of a control sample for radio sources whose lines of sights are outside the central regions of clusters. From the measured mean RM, field values of a few μG were deduced.

It should be noted that the above methods to determine the field strength yield different spatial averages. Measurement of synchrotron emission yields a volume average of the field, whereas the measurement of Faraday rotation yields a line of sight average weighted by the electron density.

2.2. NONTHERMAL X-RAY EMISSION

Compton scattering of the radio emitting (relativistic) electrons by the CMB yields nonthermal X-ray and γ-ray emission. Measurement of this radiation provides additional information that enables direct determination of the electron density and mean magnetic field, without the need to invoke equipartition (Rephaeli 1979). In the first systematic search for nonthermal X-ray emission in clusters, HEAO-1 measurements of six clusters with regions of extended radio emission were analyzed (Rephaeli et al. 1987, 1988). The search continued with the CGRO (Rephaeli et al. 1994), and ASCA (Henriksen 1998) satellites, but no significant nonthermal emission was detected, resulting in lower limits on the mean, volume-averaged magnetic fields in the observed clusters, $B_{rx} \sim 0.1$ μG.

Significant progress in the search for this emission was recently made with the RXTE and BeppoSAX satellites. Evidence for the presence of a second component in the spectrum of the Coma cluster was obtained from RXTE ($\sim$ 90 ks PCA, and $\sim$ 29 ks HEXTE) measurements (Rephaeli et al. 1999). Although the detection of the second component was not significant at high energies, its presence was deduced at energies below 20 keV. Rephaeli et al. (1999) argued that this component is more likely to be nonthermal, rather than a second thermal component from a lower temperature gas. The two spectral components and the measurements are shown in Figure 1. The best-fit power-law photon index was found to be 2.3 ± 0.45 (90% confidence), in good agreement with the radio index. The 2-10 keV flux in the power-law component is appreciable, $\sim 3 \times 10^{-11}$ erg cm^{-2} s^{-1}. The identification of the flux measured in the second component as Compton emission, when combined with the measured radio flux, yields $B_{rx} \sim 0.2$ μG, and an electron energy density of $\sim 8 \times 10^{-14}(R/1Mpc)^{-3}$ erg cm^{-3}, scaling the radial extent (R) of the emitting region to 1 Mpc.

Observations of the Coma cluster with the PDS instrument aboard the BeppoSAX satellite have led to a direct measurement of a power-law compo-

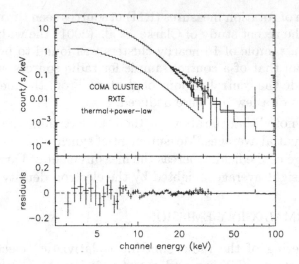

Figure 1. RXTE spectrum of the Coma cluster. Data and folded Raymond-Smith ($kT \simeq 7.51$ keV), and power-law (index = 2.34) models are shown in the upper frame; the latter component is also shown separately in the lower line. Residuals of the fit are shown in the lower frame.

nent at high energies, $25 - 80$ keV (Fusco-Femiano 1999). A best-fit power-law photon index 2.6±0.4 (90% confidence) was deduced. The measured flux in a 20-80 keV band is $\sim 2 \times 10^{-11}$ erg cm^{-2} s^{-1}, about $\sim 8\%$ of the main, lower energy 2-10 keV flux. A Compton origin yields $B_{rx} \sim 0.15$ μG. These results are in good general agreement with the RXTE results (Rephaeli et al. 1999). In addition to Coma, power-law components were measured by BeppoSAX in two other clusters, A2199 (Kaastra et al. 2000), and A2256 (Fusco-Femiano 2000). RXTE measurements of a second cluster, A2319, have also yielded evidence for a second spectral component (Gruber & Rephaeli 2001). Note that A2199 is not known to have an extended region of IC radio emission.

While the measurements of second components in the spectra of these four clusters are significant, the identification of the emission as nonthermal is not certain. To better establish the nature of this emission, spatial information is needed in order to determine the location and size of the emitting region. The large fields of view of the RXTE ($\sim 1^o$) and BeppoSAX/PDS ($\sim 1.3^o$) instruments do not allow definite identification of the origin of the observed second spectral components. What seems to have been established, however, is that there is no temporal variability in the RXTE and BeppoSAX data, so an AGN origin of the second spectral components in these clusters is unlikely.

3. Theory

3.1. MAGNETIC FIELDS

The origin of IC magnetic fields is of interest in the study of the evolution of the IC environment and of fields on cosmological scales. IC fields could possibly be of cosmological origin – intergalactic fields that had been enhanced during the formation and further evolution of the cluster, or else the fields might be mostly of galactic origin. The latter is a more likely possibility: fields were anchored to ('frozen in') the magnetized interstellar gas that was stripped from the member (normal and radio) galaxies (Rephaeli 1988). Dispersed galactic fields would have lower strengths and higher coherence scales in the IC space than is typical in galaxies. The field strength can be estimated under the assumption of flux-freezing, or magnetic energy conservation, if re-connection is ignored. Since typical galactic fields are ~few μG, mean volume-averaged field values of a few 0.1μG are expected in the IC space. It is possible that Fields may have been amplified by the hydrodynamic turbulence generated by galactic motions (Jaffe 1980). However, this process was found to be relatively inefficient (Goldman & Rephaeli 1991) in the context of a specific magneto-hydrodynamic model (Ruzmaikin et al. 1989).

Estimates of magnetic fields in extended sources are usually based on measurements of the synchrotron emission by relativistic electrons, and by Faraday rotation measurements. The estimates of B_{rx} quoted in the previous section - based on radio synchrotron and Compton emission by what was *assumed* to be the same relativistic electron population - are quite uncertain due to the need to make also other assumptions, such as the equality of the spatial factors in the expressions for the radio and X-ray fluxes (Rephaeli 1979). These are essentially volume integrations of the profiles of the electrons and fields, and since we have no information on the X-ray profile - and only rudimentary knowledge of the spatial distribution of the radio emission - this ratio was taken to be unity in the above mentioned analyses of RXTE and BeppoSAX data. The effect of this assumption is a systematically lower value of B_{rx}.

Faraday rotation measurements yield a weighted average of the field *and* gas density along the line of sight, B_{fr}. Estimates of B_{fr} are also substantially uncertain, due largely to the fact that a statistically significant value of RM can only be obtained when the data from a sample of clusters are superposed (co-added). The significantly broader distribution of RM when plotted as function of cluster-centric distance clearly establishes the presence of IC magnetic fields. However, the deduced mean value of B_{fr} is an average over all the clusters in the sample, in addition to being a weighted average of the product of a line of sight component of the field and the electron density. Both the field and density vary considerably across the cluster; in addition, the field is very likely tangled, with a wide range of coherence scales which can only be roughly estimated (probably in the range of $\sim 1 - 50$ kpc). All these make the determination of the field by Faraday rotation measurements considerably uncertain.

The unsatisfactory observational status, and the intrinsic difference between B_{rx} and B_{fr}, make it clear that these two measures of the field cannot be simply compared. Even ignoring the large observational and systematic uncertainties, the different spatial dependences of the fields, relativistic electron density, and thermal electron density, already imply that B_{rx} and B_{fr} will in general be quite different. This was specifically shown by Goldshmidt & Rephaeli (1993) in the context of reasonable assumptions for the field morphology, and the known range of IC gas density profiles. It was found that B_{rx} is typically expected to be smaller than B_{fr}. Improved measurement of the spatial profile of the radio flux, and at least some knowledge of the spatial profile of the nonthermal X-ray emission, are needed before we can more meaningfully establish the relation between B_{rx} and B_{fr} in a given cluster.

3.2. ENERGETIC PARTICLES

The radio synchrotron emitting relativistic electrons in clusters lose energy also by Compton scattering. Where the field is $B < 3$ μG, Compton losses dominate, and the characteristic loss time is $\tau_c \simeq 2.3/\gamma_3$ Gyr, where γ_3 is the Lorentz factor in units of 10^3. Such electrons have Lorentz factors $\gamma_3 > 5$, if $B \sim 1\mu$G, and therefore they lose their energy in less than 1 Gyr. If the mean field value is lower than 1μG, then the Compton loss time is even lower, because higher electron energies are needed to produce the observed radio emission. Clearly, if most of the radio-producing electrons had been injected from galaxies during a single, relatively short period, then the observed radio emission is a transient phenomenon, lasting only for a time $\sim \tau_c$, which is typically less than 1 Gyr. In this case the electron energy spectrum evolves on a relatively short timescale, shorter than typical evolutionary times of normal galaxies. It follows that electrons from an early injection period would have lost their energy by now. If the emission is indeed relatively short-lived, then it may possibly be related to a few strong radio sources.

On the other hand, if the observed radio emission has lasted more than a Compton loss time, then the relativistic electron population has perhaps reached a (quasi) steady state. This could be attained through continual ejection of electrons from radio sources and other cluster galaxies, or by re-acceleration in the IC space. The lower the mean value of the field, the shorter has the electron replenishment time to be. It is unclear whether any of these is a viable possibility, particularly so in the case of sub-μG fields. For electrons with energies $<< 1$ GeV, the main energy loss process is electronic (or Coulomb) excitations (Rephaeli 1979), and the loss time is maximal for $\gamma \sim 300$ (Sarazin 1999). In order for electrons with energies near this value to produce the observed IC radio emission, the mean magnetic field has to be at least $\sim$few μG. But in this case Compton scattering of these electrons by the CMB would only boost photon energies to the sub-keV range.

Various models have been proposed for acceleration of electrons in the IC space. All these invoke different scenarios of acceleration by shocks, resulting from galactic mergers, or shocks produced by fast moving cluster galaxies. In some of the accelerating electron models that have been proposed (Kaastra et al. 1998, Sarazin & Kempner 2000), electrons produce (nearly) power-law X-ray emission by nonthermal bremsstrahlung. Presumably, all the radio, EUV, and X-ray measurements of Coma and A2199 can be explained as emission from a population of accelerating electrons (Sarazin & Kempner 2000). However, the required electron energy density is much higher than in the relativistic electron population that produces power-law X-ray emission by Compton scattering (Rephaeli 2001. Petrosian 2001). A more plausible acceleration model has been suggested by Bykov et al. (2000). In their model, galaxies with dark matter halos moving at supersonic (and super-Alfvenic) velocities can create collisionless bow shocks of moderate Mach number $M \geq 2$. Kinetic modeling of nonthermal electron injection, acceleration and propagation in such systems seem to demonstrate that the halos are efficient electron accelerators, with the energy spectrum of the electrons shaped by the joint action of first and second order Fermi acceleration in a turbulent plasma with substantial Coulomb losses. Synchrotron, bremsstrahlung, and Compton losses of these electrons were found to produce spectra that are in quantitative agreement with current observations.

Although we do not yet have direct evidence for the presence of a substantial flux of energetic protons in the IC space, we do expect (based on the fact that protons are the main Galactic cosmic ray component) that they contribute very significantly - and even dominate - the cosmic ray energy density in clusters. Models for relativistic IC protons (Rephaeli 1987) and their γ-ray emission by neutral pion decays (produced in p-p collisions), or by the radiation from secondary electrons resulting from charged pion decays, have been proposed (Dermer & Rephaeli 1988, Blasi & Colafrancesco 1999). A substantial low-energy proton component could also cause appreciable (Coulomb) heating of the gas in the cores of clusters (Rephaeli 1987, Rephaeli & Silk 1995).

4. Conclusion

Recent RXTE and BeppoSAX measurements of power-law X-ray emission in four clusters improve our ability to characterize extragalactic magnetic fields and cosmic ray electrons. As expected, clusters of galaxies provide the first tangible basis for the exploration of nonthermal phenomena in intergalactic space. The radio and X-ray measurements strongly motivate further work on these phenomena. With the moderate spatial resolution (FWHM $\sim$ 12 arcminute) capability of the IBIS imager on the INTEGRAL satellite (which is scheduled for launch in 2002) it will be possible to determine the morphology of nonthermal (at energies > 15 keV) emission in nearby clusters. Such spatial information is crucial for a more definite identification of the observed second X-ray spectral components. Unequivocal measurement of cluster nothermal X-ray emission will greatly advance the study of nonthermal phenomena on cosmological scales.

References

Blasi, P., and Colafrancesco, S. (1999), *Astropar. Phys.*, **122**, p.169
Bykov, A. et al. (2000), *A&A*, **362**, p.29
Clarke, T. et al. (2001), *ApJ*, **547**, p.L111
Dermer, C.D, and Rephaeli, Y. (1988), *ApJ*, **329**, p.687
Fusco-Femiano, R., et al. (1999), *ApJ*, **513**, p.L21
Fusco-Femiano, R., et al. (2000), *ApJ*, **534**, p.L7
Giovannini, G., et al. (1999), *New Astron.*, 4, p.141
Giovannini, G., and Feretti, L. (2000), *New Astron.*, 5, p.335
Goldman, I., and Rephaeli, Y. (1991), *ApJ*, **380**, p.344
Goldshmidt, O., and Rephaeli, Y. (1993), *ApJ*, **411**, p.518
Gruber, D.E., and Rephaeli, Y. (2001), *ApJ*, submitted
Henriksen, M. (1998), *ApJ*, **511**, p.666
Jaffe, W. (1980), *ApJ*, **241**, p.925
Kaastra, J.S. et al. (1998), *Nuc. Phys. B*, **69**, p.567
Kaastra, J.S. et al. (2000), *ApJ*, **519**, p.L119
Kim, K.T. et al. (1991), *ApJ*, **379**, p.80
Petrosian, V. (2001), astro-ph/0101145
Rephaeli, Y. (1979), *ApJ*, **227**, p.364
Rephaeli, Y. (1987), *MN*, **225**, p.851
Rephaeli, Y. (1988), *Comm. Ap.*, **12**, p.265
Rephaeli, Y., and Gruber, D.E. (1988), *ApJ*, **333**, p.133
Rephaeli, Y. (2001), *Proceedings of the Heidelberg Gamma-Ray Meeting*, in press
Rephaeli, Y. et al. (1987), *ApJ*, **320**, p.139
Rephaeli, Y. et al. (1994), *ApJ*, **429**, p.554
Rephaeli, Y., and Silk, J. (1995), *ApJ*, **442**, p.91
Rephaeli, Y. et al. (1999), *ApJ*, **511**, p.L21
Ruzmaikin, A.A et al. (1989), *MN*, **241**, p.1
Sarazin, C.L. (1999), *ApJ*, **520**, p.529
Sarazin, C.L., and Kempner, J.C. (2000), *ApJ*, **533**, p.73

CORRELATION AND CLUSTERING
Statistical properties of galaxy large scale structures

Francesco Sylos Labini

Dpt. de Physique Thorique, Universit de Genve

24, Quai E. Ansermet, CH-1211 Genve, Switzerland

sylos@amorgos.unige.ch

Andrea Gabrielli

INFM Sezione Roma 1

Dip. di Fisica, Universita' "La Sapienza", P.le A. Moro 2 I-00185 Roma, Italy.

andrea@pil.phys.uniroma1.it

Abstract   In this lecture we clarify the basic difference between the correlation properties for systems characterized by small or large fluctuations. The concepts of correlation length, homogeneity scale, scale invariance and criticality are discussed as well. We relate these concepts to the interpretation of galaxy clsutering.

Keywords: Galaxies: general, statistics, large-scale structure of universe

1.   INTRODUCTION

The existence of large scale structures (LSS) and voids in the distribution of galaxies up to several hundreds Megaparsecs is well known from twenty years [1, 2]. The relationship of these structures with the statistics of galaxy distribution is usually inferred by applying the standard statistical analysis as introduced and developed by Peebles and coworkers [3]. Such an analysis *assumes* implicitly that the distribution is homogeneous at very small scale ($\lambda_0 \approx 5 \div 10 h^{-1} Mpc$). Therefore the system is characterized as having small fluctuations about a finite average density. If the galaxy distribution had a fractal nature the situation would be completely different. In this case the average density in finite samples is not a well defined quantity: it is strongly sample-dependent going to zero in the limit of an infinite volume. In such a situation it is not mean-

M. M. Shapiro et al. (eds.), Astrophysical Sources of High Energy Particles and Radiation, 151–160.
© 2001 *Kluwer Academic Publishers. Printed in the Netherlands.*

ingful to study fluctuations around the average density extracted from sample data. The statistical properties of the distribution should then be studied in a completely different framework than the standard one. We have been working on this problem since some time [4] by following the original ideas of Pietronero [5]. The result is that galaxy structures are indeed fractal up to tens of Megaparsecs [6]. Whether a crossover to homogeneity at a certain scale λ_0, occurs or not (corresponding to the absence of voids of typical scale larger than λ_0) is still a matter of debate [7]. At present, the problem is basically that the available redshift surveys do not sample scales larger than $50 \div 100 h^{-1} Mpc$ in a wide portion of the sky and in a complete way.

In this lecture we try to clarify some simple and basic concepts like the proper definition of correlation length, homogeneity scale, average density and scale invariance. We point out that a correct defintion and intepretation of the above concepts is necessary in order to understand phenomenologically the statistical properties of galaxy structures and to define the correct theoretical questions one would like to answer for.

2. DISTRIBUTION WITH SMALL FLUCTUATIONS

Consider a *statistically* homogeneous and isotropic particle density $n(\vec{r})$ with or without correlations with a well defined average value n_0. Let

$$n(\vec{r}) = \sum_i \delta(\vec{r} - \vec{r}_i) \tag{1}$$

be the number density of points in the system (the index i runs over all the points) and let us suppose to have an infinite system. Statistical homogeneity and isotropy refer to the fact that any n-point statistical property of the system is a function only on the scalar relative distances between these n points. The existence of a well defined average density means that

$$\lim_{R \to \infty} \frac{1}{\|C(R)\|} \int_{C(R)} d^3 r \, n(\vec{r}) = n_0 > 0 \tag{2}$$

(where $\|C(R)\| \equiv 4\pi R^3/3$ is the volume of the sphere $C(R)$) independently of the origin of coordinates. The scale λ_0, such that the *one point* average density is well-defined, i.e.

$$\left| \int_{C(R)} d^3 r \, n(\vec{r}) / \|C(R)\| - n_0 \right| < n_0 \text{ for } r > \lambda_0 , \tag{3}$$

is called *homogeneity scale*. If $n(\vec{r})$ is extracted from a density ensemble, n_0 is considered the same for each realization, i.e. it is a self-averaging quantity.

Let $\langle F \rangle$ be the ensemble average of a quantity F related to $n(\vec{r})$. If only one realization of $n(\vec{r})$ is available, $\langle F \rangle$ can be evaluated as an average over *all* the different points (occupied or not) of the space taken as origin of the coordinates. The quantity

$$\langle n(\vec{r_1})n(\vec{r_2})...n(\vec{r_l}) \rangle \, dV_1 dV_2...dV_l$$

gives the average probability of finding l particles placed in the infinitesimal volumes $dV_1, dV_2, ..., dV_l$ respectively around $\vec{r_1}, \vec{r_2}, ..., \vec{r_l}$. For this reason $\langle n(\vec{r_1})n(\vec{r_2})...n(\vec{r_l}) \rangle$ is called *complete l-point correlation function*. Obviously $\langle n(\vec{r}) \rangle = n_0$, and in a single sample such that $V^{1/3} \gg \lambda_0$, it can be estimated by

$$n_V = N/V \tag{4}$$

where N is the total number of particle in volume V.

Let us analyze the auto-correlation properties of such a system. Due to the hypothesis of statistical homogeneity and isotropy, $\langle n(\vec{r_1})n(\vec{r_2}) \rangle$ depends only on $r_{12} = |\vec{r_1} - \vec{r_2}|$. Moreover, $\langle n(\vec{r_1})n(\vec{r_2})n(\vec{r_3}) \rangle$ is only a function of $r_{12} = |\vec{r_1} - \vec{r_2}|$, $r_{23} = |\vec{r_2} - \vec{r_3}|$ and $r_{13} = |\vec{r_1} - \vec{r_3}|$. The *reduced* two-point and three correlation functions $\xi(r)$ and $\zeta(r_{12}, r_{23}, r_{13})$ are respectively defined by:

$$\langle n(\vec{r_1})n(\vec{r_2}) \rangle = n_0^2 [1 + \xi(r_{12})] \tag{5}$$
$$\langle n(\vec{r_1})n(\vec{r_2})n(\vec{r_3}) \rangle = n_0^3 [1 + \xi(r_{12}) + \xi(r_{23}) + \xi(r_{13}) + \zeta(r_{12}, r_{23}, r_{13})] \ .$$

The reduced two-point correlation function $\xi(r)$ defined in the previous equation is a useful tool to describe the correlation properties of small fluctuations with respect to the average. However we stress again that in order to perform a statistical analysis following Eqs.5, the one-point average density should be a well defined quantity and this must be carefully tested in any given sample (see below). We define

$$\sigma^2(R) \equiv \frac{\langle \Delta N(R)^2 \rangle}{\langle N(R) \rangle} \tag{6}$$

to be the mean square fluctuation normalized to the average density. From the definition of λ_0, we have

$$\sigma^2(\lambda_0) \simeq 1 \tag{7}$$

and $\sigma^2(R) \ll 1$ for $r \gtrsim \lambda_0$. Note that $\sigma^2(R)$ is again related to the *one-point* property of the distribution. We stress that the defintion of the homogeneity scale via Eq.7 can be misleading in the case where the average density is not a well-defined concept (see next section). Indeed, in such a case the quantity $\langle N(R) \rangle$ at the denominator *is not* given by $\langle N(R) \rangle = n_0 \times \|C(R)\| \sim R^3$: the scaling exponent is indeed different from the Euclidean dimension of the space $d = 3$.

In order to analyze observations from an occupied point it is necessary to define another kind of average: the *conditional* average $\langle F \rangle_p$ which characterizes the *two-point* properties of the system. This is defined as an ensemble average with the condition that the origin of coordinates is an occupied point. When only one realization of $n(\vec{r})$ is available, $\langle F \rangle_p$ can be evaluated averaging the quantity F over all the occupied points taken as origin of coordinates. The quantity

$$\langle n(\vec{r_1})n(\vec{r_2})...n(\vec{r_l}) \rangle_p \, dV_1 dV_2 ... dV_l \tag{8}$$

is the average probability of finding l particles placed in the infinitesimal volumes $dV_1, dV_2, ..., dV_l$ respectively around $\vec{r_1}, \vec{r_2}, ..., \vec{r_l}$ with the condition that the origin of coordinates is an occupied point. We call $\langle n(\vec{r_1})n(\vec{r_2})...n(\vec{r_l}) \rangle_p$ conditional l-point density. Applying the rules of conditional probability [8] one has:

$$\Gamma(r) \equiv \langle n(\vec{r}) \rangle_p = \frac{\langle n(\vec{0})n(\vec{r}) \rangle}{n_0} \tag{9}$$

$$\langle n(\vec{r_1})n(\vec{r_2}) \rangle_p = \frac{\langle n(\vec{0})n(\vec{r_1})n(\vec{r_2}) \rangle}{n_0}.$$

where $\Gamma(r)$ is called the conditional average density [5].

However, in general, the following convention is assumed in the definition of the conditional densities: the particle at the origin does not observe itself. Therefore $\langle n(\vec{r}) \rangle_p$ is defined only for $r > 0$, and $\langle n(\vec{r_1})n(\vec{r_2}) \rangle_p$ for $r_1, r_2 > 0$. In the following we use this convention as corresponding to the experimental data in galaxy catalogs.

We have defined above the homogeneity scale by means of the one-point properties of the distribution. Here we may define it in another wa by looking at the two-point properties: If the presence of an object at the point $\vec{r_1}$ influences the probability of finding another object at $\vec{r_2}$, these two points are correlated. Hence there is a correlation at the scale distance r if

$$G(r) = \langle n(\vec{0})n(\vec{r}) \rangle \neq \langle n \rangle^2 . \tag{10}$$

On the other hand, there is no correlation if

$$G(r) = \langle n \rangle^2 > 0 . \tag{11}$$

Therefore the proper definition of λ_0, the *homogeneity scale*, is the length scale beyond which $G(r)$ or equivalently $\Gamma(r)$ become nearly constant with scale and show a well-defined flattening. If this scale is smaller than the sample size then one may study for instance the behaviour of $\sigma^2(R)$ with scale (Eq.6) in the sample.

The length-scale λ_0 represents the typical dimension of the voids in the system. On the other hand there is another length scale which is very important for the characterization of point spatial distributions: the *correlation length* r_c. The length r_c separates scales at which density fluctuations are correlated (i.e. probabilistically related) to scales where they are uncorrelated. It can be defined only if a crossover towards homogeneity is shown by the system, i.e. if λ_0 exists [9]. In other words r_c defines the organization in geometrical structures of the fluctuations with respect to the average density. Clearly $r_c > \lambda_0$: only if the average density can be defined one may study the correlation length of the fluctuations around it. Note that r_c *is not* related to the absolute amplitude of fluctuations, but to their probabilistic correlation. In the case in which λ_0 is finite and then $\langle n \rangle > 0$, in order to study the correlations properties of the fluctuations around the average and then the behaviour of r_c, we can study the reduced two-point correlation function $\xi(r)$ defined in Eq.5.

The *correlation length* can be defined through the scaling behavior of $\xi(r)$ with scale. There are many definitions of r_c, but in any case, in order to have r_c finite, $\xi(r)$ must decay enoughly fast to zero with scale. For instance, if

$$|\xi(r)| \to \exp(-r/r_c) \text{ for } r \to \infty, \tag{12}$$

this means that for $r \gg r_c$ the system is structureless and density fluctuations are weakly correlated. The definition of the correlation length r_c by Eq.12 is equivalent to the one given by [9].

2.1. SOME EXAMPLES

Let us consider some simple examples. The first one is a Poisson distribution for which there are no correlation between different points. In such a situation [10] the average density is well defined, and

$$\xi(r_{12}) = \delta(\vec{r_1} - \vec{r_2})/n_0. \tag{13}$$

Analogously, one can obtain the three point correlation functions:

$$\zeta(r_{12}, r_{23}, r_{13}) = \delta(\vec{r_1} - \vec{r_2})\delta(\vec{r_2} - \vec{r_3})/n_0^2. \tag{14}$$

The two previous relations say only that there is no correlation between different points. That is, the reduced correlation functions ξ and ζ have only the so called *"diagonal"* part. This diagonal part is present in the reduced correlation functions of any statistically homogeneous and isotropic distribution with correlations. For instance [11] $\xi(r)$ in general can be written as $\xi(r) = \delta(\vec{r})/n_0 + h(r)$, where $h(r)$ is the non-diagonal part which is meaningful only for $r > 0$. Consequently, we obtain for the purely Poisson case (remember that conditional densities are defined only for points out of the origin):

$$\langle n(\vec{r}) \rangle_p = n_0 \tag{15}$$
$$\langle n(\vec{r_1})n(\vec{r_2}) \rangle_p = n_0^2 [1 + \delta(\vec{r_1} - \vec{r_2})/n_0] \ .$$

The second example is a distribution which is homogeneous but with a finite correlation lenght r_c. In such a situation $\Gamma(r)$ has a well-defined flattening and one may study the properties of $\xi(r)$. The correlation length r_c is usually defined as the scale beyond which $\xi(r)$ is exponentially damped. It measures up to which distance density fluctuations density are correlated. Note that while λ_0 refers to an one-point property of the system (the average density), r_c refers to a two-points property (the density-density correlation) [9, 12]. In such a situation $\xi(r)$ is in general represented by

$$\xi(r) = A \exp(-r/r_c) \tag{16}$$

where A is a prefactor which basically depends on the homoegeneity scale λ_0. We remind that λ_0 gives the scale beyond which $\sigma^2(R) \ll 1$ (Eq.6), and not the scale beyond which density fluctuations are not correlated anymore. This means that the typical dimension of voids in the system is not larger than λ_0, but one may find structures of density fluctuations of size up to $r_c \geq \lambda_0$.

Finally, let us now consider a mixed case in which the system is homogeneous (i.e. λ_0 is finite), but it has long-range power-law correlations. This means that fluctuations around the average, independently on their amplitude, are correlated at all scales, i.e. one finds structures of all scales. However, we stress again these are structures of fluctuations with respect to a mean which is well defined. This last event is in general described by the divergence of the correlation length r_c. Therefore let us consider a system in which $\langle n(\vec{r}) \rangle = n_0 > 0$ and $\xi(r) = [\delta(\vec{r})]/n_0 + h(r)$, with

$$|h(r)| \sim r^{-\gamma} \quad \text{for } r \gg \lambda_0 , \tag{17}$$

and $0 < \gamma \leq 3$. For $\gamma > 3$, despite the power law behavior, $\xi(r)$ is integrable for large r, and depending on the studied statistical quantity of the point distribution, we can consider the system as having a finite r_c (i.e. behaving like an exponentially damped $\xi(r)$) or not. Eq. 17 characterizes the presence of scale-invariant structures of fluctuations with long-range correlations, which, in Statistical Physics is also called "critical" [13].

3. DISTRIBUTION WITH LARGE FLUCTUATIONS

A completely different case of point distribution with respect the homogeneous one with or without correlations is the fractal one. In the case of a fractal distribution, the average density $\langle n \rangle$ in the infinite system is zero, then $G(r) = 0$ and $\lambda_0 = \infty$ and consequently $\xi(r)$ is not defined. For a fractal point distribution with dimension $D < 3$ the conditional one-point density $\langle n(\vec{r}) \rangle_p$ (which is hereafter called $\Gamma(r)$) has the following behavior [4]

$$\langle n(\vec{r}) \rangle_p \equiv \Gamma(r) = Br^{D-3} , \tag{18}$$

for enough large r. The intepretation of this behavior is the following. We may compute the average mass-length relation from an occupied point which gives the average numebr of points in a spherical volume of radius R centered on an occupied point: this gives

$$\langle N(R) \rangle_p = (4\pi B)/D \times R^D , \tag{19}$$

The constant B is directly related to the lower cut-off of the distribution: it gives the mean number of galaxies in a sphere of radius $1h^{-1}Mpc$. Eq.19 implies that the average density in a sphere of radius R around an occupied point scales as $1/R^{3-D}$. Hence it depends on the sample size R, the fractal is asymptotically empty and thus $\lambda_0 \to \infty$. We have two limiting cases for the fractal dimension: (1) $D = 0$ means that there is a finite number of points well localized far from the boundary of the sample (2) $D = 3$ the distribution has a well defined positive average density, i.e. the conditional average density does not depend on scale anymore. Given the metric interpretation of the fractal dimension, it is simple to show that $0 \leq D \leq 3$. Obviously, in the case $D = 3$ for which λ_0 is finite $\Gamma(r)$ provides the same information of $G(r)$, i.e. it characterizes the crossover to homogeneity.

A very important point is represented by the kind of information about the correlation properties of the infinite system which can be extracted from the analysis of a finite sample of it. In [5] it is demonstrated that, in the hypothesis of statistical homogeneity and isotropy, even in the super-correlated case of a fractal the estimate of $\Gamma(r)$ extracted from the finite sample of size R_s, is not dependent on the sample size R_s, providing a good approximation of that of the whole system. Clearly this is true a part from statistical fluctuations [10] due to the finiteness of the sample. In general the $\Gamma(r)$ extracted from a sample can be written in the following way:

$$\Gamma(r) = \frac{1}{N} \sum_{i=1}^{N} \frac{1}{4\pi r^2 \Delta r} \int_{r}^{r+\Delta r} n(\vec{r}_i + \vec{r}\,')d^3r', \tag{20}$$

where N is the number of points in the sample, $n(\vec{r}_i + \vec{r}\,')$ is the number of points in the volume element d^3r' around the point $\vec{r}_i + \vec{r}\,'$ and Δr is the thickness of the shell at distance r from the point at $\vec{r}_i$. Note that the case of a sample of a homogeneous point distribution of size $V \ll \lambda_0^3$, must be studied in the same framework of the fractal case.

4.   PROBLEMS OF THE STANDARD ANALYSIS

In the fractal case ($V^{1/3} \ll \lambda_0$), the sample estimate of the homogeneity scale, through the value of r for which the sample-dependent correlation function $\xi(r)$ (given by Eq.21) is equal to 1, is meaningless. This estimate is the so-called *"correlation length"* r_0 [3] in the standard approach of cosmology. As we discuss below, r_0 has nothing to share with the *true* correlation length r_c. Let us see why r_0 is unphysical in the case $V^{1/3} \ll \lambda_0$. The length r_0 [3] is defined by the relation $\xi(r_0) = 1$, where $\xi(r)$ is given operatively by

$$\xi(r) = \frac{\Gamma(r)}{n_V} - 1 . \tag{21}$$

where n_V is given by Eq.4. What does r_0 mean in this case ? The basic point in the present discussion [5], is that the mean density of the sample, n_V, used in the normalization of $\xi(r)$, is not an intrinsic quantity of the system, but it is a function of the finite size R_s of the sample.

Indeed, from Eq.18, the expression of the $\xi(r)$ of the sample in the case of fractal distributions is [5]

$$\xi(r) = \frac{D}{3} \left(\frac{r}{R_s} \right)^{D-3} - 1 . \tag{22}$$

being R_s the radius of the assumed spherical sample of volume V. From Eq.22 it follows that r_0 (defined as $\xi(r_0) = 1$) is a linear function of the sample size R_s

$$r_0 = \left(\frac{D}{6} \right)^{\frac{1}{3-D}} R_s \tag{23}$$

and hence it is a spurious quantity without physical meaning but it is simply related to the sample's finite size. In other words, this is due to the fact that n_V in the fractal case is in any finite sample a *bad* estimate of the asymptotic density which is zero in this case

We note that the amplitude of $\Gamma(r)$ (Eq.18) is related to the lower cut-off of the fractal, while the amplitude of $\xi(r)$ is related to the upper cut-off (sample size R_s) of the distribution. This crucial difference has never been appreciated appropriately.

Finally, we stress that in the standard analysis of galaxy catalogs the fractal dimension is estimated by fitting $\xi(r)$ with a power law, which instead, as one can see from Eq.22, is power law only for $r \ll r_0$ (or $\xi \gg 1$). For distances around and beyond r_0 there is a clear deviation from the power law behavior due to the definition of $\xi(r)$. Again this deviation is due to the finite size of the observational sample and does not correspond to any real change in the correlation properties. It is easy to see that if one estimates the exponent at distances $r \lesssim r_0$, one systematically obtains a higher value of the correlation exponent due to the break in $\xi(r)$ in a log-log plot.

5.   DISCUSSION AND CONCLUSION

From an operative point of view, having a finite sample of points (e.g. galaxy catalogs), the first analysis to be done is the determination of $\Gamma(r)$ of the sample itself. Such a measurement is necessary to distinguish between the two cases: (1) a crossover towards homogeneity in the sample with a flattening of $\Gamma(r)$, and hence an estimate of $\lambda_0 < R_s$ and $\langle n \rangle$; (2) a continuation of the fractal behavior. Obviously only in the case (1), it is physically meaningful to introduce an estimate of the correlation function $\xi(r)$ (Eq.21), and extract from it the length scale r_0 ($\xi(r_0) = 1$) to estimate the intrinsic homogeneity scale λ_0. In this case, the functional behavior of $\xi(r)$ with distance gives instead information on the correlation length of the density fluctuations. Note that there are always subtle finite size effects which perturb the behvaiour of $\xi(r)$ for $r \sim V1/3$, and which must be properly taken into account. These same arguments apply to the estimation of the power spectrum of the density fluctuations, which is just the fourier conjugate of the correlation function [4]. The application of these concepts to the case of real galaxy data can be found in [4, 6, 10].

Acknowledgments

We thank Y.V. Baryshev, R. Durrer, J.P.Eckmann, P.G. Ferreira, M. Joyce, M. Montuori and L. Pietronero for useful discussions. This work is partially supported by the EC TMR Network "Fractal structures and self-organization" ERBFMRXCT980183 and by the Swiss NSF.

160

References

[1] De Lapparent V., Geller M. & Huchra J., Astrophys.J., 343, (1989) 1

[2] Tully B. R. Astrophys.J. 303, (1986) 25

[3] Peebles, P.J.E., (1980) *Large Scale Structure of the Universe*, Princeton Univ. Press

[4] Sylos Labini F., Montuori M., Pietronero L., Phys.Rep.,293, (1998) 66

[5] Pietronero L., Physica A, 144, (1987) 257

[6] Joyce M., Montuori M., Sylos Labini F., Astrophys. Journal 514,(1999) L5

[7] Wu K.K., Lahav O.and Rees M., Nature, 225, (1999) 230

[8] Feller W., *An Introduction to Probability Theory and its Applications*, vol. 2, ed. Wiley & Sons

[9] Gaite J., Dominuguez A. and Perez-Mercader J. Astrophys.J.Lett. 522, (1999) L5

[10] Gabrielli A. & Sylos Labini F., Preprint (astro-ph/0012097)

[11] Salsaw W.C. "The distribution of galaxies" Cambridge University Press (2000)

[12] Gabrielli A., Sylos Labini F., and Durrer R. Astrophys.J. Letters 531, (2000) L1

[13] Huang K., "Statistical Mechanics" John Wiley & Sons (1987) Chapetr 16.

THE SUNYAEV-ZELDOVICH EFFECT AND ITS COSMOLOGICAL SIGNIFICANCE

YOEL REPHAELI

School of Physics & Astronomy
Tel Aviv University, Tel Aviv, Israel

Abstract.

Comptonization of the cosmic microwave background (CMB) radiation by hot gas in clusters of galaxies - the Sunyaev-Zeldovich (S-Z) effect - is of great astrophysical and cosmological significance. In recent years observations of the effect have improved tremendously; high signal-to-noise images of the effect (at low microwave frequencies) can now be obtained by ground-based interferometric arrays. In the near future, high frequency measurements of the effect will be made with bolomateric arrays during long duration balloon flights. Towards the end of the decade the PLANCK satellite will extensive S-Z surveys over a wide frequency range. Along with the improved observational capabilities, the theoretical description of the effect and its more precise use as a probe have been considerably advanced. I review the current status of theoretical and observational work on the effect, and the main results from its use as a cosmological probe.

1. Introduction

The cosmological significance of the spectral signature imprinted on the CMB by Compton scattering of the radiation by electrons in a hot intergalactic medium was realized early on by Zeldovich & Sunyaev (1969). A more important manifestation of this effect occurs in clusters of galaxies (Sunyaev & Zeldovich 1972). Measurements of this S-Z effect with single-dish radio telescopes were finally successful some 15 years later (for general reviews, see Rephaeli 1995a, Birkinshaw 1999). Growing realization of the cosmological significance of the effect has led to major improvements in observational techniques, and to extensive theoretical investigations of its many facets. The use of interferometric arrays, and the substantial progress

M. M. Shapiro et al. (eds.), Astrophysical Sources of High Energy Particles and Radiation, 161–171.

in the development of sensitive radio receivers, have led to first images of the effect (Jones 1993, Carlstrom et al. 1996). Some 40 clusters have already been obtained with the OVRO and BIMA arrays (Carlstrom et al. 1999, Carlstrom et al. 2001). Theoretical treatment of the S-Z effect has also improved, starting with the work of Rephaeli (1995b), who performed an exact relativistic calculation and demonstrated the need for such a more accurate description.

The S-Z effect is essentially independent of the cluster redshift, a fact that makes it unique among cosmological probes. Measurements of the effect yield directly the properties of the hot intracluster (IC) gas, and the *total* dynamical mass of the cluster, as well as indirect information on the evolution of clusters. Of more basic importance is the ability to determine the Hubble (H_0) constant and the density parameter, Ω, from S-Z and X-ray measurements. This method to determine H_0, which has clear advantages over the traditional galactic distance ladder method based on optical observations of galaxies in the nearby universe, has been yielding increasingly higher quality results. But substantial systematic uncertainties, due largely to modeling of the thermal and spatial distributions of IC gas, need still to be reduced in order to fully exploit the full potential of this method. Sensitive spectral and spatial mapping of the effect, and the minimization of systematic uncertainties in the S-Z and X-ray measurements, constitute the main challenges of current and near future work in this rapidly progressing research area.

I describe the effect and review some of the recent observational and theoretical work that significantly improved its use as a precise cosmological probe. Progress anticipated in the near future is briefly discussed.

2. Exact Description of the Effect

The scattering of the CMB by hot gas heats the radiation, resulting in a systematic transfer of photons from the Rayleigh-Jeans (R-J) to the Wien side of the (Planckian) spectrum. An accurate description of the interaction of the radiation with a hot electron gas necessitates the calculation of the exact frequency re-distribution function in the context of a relativistic formulation. The calculations of Sunyaev & Zeldovich (1972) are based on a solution to the Kompaneets (1957) equation, a *nonrelativistic* diffusion approximation to the exact kinetic (Boltzmann) equation describing the scattering. The result of their treatment is a simple expression for the intensity change resulting from scattering of the CMB (temperature T) by electrons with *thermal* velocity distribution (temperature T_e),

$$\Delta I_t = i_o y g(x) , \qquad (1)$$

where $i_o = 2(kT)^3/(hc)^2$. The spatial dependence is contained in the Comptonization parameter, $y = \int (kT_e/mc^2) n \sigma_T dl$, a line of sight integral (through the cluster) over the electron density (n); σ_T is the Thomson cross section. The spectral function,

$$g(x) = \frac{x^4 e^x}{(e^x - 1)^2} \left[\frac{x(e^x + 1)}{e^x - 1} - 4 \right], \qquad (2)$$

where $x \equiv h\nu/kT$ is the non-dimensional frequency, is negative in the R-J region and positive at frequencies above a critical value, $x = 3.83$, corresponding to ~ 217 GHz. Typically in a rich cluster $y \sim 10^{-4}$ along a line of sight through the center, and the magnitude of the relative temperature change due to the thermal effect is $\Delta T_t/T = -2y$ in the R-J region.

The above thermal intensity change is the full effect only if the cluster is at rest in the CMB frame. Generally, the effect has a second component when the cluster has a finite (peculiar) velocity in the CMB frame. This *kinematic* (Doppler) component is

$$\Delta I_k = \frac{x^4 e^x}{(e^x - 1)^2} \frac{v_r}{c} \tau , \qquad (3)$$

where v_r is the line of sight component of the cluster peculiar velocity, τ is the Thomson optical depth of the cluster. The related temperature change is $\Delta T_k/T = -(v_r/c)\tau$ (Sunyaev & Zeldovich 1980).

The quantitative nonrelativistic description of the two components of the S-Z effect by Sunyaev & Zeldovich (1972) is generally valid at low gas temperatures and at low frequencies. Rephaeli (1995b) has shown that this approximation is insufficiently accurate for use of the effect as a precise cosmological probe: Electron velocities in the IC gas are high, and the relative photon energy change in the scattering is sufficiently large to require a relativistic calculation. Using the exact probability distribution in Compton scattering, and the relativistically correct form of the electron Maxwellian velocity distribution, Rephaeli (1995b) calculated ΔI_t in the limit of small τ, keeping terms linear in τ (single scattering). Results of this semi-analytic calculation, shown in Figure 1, demonstrate that the relativistic spectral distribution of the intensity change is quite different from that derived by Sunyaev & Zeldovich (1972). Deviations from their expression increase with T_e and can be quite substantial. These are especially large near the crossover frequency, which shifts to higher values with increasing gas temperature.

The results of the semi-analytic calculations of Rephaeli (1995b) generated considerable interest which led to various generalizations and extensions of the relativistic treatment. Challinor & Lasenby (1998) generalized the Kompaneets equation and obtained analytic approximations to its solution for the change of the photon occupation number by means of a power

164

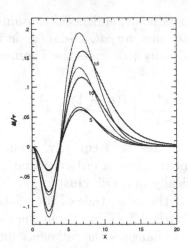

Figure 1. The spectral distribution of $\Delta I_t/\tau$ (in units of $h^2c^2/2k^3T^3$). The pairs of thick and thin lines, labeled with $kT_e = 5$, 10, and 15 keV, show the relativistic and nonrelativistic distributions, respectively.

series in $\theta_e = kT_e/mc^2$. Itoh et al. (1998) adopted this approach and improved the accuracy of the analytic approximation by expanding to fifth order in θ_e. Sazonov & Sunyaev (1998) and Nozawa et al. (1998) have extended the relativistic treatment also to the kinematic component obtaining – for the first time – the leading cross terms in the expression for the total intensity change $(\Delta I_t + \Delta I_k)$ which depends on both T_e and v_r. An improved analytic fit to the numerical solution, valid for $0.02 \leq \theta_e \leq 0.05$, and $x \leq 20$ ($\nu \leq 1130$ GHz), has recently been given by Nozawa et al. (2000). In view of the possibility that in some rich clusters $\tau \sim 0.02 - 0.03$, the approximate analytic expansion to fifth order in θ_e necessitates also the inclusion of multiple scatterings, of order τ^2. This has been accomplished by Itoh et al. (2000), and Shimon & Rephaeli (2001).

The more exact relativistic description of the effect should be used in all high frequency S-Z work, especially when measurements of the effect are used to determine precise values of the cosmological parameters. Also, since the ability to determine peculiar velocities of clusters depends very much on measurements at (or very close to) the crossover frequency, its exact value has to be known. This necessitates knowledge of the gas temperature since in the exact relativistic treatment the crossover frequency is no longer independent of the gas temperature, and is approximately given by $\simeq 217[1 + 1.167kT_e/mc^2 - 0.853(kT_e/mc^2)^2]$ GHz (Nozawa et al. 1998a). Use of the S-Z effect as a cosmological probe necessitates also X-ray measurements to determine (at least) the gas temperature. Therefore, a relativistically correct expression for the (spectral) bremsstrahlung emissivity must be used (Rephaeli & Yankovitch 1997). In the latter paper first order relativistic corrections to the velocity distribution, and electron-electron bremsstrahlung, were taken into account in correcting values of H_0 that were previously derived using the nonrelativistic expression for the emissivity (see also Hughes & Birkinshaw 1998). Nozawa et al. (1998b) have performed a more exact calculation of the relativistic bremsstrahlung Gaunt factor.

Scattering of the CMB in clusters affects also its polarization towards the cluster. Net polarization is induced due to the quadrupole component in the spatial distribution of the radiation, and when the cluster peculiar velocity has a component transverse to the line of sight, $v_\perp$ (Sunyaev & Zeldovich 1980, Sazonov & Sunyaev 1999). The leading contributions to the latter, kinematically-induced polarization, are proportional to $(v_\perp/c)\tau^2$ and $(v_\perp/c)^2\tau$. Itoh et al. (2000) have included relativistic corrections in the expression they derived for the kinematically induced polarization.

3. Recent Measurements

The quality of S-Z measurements has increased very significantly over the last seven years mainly as a result of the use of interferometric arrays. Telescope arrays have several major advantages over a single dish, including insensitivity of the measurements to changes in the atmospheric emission, sensitivity to specific angular scales and to signals which are correlated between the array elements, and the high angular resolution that enables nearly optimal subtraction of signals from point sources. With the improved sensitivity of radio receivers it became feasible to use interferometric arrays for S-Z imaging measurements (starting with the use of the Ryle telescope by Jones et al. 1993). Current state-of-the-art work is done with the BIMA and OVRO arrays; images of some 35 moderately distant clusters (in the redshift range $0.17 - 0.89$) have already been obtained at ~ 30 GHz (Carlstrom et al. 1999, 2001). The S-Z image of the cluster CL0016+16, observed at 28 GHz with the BIMA array (Carlstrom et al. 1999), is shown in the upper frame in Figure 2. The ROSAT X-ray image is superposed on the contour lines of the S-Z profile in the lower frame. These images nicely demonstrate the good agreement between the orientation of the X-ray and S-Z brightness distributions, as well as the relative smallness of the X-ray size ($\propto n^2$) in comparison with the S-Z size ($\propto n$) of the cluster. Work has begun recently with the CBI, a new radio (26-36 GHz) interferometric array (in the Chilean Andes) of small (0.9m) dishes on a platform with baselines in the 1m to 6m range. The spatial resolution of the CBI is in the $3' - 10'$ range, so (unlike the BIMA and OVRO arrays) it is suitable for S-Z measurements of nearby clusters. Some 9 clusters have already been

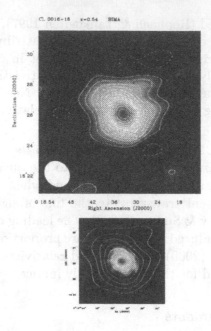

Figure 2. S-Z and X-ray images of the cluster CL0016+16. The S-Z image of the cluster, obtained with the BIMA array, is shown in the upper frame. In the lower frame, contours of the S-Z effect in the cluster are superposed on the *ROSAT* X-ray image (from Carlstrom et al. 1999).

observed with the CBI (Udomprasert et al. 2000).

S-Z observations at higher frequencies include measurements with the SuZIE array, and the PRONAOS and MITO telescopes. Three moderately distant clusters were measured with the small (2X3) SuZIE array: A1689 & A2163 (Holzapfel et al. 1997a, 1997b), and recently the cluster A1835 was observed at three spectral bands centered on 145, 221, 279 GHz (Mauskopf et al. 2000). Observations at four wide spectral bands (in the overall range of 285-1765 GHz) were made of A2163 with the PRONAOS atmospheric 2m telescope (Lamarre et al. 1998), leading to what seems to be the first detection of the effect by a balloon-borne experiment. The new 2.6m MITO telescope – which currently operates at four spectral bands and has a large $\sim 17'$ beam – was used to observe the effect in the Coma cluster (D'Alba et al. 2001).

The sample of observed clusters now includes the distant (z=0.45) cluster RXJ 1347 which was measured to have the largest determined Comptonization parameter, $y = 1.2 \times 10^{-3}$ (Pointecouteau et al. 1999). The observations were made with the Diabolo bolometer operating at the IRAM 30m radio telescope. The Diabolo has a $0.5'$ beam, and a dual channel bolometer (centered on 2.1 and 1.2 mm). Four other clusters were also observed with the Diabolo bolometer (Desert et al. 1998).

4. The S-Z Effect as a Cosmological Probe

The fact that the S-Z effect is (essentially) independent of the cluster redshift makes it a unique cosmological probe. This feature, its simple well understood nature, and the expectation that the inherent systematic uncertainties associated mainly with modeling of the IC gas can be reduced, have led to ever growing observational and theoretical interest in using the effect for the determination of cluster properties and global cosmological parameters. (For extensive discussions of use of the effect as a probe, see the reviews by Rephaeli 1995a, and Birkinshaw 1999.)

In principle, high spatial resolution measurements of the S-Z effect yield the gas temperature and density profiles across the cluster. Cluster gas density and temperature profiles have so far been mostly deduced from X-ray measurements. S-Z measurements can more fully determine these profiles due to the linear dependence of ΔI_t on n (and T), as compared to the n^2 dependence of the (thermal bremsstrahlung) X-ray brightness profile. This capability has been reached for the first time in the analysis of the interferometric BIMA and OVRO images (Carlstrom et al. 2001).

The cluster full mass profile, $M(r)$, can be derived directly from the gas density and temperature distributions by solving the equation of hydrostatic equilibrium (assuming, of course, the gas has reached such a state in the underlying gravitational potential). This method has already been employed in many analyses using X-ray deduced gas parameters (e.g. Fabricant et al. 1980). Grego et al. (2001) have recently used this method to determine total masses and gas mass fractions of 18 clusters based largely on the results of their interferometric S-Z measurements. Isothermal gas with the familiar density profile, $(1 + r^2/r_c^2)^{-3\beta/2}$, was assumed. The core radius, r_c, and β were determined from analysis of the S-Z data, whereas the X-ray value of the temperature was adopted. Mean values of the gas mass fraction were found to be in the range $(0.06 - 0.09)h^{-1}$ (where h is the value of H_0 in units of 100 km s^{-1} Mpc^{-1}) for the currently popular open and flat, Λ-dominated, CDM models. Note that the deduction of an approximate 3D density and temperature distributions from their sky-projected profiles requires use of a deprojection algorithm. Such an algorithm was developed by Zaroubi et al. (1998).

Measurement of the kinematic S-Z effect yields the line of sight component of the cluster peculiar velocity (v_r). This is observationally feasible only in a narrow spectral band near the critical frequency, where the thermal effect vanishes while the kinematic effect – which is usually swamped by the much larger thermal component – is maximal (Rephaeli & Lahav 1991).

SuZIE is the first experiment with a spectral band centered on the crossover frequency. Measurements of the clusters A1689 and A2163 (Holzapfel et al. 1997b) and A1835 (Mauskopf et al. 2000) yielded substantially uncertain results for v_r (170^{+815}_{-630}, 490^{+1370}_{-880}, and 500 ± 1000 km s^{-1}, respectively). Balloon-borne measurements of the effect with PRONAOS have also yielded insignificant determination of the peculiar velocity of A2163 (Lamarre et al. 1998).

Most attention so far has been given to the determination of the Hubble constant, H_0, and the cosmological density parameter, Ω, from S-Z and X-ray measurements. Briefly, the method is based on determining the angular diameter distance, d_A, from a combination of ΔI_t, the X-ray surface brightness, and their spatial profiles. Averaging over the first eight determinations of H_0 (from S-Z and X-ray measurements of seven clusters) yielded $H_0 \simeq 58 \pm 6$ km s^{-1} Mpc^{-1} (Rephaeli 1995a), but the database was then very non-uniform and the errors did not include systematic uncertainties. Repeating this with a somewhat updated data set, Birkinshaw (1999) deduced essentially a similar mean value (60 km s^{-1} Mpc^{-1}), but noted that the individual measurements are not independent, so that a simple error estimation is not very meaningful. A much larger S-Z data set is now available from the interferometric BIMA and OVRO observations, and since the redshift range of the clusters in the sample is substantial, the dependence on Ω is appreciable. A fit to 33 cluster distances gives $H_0 = 60$ km s^{-1} Mpc^{-1} for $\Omega = 0.3$, and $H_0 = 58$ km s^{-1} Mpc^{-1} for $\Omega = 1$, with direct observational errors of $\pm 5\%$ (Carlstrom et al. 2001). The main known sources of systematic uncertainties (see discussions in Rephaeli 1995a, and Birkinshaw 1999) are presumed to introduce an additional error of $\sim 30\%$ (Carlstrom et al. 2001). The current number of clusters with S-Z determined distances is sufficiently large that a plot of d_A vs. redshift (a Hubble diagram) is now quite of interest, but with the present level of uncertainties the limits on the value of Ω are not very meaningful (as can be seen from figure 11 of Carlstrom et al. 2001).

The S-Z effect induces anisotropy in the spatial distribution of the CMB (Sunyaev 1977, Rephaeli 1981); this is the main source of secondary anisotropy on angular scales of few arcminutes. The magnitude of the temperature anisotropy, $\Delta T/T$ can be as high as $few \times 10^{-6}$ on angular scales of a few arcminutes, if the gas evolution in clusters is not too strong (Colafrancesco et al. 1994). Because of this, and the considerable interest in this range of angular scales – multipoles (in the representation of the CMB temperature in terms of spherical harmonics) $\ell \geq 1000$ – the S-Z anisotropy has been studied extensively in the last few years. The basic goal has been to determine the S-Z anisotropy in various cosmological, large scale structure, and IC gas models. The anisotropy is commonly characterized by the

ℓ dependence of its power spectrum. The strong motivation for this is the need to accurately calculate the power spectrum of the full anisotropy in order to make precise global parameter determinations from the analysis of large stratospheric and satellite databases. In addition, mapping the S-Z anisotropy will yield direct information on the evolution of the cluster population.

Results from many calculations of the predicted S-Z anisotropy are not always consistent even for the same cosmological and large scale structure parameters. This is simply due to the fact that the calculation involves a large number of input parameters in addition to the global cosmological parameters (*e.g.* the present cluster density, and parameters characterizing the evolutionary history of IC gas), and the sensitive dependence of the anisotropy on some of these. The anisotropy can be more directly estimated from simulations of the S-Z sky, based largely on results from cluster X-ray surveys and the use of simple scaling relations (first implemented by Markevitch et al. 1992).

The observational capabilities of upcoming long duration balloon-borne experiments and satellites are expected to result in detailed mapping of the small angular scale anisotropy. These have motivated many recent works; Colafrancesco et al. (1997), and Kitayma et al. (1998), have calculated the S-Z cluster number counts in an array of open and flat cosmological and dark matter models, and da Silva et al. (1999) have carried out hydrodynamical simulations in order to generate S-Z maps and power spectra. Cooray et al. (2000) have, in particular, concluded that the planned multi-frequency survey with the Planck satellite should be able to distinguish between the primary and S-Z anisotropies, and measure the latter with sufficient precision to determine its power spectrum and higher order correlations.

The main characteristics of the predicted power spectrum of the induced S-Z anisotropy are shown in Figure 3. Plotted are the angular power spectra, $C_\ell(\ell + 1)/2\pi$, vs. the multipole, ℓ, for both the primary and S-Z induced anisotropies in the (currently fashionable) flat cosmological model with $\Omega_\Lambda = 0.7$ (where Λ is the cosmological constant) and with a CDM density parameter $\Omega_M = 0.3$. The figure is from the work of Sadeh and Rephaeli (2001), who studied the S-Z anisotropy in the context of treatment (which is an extension of the approach adopted by Colafrancesco et al. 1997) which is based on a Press & Schechter cluster mass function, normalized by the observed X-ray luminosity function (see also Colafrancesco et al. 1994). IC gas was assumed to evolve in a simple manner consistent with the results of the EMSS survey carried out with the Einstein satellite. The primary anisotropy was calculated using the CMBFAST code of Seljak & Zaldarriaga (1996). The solid line shows the primary anisotropy which

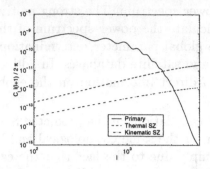

Figure 3. Primary and S-Z power spectra in the flat CDM model (Sadeh & Rephaeli 2001). The solid line shows the primary anisotropy as calculated using the CMBFAST computer code of Seljak & Zaldarriaga (1996). The dashed line shows the thermal S-Z power spectrum, and the dotted-dashed line is the contribution of the kinematic component.

dominates over the S-Z anisotropy for $\ell < 3000$. The S-Z power is largely due to the thermal effect; this rises with ℓ and is maximal around $\ell \sim 1000$. In this model, the S-Z power spectrum contributes a fraction of 5% (10%) to the primary anisotropy at $\ell \simeq 1520$ ($\ell \simeq 1840$). It can be concluded from this (and other studies) that the S-Z induced anisotropy has to be taken into account, if the extraction of the cosmological parameters from an analysis of measurements of the CMB power spectrum at $\ell > 1500$ is to be precise.

5. Future Prospects

The S-Z effect is a unique cosmological and cluster probe. In the near future sensitive observations of the effect with ground-based and balloon-borne telescopes, equipped with bolometric multi-frequency arrays, are expected to yield high-quality measurements of its spectral and spatial distributions in a The capability to measure the effect at several high-frequency bands (in the range $150 - 450$ GHz) will make it possible to exploit the S-Z characteristic spectrum as a powerful diagnostic tool. The main limitation on the accuracy of the cosmological parameters will continue to be due to systematic uncertainties. Therefore, the most significant results will be obtained from measurements of the effect in *nearby* clusters, $z \leq 0.1$, where systematic uncertainties can be most optimally reduced. The combination of S-Z measurements of a large number of clusters with the new generation of bolometric arrays, and the improved X-ray data which are currently available from observations of clusters with the XMM and *Chandra* satellites, will greatly improve the accuracy of the derived values of cluster masses, and of the Hubble constant. In particular, it will be possible to measure H_0

with an overall uncertainty of just $\sim 5\%$.

References

Birkinshaw, M. (1999) Title, *Phys.Rep.*, **310**, p.97
Carlstrom, J.E., Joy, M. and Grego, L. (1996), *ApJ*, **456**, p. L75
Carlstrom, J.E. et al. (1999), *Physica Scripta*, **60**
Carlstrom, J.E. et al. (2001), astro-ph/0103480
Colafrancesco, S., Mazzotta, P., Rephaeli, Y. and Vittorio, N. (1994), *ApJ*, *433*, p.454
Colafrancesco, S., Mazzotta, P., Rephaeli, Y. and Vittorio, N. (1997), *ApJ*, *479*, p.1
Challinor, A. and Lasenby, A. (1998), *ApJ*, *510*, p.930
Cooray, L., Hu, W. and Tegmark, M. (2000), astro-ph/0002238.
D'Alba, L. et al. (2001), astro-ph/0010084
da Silva, A.C. et al. (1999), astro-ph/9907224
Desert, F.X. et al. (1998), *New Astron.*, **3**, p.655
Fabricant, D.M., Lecar, M. and Gorenstein, P. (1980), *ApJ*, *241*, p.552
Grego, L. et al. (2001), *ApJ*, in press
Holzapfel, W.L. et al. (1997a), *ApJ*, **480**, p.449
Holzapfel, W.L. et al. (1997b), *ApJ*, **481**, p.35
Hughes, J.P. and Birkinshaw, M. (1998), *ApJ*, **501**, p.1
Itoh, N., Kohyama, Y. and Nozawa, S. (1998), *ApJ*, **502**, p.7
Itoh, N., Nozawa, S. and Kohyama, Y. (2000), astro-ph/0005390
Jones, M. et al. (1993), *Nature*, **365**, p.320
Kitayama, T et al. (1998), *PASJ*, **50**, p.1
Kompaneets, A.S. (1957), *Soviet Phys.-JETP*, **4**, p.730
Lamarre, J.M. et al. (1998), *ApJ*, **507**, p.L5
Markevitch, M. et al. (1998), *ApJ*, **395**, p.326
Mauskopf, P.D. et al. (2001), *ApJ*, **538**, p.505
Nozawa, S., Itoh, N. and Kohyama, Y. (1998a), *ApJ*, **507**, p.530
Nozawa, S., Itoh, N. and Kohyama, Y. (1998b), *ApJ*, **508**, p.17
Nozawa, S. et al. (2000), *ApJ*, **536**, p.31
Pointecouteau, E. et al. (1999), *ApJ*, **519**, p.L115
Rephaeli, Y. (1981), *ApJ*, **351**, p.245
Rephaeli, Y. (1995a), *ARAA*, **33**, p.541
Rephaeli, Y. (1995b), *ApJ*, **445**, p.33
Rephaeli, Y. and Lahav, O. (1991), *ApJ*, **372**, p.21
Rephaeli, Y. and Yankovitch, D. (1997), *ApJ*, **481**, p.L55
Sadeh, S. and Rephaeli, Y. (2001), preprint
Sazonov, S.Y. and Sunyaev, S.Y. (1998), *ApJ*, **508**, p.1
Sazonov, S.Y. and Sunyaev, S.Y. (1999), *MN*, **310**, p.765
Seljak, U. and Zaldarriaga, M. (1996), *ApJ*, **469**, p.437
Shimon, M. and Rephaeli, Y. (2001), preprint
Sunyaev, R.A. (1977), *Comm.Ap.Sp.Phys.*, **7**, p.1
Sunyaev, R.A. and Zeldovich, Y.B. (1972), *Comm.Ap.Sp.Phys.*, **4**, p.173
Sunyaev, R.A. and Zeldovich, Y.B. (1980), *MN*, **190**, p.413
Udomprasert, P.S., Mason, B.S. and Readhead, A.C.S. (2000), astro-ph/0012248
Zaroubi, S., et al. (1998), *ApJ*, **500**, p.L87
Zeldovich, Y.B. and Sunyaev, R.A., (1969), *Comm.Ap.Sp.Phys.*, **4**, p.301

AN OVERVIEW ON GEV-TEV GAMMA RAY ASTRONOMY: PAST-PRESENT-FUTURE

V. FONSECA

Dpto. Fisica Atomica, Fac. Ciencias Fisicas
Universidad Complutense, 28040 Madrid, Spain

Abstract. High energy gamma astronomy is nowadays a very exciting research field. The results obtained by detectors placed on board of satellites or by ground-based instruments have shown that the universe is full of gamma-ray sources accessible to measurements. For the first time in history, present technological developments make possible to open a new window to the study of the universe in the GeV - TeV energy range. Very sensitive detectors to be launched on satellites or to be placed on ground are under construction. In this paper an overview of the time evolution and present status of gamma-ray detectors is presented.

1. Introduction

Progress in astronomy has often been the direct result of the development of a new observational technique, either the study of a previously unexplored wavelength region of the electromagnetic spectrum or a great enhancement in sensitivity compared with earlier observations. The development of new techniques in high energy gamma ray astronomy holds the promise of similar progress.

Victor Hess discovered in 1912 that the Earth is constantly bombarded by cosmic radiation coming from all directions. The vast majority of cosmic radiation are charged particles, mainly protons, and a very small amount ($\approx 10^{-4}$) are gamma rays. Due to the weak interstellar magnetic fields, only gamma rays and other neutral cosmic particles arrive to Earth keeping the information about its origin of emission.

Until the advent of high-energy accelerators in the early 1950s, experiments at the high energy frontier were conducted with cosmic rays. The muon, pion and kaon were all discovered in cosmic ray experiments. It was

173

M. M. Shapiro et al. (eds.), Astrophysical Sources of High Energy Particles and Radiation, 173–190.

only natural that questions about the origin of cosmic rays and how they are accelerated were important to nuclear and particle physicists. Even after most of the experimental work in particle and nuclear physics had moved to terrestrial accelerators, particle physics research continued with cosmic rays because they could be used to study particle interactions at energies above those available at these accelerators.

High energy gamma ray astronomy resulted from two distinct fields of study. In the early 1970s, space-based detectors extended the field of X-ray astronomy upwards in energy and observed a number of discrete sources of 100 MeV photons. The study of cosmic photons above 1 TeV (10^{12} eV) was done with ground-based instruments, which were an outgrow of cosmic ray studies. Here, however, the development of new detection techniques did not immediately reveal new sources or phenomena. It was only recently, after the development of second and third generation telescopes, that the ground based observations made important contributions to our understanding of the cosmos.

The study of astrophysical sources of high energy gamma rays provides a unique insight into the production mechanism. As the observed spectra are non thermal, one can study acceleration mechanisms in sources such as supernova remnants and active galactic nuclei. Extra-galactic sources serve as beacons that allow us to probe the intervening intergalactic medium and thus may give us clues to the early universe. The production of high energy gamma rays may serve as a marker for Fermi acceleration of charged particles and provide a means of determining the origin of the cosmic rays.

The energy of cosmic radiation covers many orders of magnitude. The energy term definition used here is:

- High Energy (HE) gamma rays: 30 MeV-30 GeV (10^9 eV)
- Very High energy (VHE) gamma rays: 30 GeV-30 TeV (10^{12} eV)
- Ultra High Energy (UHE) gamma rays: 30 TeV-30 PeV (10^{15} eV)
- Extremely High Energy (EHE) gamma rays: 30 PeV and up.

2. Detection techniques

The two types of high energy gamma-ray detectors are satellite-based detectors that convert the detected photon and track the resulting electron-positron pair to determine the photon direction, and ground based experiments that detect the extensive air showers produced when high energy photons interact with the atmosphere of the earth. Therefore detectors measure directly gamma radiation above the atmosphere and indirectly at ground level.

Three important facts govern the techniques that are used in high energy gamma astronomy.

1) The flux of high energy charged cosmic rays is much larger than the gamma ray flux. These charged particles since are bent in the interstellar magnetic fields, they form an essentially isotropic background. Rejection of the large cosmic ray background is extremely important in gamma ray astronomy.

2) The fluxes of high energy gamma rays from astrophysical sources are quite low and decrease rapidly with increasing energy. For example, Vela, the strongest gamma ray source in the sky can be detected by a 1000 cm^2 detector in a satellite at a rate of one photon/minute above 100 MeV and one photon every two hours above 10 GeV. As energy increases larger instruments on satellites and earth based are required.

3) The earth atmosphere is opaque to high energy photons. At sea level the atmosphere is 1030 g/cm^2 thick, which corresponds to 28 radiation lengths, this implies that the probability that a high energy photon incident from the zenith angle reach ground level without interacting electromagnetically is about $3x10^{-10}$. Thus only a detector above the earth's atmosphere, in a balloon or a satellite, can detect primary cosmic rays.

3. Direct Measurements with Detectors on Satellites

Two gamma ray satellites were launched in the 1970s: SAS-2 in 1972 [1] and COS-B in 1975 [2]. The most sensitive high energy gamma ray telescope was the EGRET instrument aboard the Compton Gamma Ray Observatory (CGRO) [3]. CGRO was launched in April 5, 1991, after a long delay due to the explosion of the Challenger space shuttle. The performance of EGRET was better than that of previous detectors, as a comparison for photon energies of 1 GeV the energy resolution, the angular resolution and the effective area of EGRET were: 19%, 1.2° and 1300 m^2 respectively and for COS-B was: 67%, 2.4° and 48 m^2. EGRET detected more than 270 gamma ray sources before it was decommissioned in June, 2000.

Several future satellites are under development. The Alpha Magnetic Spectrometer (AMS) is designed primarily to detect cosmic anti-nuclei on the International Space Station Alpha [4]. AMS can also function as a gamma-ray detector sensitive to photons between 300 MeV and 300 GeV. The point source sensitivity, expressed as the minimum detectable flux from a point source over a year, is expected to be comparable to that of EGRET. However the mass of the space station constitutes a target for cosmic ray interactions and a source of noise. In addition, AMS cannot be pointed to specific objects as its orientation is fixed with respect to the space station.

The Gamma Ray Large Area Space Telescope (GLAST) detector is under development by an international collaboration [5], [6]. GLAST will have a large field of view (about 2π sr) which will enable it to observe a larger part of the sky at once. The point source sensitivity of GLAST should be nearly 100 times better than that of EGRET. It is expected that GLAST will be in orbit in year 2005.

The AGILE (Astro-rivelatore Gamma a Immagini LEggero) gamma ray telescope will be launched in year 2002 [7]. It will be physically smaller than the EGRET detector but it will achieve a similar sensitivity due to its wide field of view.

4. Extensive Air Showers

Gamma rays striking the earth's atmosphere interact with air molecules high in the atmosphere producing Extensive Air Showers (EAS). Earth based gamma ray astronomy detectors measure the products of these interactions.

The total cross section for photon-proton collisions has been measured for center of mass energies up to 200 GeV [8] which is equivalent to 20 TeV photon colliding with a proton at rest. The predominant interactions are electromagnetic; the cross section for the production of hadron and muon pairs are several orders of magnitude smaller than that for electron pair production. In the electromagnetic shower, photons produce electron-positron pairs, and electron and positrons produce photons via bremsstrahlung. The resulting electromagnetic cascade grows nearly exponential as it propagates through the atmosphere; the primary energy is divided among more and more particles until the mean energy of the electrons and positrons approaches the critical energy ϵ(about 80 MeV in air). At this point, the ionization energy loss mechanism, which does not produce additional shower particles, becomes more important than bremsstrahlung. Thus energy is lost from the shower and the number of particles decreases as the shower continues to propagate.

The longitudinal development (the number of electron and positrons as a function of atmospheric depth) of an EAS has a maximum which occurs approximately $\ln(E/\epsilon)$ radiation lengths into the atmosphere, well above ground for all but the very highest primary energies. Nevertheless a large number of shower particles may reach the ground, especially at mountain altitudes. In an electromagnetic air shower the particles are ultrarelativistic and the dominant physical processes are sharply peaked forward. The cascade arrives at the ground in a thin front only a few meters thick. The lateral extent of a shower (or lateral development) is due mainly to multiple scattering of the electrons and positrons.

The more plentiful incident high energy cosmic rays, protons and nuclei, also interact high in the atmosphere, producing extensive air showers; the nuclear collision length is 62 g/cm^2 for protons and less for heavier nuclei.

The initial interaction generates a hadronic cascade; these interactions tend to quickly divide the primary energy among a large number of particles, so relatively few hadrons survive to the ground. Some charged pions and other hadrons decay before interacting, producing muons (and neutrinos); the weakly interacting muons generally lose energy only by ionization and so have a relatively high probability of reaching the ground. High energy neutral pions produced in the hadronic interactions decay rapidly into photons; these photons produce electromagnetic cascades in the same manner as discussed above. Thus the particles reaching the ground in a hadron induced EAS are mostly electrons, positrons, photons and muons (plus neutrinos); except for the presence of muons, a cosmic ray shower at ground level is not very dissimilar to a gamma ray shower. There are roughly 20 times more muons in a hadron induced shower than in a photon induced shower of the same energy.

In addition to shower particles, Cherenkov photons are produced in EAS in a much greater amount. Cherenkov light in the atmosphere is produced by charged particles traveling faster than the speed of light in air. The maximum of the shower development occurs at a height of between 10 and 7 Km above sea level for gamma rays of energies between 100 GeV and 10 TeV. The median altitude for Cherenkov emission from a 1 TeV gamma shower is 8 Km. Half of the emission occurs within 21 m of the shower axis (70 m for a proton shower [9]). The Cherenkov light from an 1 TeV gamma EAS illuminates a light pool on the ground of radius about 130 m (at 2 Km a.s.l.) with an average photon density of about 200 photon/m^2 in the visible. The Cherenkov light is a thin front resulting from a pulse of duration 2-3 ns and an average angular extent on the sky of about 0.5°. The total distance traveled by all particles above the Cherenkov threshold is directly related to the energy of the primary particle, so the Cherenkov yield gives a measurement of the initial gamma ray energy. For a primary gamma ray of 1 TeV, the radial distribution of the density of Cherenkov photons is flat up to about 130 m, therefore a single sample at any point within this distance of the impact point gives a good estimate of the primary energy.

The large area of illumination on the ground implies that a single atmospheric Cherenkov light telescope (ACT) anywhere within the light pool can detect the shower, giving a large effective detection area of above 5x10^4 m^2 for vertically incident showers.

An extensive air shower can be detected on ground by using two methods, the sampling and the imaging methods. In the first case an array of regularly distributed counters samples the shower front particles (lateral/radial distribution measurement of the EAS) and/or the Cherenkov photons. In the second case a single large imaging Cherenkov telescope (IACT) (or an array of several identical IACTs working in stereoscopic mode) gets an image of the Cherenkov light produced in the shower maximum of the EAS (longitudinal distribution measurement of the EAS). In both cases, one uses the properties of the EAS to infer information about the primary particle that initiated it.

5. The Sampling Technique of Ground-Based Gamma Ray Astronomy: Particle Detector Arrays

Extensive air showers particle detector arrays (EAS-PDA) are used to detect EAS with energies above 20 TeV. Fluxes at these energies are small so a detection area in excess 10^4 m^2 is needed. The cost of a full coverage detector with this area is prohibitive. However one can use the fact that a large number of particles reach the ground at shower energies above TeV. A typical 100 TeV photon shower has about 50000 e$^+$, e$^-$ and about five times as many low energy gamma rays, spread out over an area in excess of 10^4 m^2 at mountain altitudes. Because of the large amount of particles reaching ground, an EAS-PDA needs sample only a relatively small fraction of these particles. An EAS-PDA consists of a number of charged particle detectors spread over a large area. Typical arrays have from 50 to 1000 scintillation counters, each about 1 m^2 in size, spread over an area of 10^4 - 2×10^5 m^2; the actual sensitive detector area is less than 1% of the total enclosed area of the array. The performance of the array can be improved by placing lead, one radiation length thick, above each counter to convert shower photons into charged particles ([14], [15], [16]). Table 1 lists the major extensive air shower particle detector arrays.

The direction of the primary is reconstructed by measuring the relative times at which the individual counters in the array are struck by the shower front. The angular resolution depends upon properties of both the EAS and the detector. A good angular resolution is obtained by fitting the shower front which is curved by an amount that is a function of position from the shower core. The angular resolution is determined in a number of ways. The most powerful method uses the measured shape of the shadow of the sun and the moon in the otherwise isotropic flux of charged cosmic rays [17]. Another method to determine the angular resolution is a function of the number of struck counters, as determined by dividing an array into two overlapping parts and comparing the angles reconstructed separately from these two arrays [16].

The energy threshold of an EAS-PDA depends upon the minimum number of counters that must be struck to allow a shower to be reconstructed, the altitude of the array and spacing of the counters. The energy threshold of an array is not a well defined quantity because the number of shower particles reaching the ground fluctuates greatly from shower to shower for identical primaries; the principal source of these fluctuations is the variation in the altitude and nature of the first interaction. A particle array does not measure the energy of the primary particle well, since the detectors sample the shower front covering only 1%. The energy response can be determined with the aid of Monte Carlo simulations of air showers and the array.

Signals in ground based gamma astronomy come in the presence of noise from the isotropic (hadronic) cosmic rays. For example, the strong signal observed by IACTs from the Crab constitutes an increase in raw counting rate of less than 1%. The background rejection, achievable with EAS-APDs is quite low and the resulting signal to noise ratio is rather poor.

TABLE 1. Recent major extensive air shower particle detector arrays. The energy threshold corresponds to the median energy for gamma rays.

group	site	area $(\times 10^3 m^2)$	No detectors	Eth (TeV)	rate (Hz)	years operational
CASA-MIA	Dugway,USA	230	1089	90	18	1991-96
CYGNUS	Los Alamos,USA	86	204	50	5	1986-96
HEGRA	La Palma,Spain	41	257	40	18	1987-00
HEGRA /AIROBICC		41	97	15	18	1992-00
SPASE	South Pole	10	24	100	1	1987-92
EAS-TOP	G.Sasso,Italy	50	35	120	2	1993-99
TIBET	Yangbajing,Tibet	8	49	8	5	1990-93
		44	221	8	230	1995-
		5	109	2	120	1996-
Milagrito	N.Mexico,USA	1.5	225	1	300	1997-99
Milagro	N.Mexico,USA	5	723	1	1000	1999-

Ground based high energy gamma ray observations in the 1970s and 1980s produced many claimed source detections in the TeV and PeV regions. These are summarized in a review article by Weekes (1988)[10], which presented a source catalog containing 13 sources. The most widely known results involved detections of TeV emission from Cygnus X-3 (a 4σ UHE steady excess [11], [12]) and Hercules X-1 (episodic VHE and UHE emissions with periodicity at the X ray period). The EAS-PDA detectors built in the 1990s to improve the early claimed detections in the VHE and UHE regions gave no positive results. Following the null results from the galactic and extra-galactic sources reported by the EAS-PDA detectors with energy thresholds higher than about 40 TeV, much of the motivation for improving these arrays has disappeared.

Improvements in EAS-PDA follow the IACTs approach, that is to decrease the detection energy threshold. This can be achieved in several ways: a) using a matrix of wide angle Cherenkov counter open to the night sky, as the AIROBICC array [19] build in the HEGRA experiment at 2200 a.s.l. with an energy threshold of 15 TeV; b) building EAS-PDA with a dense matrix of scintillation counters placed at higher altitude as the Tibet array at 4300 m a.s.l. with an energy threshold of 2 TeV [22]; c) building an EAS-PDA based on an increased EAS sampling area with large water Cherenkov detector as the Milagrito/Milagro detectors with an energy threshold of about 2 TeV [20], [21].

6. The Cherenkov Technique of Ground-Based Gamma Ray Astronomy: Imaging Telescopes and Solar Plants

In 1948 Blackett [23] predicted that there should be Cherenkov light emission in the atmosphere from the relativistic particles produced in the extensive air showers due to cosmic ray interactions. Confirmation of that prediction and the beginning of the atmospheric air Cherenkov technique was the discovery of Galbraith and Jelley which goes back into 1953. They used a single photomultiplier in the focus of a simple parabolic mirror of 25 cm diameter and counted Cherenkov light flashes from extended air showers, initiated by energetic cosmic ray particles, at a rate of 1/min [24]. This first type of small telescopes are called first generation. Improvements followed by the increase of the instrumental scale by using larger reflectors (1-2 m diameter) for individual telescopes and by using more than one telescope located either close to each other or separated by some distance, and by setting them into coincidence in time. The next milestone was set in the 1980s by the use of a coordinate sensitive imaging camera (instead of a single photomultiplier) in the focal plane of a large reflector. Today these detectors are known under the name of imaging atmospheric Cherenkov telescopes (IACTs) and are also called second generation telescopes.

The development of the imaging atmospheric Cherenkov telescope technique was due to the intuition of Weekes and Turver [25]. They believed that recording images of the Cherenkov light pool would improve the angular resolution of an air Cherenkov telescope. The shape of the Cherenkov images could also be used to differentiate between hadronic and electromagnetic showers, leading to additional background rejection. The original Whipple telescope had a 10 m tessellated mirror with a camera in its focal plane with 7 photomultipliers (pixels) of 0.5^o diameter each. The camera was further improved to 39 pixels and from 1988-1996 the telescope had a camera with 109 pixels. The field of view has since been increased.

The breakthrough in the imaging technique came with the introduction

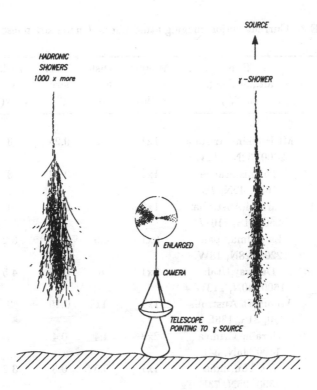

Figure 1. Simplified observation of a hadron and an electromagnetic shower seen by a Cherenkov telescope. The shower images in the field of view are right-left exchanged.

of the second moment parameterization of the Cherenkov images by Hillas [26] which was applied to the Whipple data. They got the first highly significant measurement (9 σ) of gamma rays from the Cab nebula in 1988 [27]. A new era in experimental gamma astronomy started with this measurement. The Whipple collaboration also discovered the first VHE extragalactic source, the Markarian 421 Active Galactic Nucleus, in 1992. The second similar source, the Markarian 501, was discovered in 1995. All the three sources were confirmed first by the HEGRA, then by the CAT and later by other collaborations.

Table 2 lists some of the major existing IACTs. In figure 1 the concept of imaging an extensive air shower by a telescope is shown.

Two problems in using IACT are the night sky background and the large isotropic background from cosmic ray showers. IACTs can only measure during clear, moonless nights with a duty cycle of about 10%. IACTs take advantage of the fact that hadron showers are more chaotic than gamma-showers, since the development of a hadron shower (at the same energy than a gamma shower) is governed by a relatively smaller number of particles in the hadronic core. The larger transverse momentum of hadronic interactions

TABLE 2. Current major imaging atmospheric Cherenkov telescopes

Group - experiment	Site alt.,lat.,long. (m,°,°)	Mounts x dishes	Dish area m²	Pixel size (°)	Field of view(°)	Energy threshold (GeV)
Whipple-10m	Mt.Hopkins,Arizona 2300, 32N, 111W	1x1	72	0.25	3	250
CAT	Themis,France 1650, 42N, 2E	1x1	17	0.12	3	250
Durham-Mk VI	Narrabi,Australia 250, 31S, 110W	1x3	42	0.25	4	300
HEGRA	La Palma,Spain 2200, 28N, 18W	6x1	8.5	0.25	3.2	500
Telescope Array	Dugway,Utah 1600, 40N, 113W	3x1	6	0.25	4.5	600
Cangaroo	Woomera,Australia 0, 31S, 136E	1x1	11	0.19	3	1000
CAO-GT-48	Ukraine,Crimea 1100, 45N, 34E	2x3	4.4	0.4	3	1000
TACTIC	Mt.Abu,India 1300, 25N, 73E	1x1	9.5	0.3	3.2	1000
Lebedev-SHALON	T.Shan,Kazahkstan 3338, 43N,77E	1x1	10	0.6	8	1000

gives a broader lateral distribution.

The longitudinal development of gamma ray showers of a given energy is well defined with small fluctuations. Consequently the image in the focal plane of an imaging Cherenkov telescope has a well defined longitudinal profile that depends only on the impact parameter and the angular origin of the shower. In a fine pixel imaging Cherenkov telescope, this profile can be compared to theoretical gamma ray shower profiles, as a function of the shower energy, impact parameter and angular origin. In the case of the CAT telescope [28], this gives an angular resolution for individual gamma ray showers of about 0.1°. For an array of several IACTs viewing a shower simultaneously, the angular origin of the shower on the sky is given by the point of intersection of the shower axis as seen by the telescopes. An angular resolution of 0.1° is achieved with the HEGRA IACT system [29], [30]. In such second generation Cherenkov telescopes, the cosmic ray background can be reduced to a level comparable to, or below, the signal from strong gamma ray sources. In figure 2, the concept for stereoscopic observation of

EAS is shown. The existing IACTs have an energy threshold of ≈ 250 GeV.

Another type of detectors, Cherenkov front sampling detectors are being developed to achieve very low energy thresholds (20-40 GeV). In this detection technique, the arrival time of the Cherenkov pulse is measured at a number of mirrors distributed on the ground within the light pool. The Cherenkov photons (produced in the atmosphere at different altitudes) arrive to ground in a cone with a wide opening angle centered on the shower axis. A fit of the arrival times of the Cherenkov photons at the ground front of this cone gives the shower axis direction due to the primary gamma. The cosmic ray rejection power of this technique is poor due to the lack of information on the longitudinal shower development.

Modern wave-front detectors make use of the enormous mirror areas available at solar energy plants. These plants consist of a large array of heliostats, large area mirrors that track the sun and focus the sun's energy onto a heat exchanger located on a central tower. By using a secondary mirror on the central tower, each heliostat can be imaged onto a single photomultiplier tube [31]. In Table 3 a list of the existing solar plants is shown.

TABLE 3. Atmospheric Cherenkov Solar Array Telescopes

Group/	Site	Heliostats Now(future)	E.Threshold (GeV)
STACEE	Albuquerque, USA	32(48)	180
CELESTE	Themis, France	40(54)	50
SOLAR-2	Barstow, USA	32(64)	20
GRAAL	Almeria, Spain	(13-18)x4	200

7. Improvements and future IACT project

There is an observational gap in the electromagnetic spectrum in the range from 20 GeV (upper limit of EGRET) to 300 GeV (lower limit of the existing IACTs). New IACTs are under development to cover this unknown energy range. Improvements in air Cherenkov telescopes are proceeding along two paths: improved flux sensitivity and lower energy thresholds. Improved flux sensitivity allows one to detect weaker sources in a shorter amount of time (and thereby scan larger regions of the sky) and to study the time variability of the sources. A lower energy threshold allows the study of the energy region between space-based and ground-based instruments.

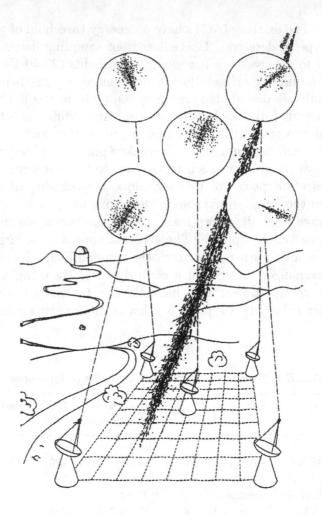

Figure 2. Symbolic arrangement of 5 telescopes in the stereo mode.

This should allow the measurement of cutoffs in the pulsed emission energy spectra of many of the pulsars detected by EGRET, the detection of more distant AGNs, and the measurement of the cutoffs in their energy spectra.

Recent progress in IACTs has made possible better angular and time resolution and the use of several IACTs either on the same mount [32] or on neighboring mounts about 60 m apart, simultaneously observing the same showers, as shown by the HEGRA experiment [30]. Further progress will come increasing the number of detected photons with larger mirrors and high quantum efficiency detectors above the current 20-25% of conventional photomultiplier tubes. New photocathodes such as GaAsP yield twice as many photon conversions, reaching about 50% quantum efficiency and an extended to the red wavelength response [33]. Solid state detectors with about 100% efficiency for visible photons are in a research and development stage [34, 35] This improvement in the quantum efficiency would have a major impact in many sectors of professional and research activities.

The future IACTs (third generation IACTs) under construction aiming either 1) to lower the energy threshold to 10-20 GeV or 2) to improve the flux sensitivity and make extensive observations are briefly described in what follows.

1) In the first approach, the MAGIC project [36] is based on a large parabolic dish of 17 m, using aluminum mirror elements with permanent and automatic alignment control. This project makes progress in many technical matters: mirrors, light mechanical structures for the telescope frame, automation, new high quantum efficiency photodetectors, massive data flow and new strategies for event reduction, which will be of benefit to the entire field.

2) In the second approach, several projects are under construction. The VERITAS [37] and HESS [38] take full advantage of the recent progress in the IACTs technique, merely by extending to IACT arrays of 10 to 15 telescopes of 10 m diameter, each equipped with high resolution cameras of about 500 pixels. The CANGAROO project [39] will also build 5 IACTs of 10 m each in Australia. These arrays will permit very flexible observation strategies, either turning each telescope toward a different source or using the enhanced power of several telescopes working together for chosen sources. The future IACT projects will be located two in the northern hemisphere (MAGIC and VERITAS) and two in the southern hemisphere (HESS and CANGAROO) what permits coverage of the whole sky. In table 4, the next third generation of IACTs is shown. In figure 3, a comparison of the sensitivities of several future gamma ray instruments is shown.

TABLE 4. Next generation imaging atmospheric Cherenkov telescopes

Group/ Instrument	Site	Reflector(s) Number xaperture	Camara pixels	Threshold (GeV)	Epoch Begin
MAGIC	LaPalma,Spain	17m	1x800	30	2001
HESS	Namibia	4x10	4x700	50	2002
CANGAROO III	Woomera,Austr.	4x10	4x512	75	2003
VERITAS	Arizona,USA	7x10	7x499	75	2004

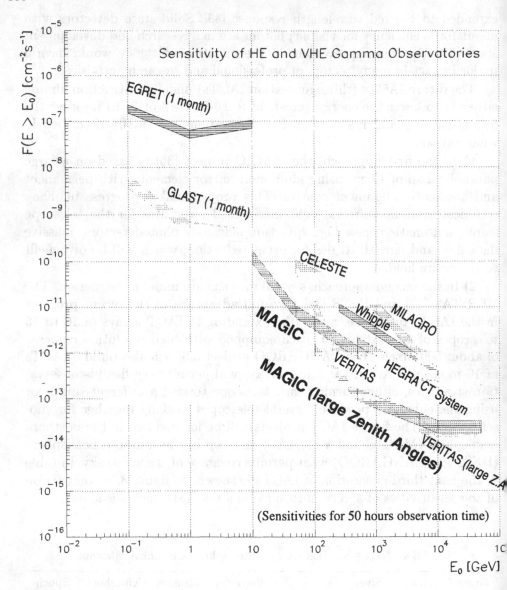

Figure 3. Sensitivity of several gamma ray experiments.

8. OBSERVATIONS

At the end of the 1980s, results in high energy gamma astronomy were sparse and often somewhat confusing. The launch of the CGRO satellite

THIRD EGRET CATALOGUE OF GAMMA-RAY POINT SOURCES

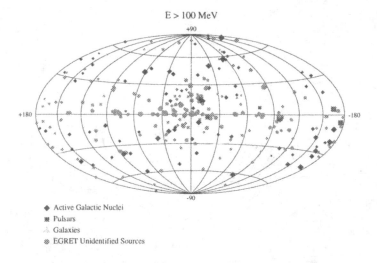

Figure 4. Third EGRET catalog

was delayed by the Challenger disaster, so there were no satellite based observations in the 1980s. With the exception of the widely accepted detection of TeV gamma rays from Crab, all other claimed observations of sources emitting gammas with energies above TeV were disputed.

The 1990s began inauspiciously when the spark chamber failed on the Gamma-1 telescope aboard the Russian Gamma Observatory, launched in July 1990. Gamma-1 was designed to detect gamma rays from 50 MeV to 6 GeV, but the spark chamber failure eliminated the pointing ability of Gamma-1 and severely reduced the capabilities of the mission. Nevertheless the first seven years of the 1990s brought a remarkably rich view of the gamma ray sky. This was largely due to the new detections made by the EGRET telescope [40], [41], [42], [43]. aboard the CGRO satellite. EGRET has shown that there are many gamma ray sources (in the energy range from 100 MeV to 20 GeV), both galactic and extragalactic, as well as a strong component of diffuse emission from the galactic plane (Figure 4).

The gamma ray sky above 300 GeV, is presently much sparser, although only a small fraction has been studied. However, a few steady galactic sources and several sporadic emissions from active galaxies have been observed. These AGNs show remarkable variability, which has severely constrained models of very high energy emission. The discovery of the AGN Mrk 501 was the first time that a gamma ray source was found by ground

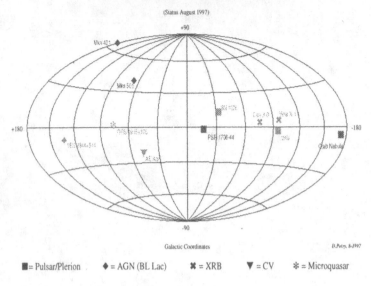

Figure 5. Very High Energy sky map.

based observations before observations in space. In addition, photons up to 50 TeV have been detected from the Crab by the CANGAROO collaboration , using the IACT technique at large zenith angle [13]. These results came only after about 30 years of refinement of the air Cherenkov technique with IACTs. In figure 5, the present very high energy sky map is shown. In contrast, observations, with the present EAS-PDA, in the 1990s have shown that there are no bright sources emitting gammas at energies above 50 TeV. For more detailed information see recent reviews. [44], [45]

In the last International Cosmic Ray Conference which took place at Utah, USA (1999), The Tibet collaboration and the Milagro/Milagrito water Cherenkov detectors presented results of > 2 TeV gamma emissions from several sources [46], [47]. The improved Tibet EAS-PAD at 4300 m a.s.l, has collected a 5.5 σ steady excess from the Crab nebula and also the blazar Mrk501 was significantly detected. The Milagrito detector (a prototype of the Milagro detector) at 2650 a.s.l. has detected the blazar Mrk501 and a possible VHE emission from the gamma ray burst 970417a observed by the BATSE detector on board of the CGRO.

The 1990s have seen gamma ray astronomy develop so that it now plays an important role in our understanding of the cosmos. A number of unexpected discoveries have been made and many more are to be expected. High energy gamma ray astronomy should flower in the next millennium.

References

1. Derdeyn, S. M, Ehrmann C.H., Fichtel C. E., Kniffen D. A. and Ross R. W., (1972), *Nucl. Instrum. Methods* **98**, 557
2. Bignami, G. F. et al.,(1975), *Space Sci.Intrum.* **1**, 245
3. Thompson, D. J., et al.,(1993), *Astrophys. J. Suppl. Ser.* **86**, 629
4. Ahlen, S., et al.,(1994), *Nucl. Instrum. Methods Phys. Res. A*, **350**, 351
5. Michelson, P. F.,(1994), *Proceedings of the Workshop on Towards a Major Atmospheric Detector III, ed. T. Kifune (Universal Academy Press, Tokyo),* 221.
6. Bloom, E. D. (1996), *Space Sci. Rev.,* **75**, 109.
7. Battiston, R. (1999) *Proceedings of the Workshop GeV-TeV Gamma Ray Astrophysics, Snowbird, Utah.* See also *AIP Proc. Conf.,* **515**, 474
8. Aid, S. et al.,(1995) *Z. Phys. C,* **69**, 27.
9. Hillas, A. M., (1996) *Space Sci. Rev.,***75**, 17.
10. Weekes, T. C., (1988), *Phys. Rep.* **160**, 1 and references therein.
11. Samorski, M. and Stamm, W., (1983), *Astrophys. J.,***268**, L17.
12. Lloyd-Evans, J., et al., (1983), *Nature (London,* **305**, 784.
13. Tanimori, T. et al., (1998), *Astrophys. J. Letters,* **492**, L33-L36.
14. Linsley, J., (1987) *20th ICRC, Moscow, edited by V. A. Kozyarivsky et al.,* **2**, 442.
15. Alexandreas, D. E., et al., (1992), *Nucl. Instrum. Methods Phys. Rev. A,* **311**, 350.
16. Merck, M., et al., (1996) *Astropart. Phys.,* **5**, 379.
17. Amenomory, M., et al.,(1996) *Astrophys. J.,* **464**, 954.
18. Alexandreas, D. E., et al., (1993) *Nucl. Instrum. Methods Phys. Rev. A,* **328**, 570.
19. Karle, A., et al., (1996) *Astropart. Phys.,* **4**, 1.
20. Williams, D. A., et al., (1999) *Proc. of GeV-TeV Gamma-Ray Astrophysics: Towards a Major Atmospheric Cherenkov Telescope VI. Ed. B. L. Dingus, M. H. Salamon and D. B. Kieda, AIP Conference Proc., N.Y.,***515**.
21. Smith, A., et al., (1999) *Proc. of GeV-TeV Gamma-Ray Astrophysics: Towards a Major Atmospheric Cherenkov Telescope VI. Ed. B. L. Dingus, M. H. Salamon and D. B. Kieda, AIP Conference Proc., N.Y.,***515**.
22. Amenomori, M., et al., (1998) *25th ICRC, Durban, (World Scientific Singapore),* **5**, 245.
23. Blackett, P. M. S., (1948),*Rep. Conf. Gassiot Comm. Phys. Soc.*, Physical Society, London, 34.
24. Galbraith, W. and Jelley, J. V. (1953),*Nature* ,**171**, 349.
25. Weekes, T. C. and Turver, K. E., (1997) *Proceedings of the 12th ESLAB Symposium, Frascati, ESA, SP-124,* 279.
26. Hillas, A. M. (1985) *Proc. 19th ICRC, La Jolla,* **3**, 445.
27. Weekes, T. C. et al.(1989) *Ap. J,* **342**, 379.
28. Barrau, A., et al., (1998) *Nucl. Instrum. Methods Phys. Res. A,* **416**, 278.
29. Daum, A., et al., (1997), *Astropart. Phys.,*8, 1.
30. Puhlhofer, G., et al., (1997) *Astropart. Phys.,*8, 101.
31. Tumer, O. T.,(1990), *Nucl. Phys. B (Proc. Suppl.),* **14A**, 351.
32. Chadwick, P. M., et al., (1998) *25th ICRC, Durban (World Scientific, Singapore,* **5**, 101.
33. Bradbury, S., et al., (1997) *Nucl. Instrum. Methods Phys. Res. A,* **387**, 45.
34. Wayne, M. R., (1997) *Nucl. Instrum. Methods Phys. Res. A,* **387**, 278.
35. Ruchti, R. C., (1996) *Annu. Rev. Nucl. Paart. Sci.,* **46**, 281.
36. Barrio, J. A., et al., (1998) *The MAGIC telescope, design study, Max- Planck-Institute for Physics, Munich, Internal Report MPI-PhE/98,* **5**, 1.
37. Weekes, T. C., (1998) *Phys. Rep.,*160, 1, and references therein.
38. Hermann, G., (1997) *Proc. of the XXVII Recontres de Moriond. (Editions Fron-*

190

tieres, *Gif-sur-Yvette, France*, 141.

39. Mori M., et al., (1999) *Proc. of GeV-TeV Gamma-Ray Astrophysics: Towards a Major Atmospheric Cherenkov Telescope VI. Ed. B. L. Dingus, M. H. Salamon and D. B. Kieda, AIP Conference Proc., N.Y.,***515**.

40. Thompson, D. J., et al., (1995) *Astron. Astrophys., Suppl. Ser.*, **101**, 259.

41. Fichtel, C. E., (1996), *Astron. Astrophys., Suppl. Ser.*,**120**, 23.

42. Dingus, B. L., (1994), *Proc. of the 1994 Snowmass Summer School on Particle, Nuclear Astrophysics and Cosmology in the Next Millenium, ed. Kolb, E. W. and Peccei, R. D. (World Scientific, Singapore)*, 99.

43. Michelson, P. F., (1994) *Proc.of the Workshop: Towards a Major Atmospheric Detector III, ed. T. Kifune (Universal Academy Press, Tokyo)*, 257.

44. Ong, R. A., (1998), *Physics Reports*, **305, No. 3-4**, 93.

45. Hoffman, C. M. et al., (1999), *Rev. of Modern Physics*, **71, No. 4**, 897.

46. Buckley, J. H., (2000), *26th International Cosmic Ray Conference, ed. by B. L. Dingus et al., American Institute of Physics*, **516**, 195.

47. Dingus, B. L., (2000), *26th International Cosmic Ray Conference, ed. by B. L. Dingus et al., American Institute of Physics*, **516**, 351.

A BRIEF INTRODUCTION TO THE STUDY OF AFTERGLOWS OF GAMMA RAY BURSTS

RAMANATH COWSIK

Indian Institute of Astrophysics Bangalore, 560 034, India
and
Tata Institute of Fundamental Research, Mumbai, 400 005, India.

1. Preamble

Gamma Ray Burst or GRB for short, is the term used to describe intense pulse of astronomical gamma rays of mean energy ~ 1 MeV, and such events were discovered in the late 1960's by American satellites *(VELA)* serendipitously. In the last 30 odd years since their discovery more than 3000 papers have been written about them (including my own unpublished work with Richard Lingenfelter even before the formal discovery) and yet it is only in the last few years there has been any real progress in their understanding. This recent progress was made possible by the observations made with the Wide Field Camera (WFC) aboard the Italian-Dutch Satellite *BeppoSAX*, which operating in the soft X-ray band of 2-30 keV is capable of providing speedily the direction of the sources of the GRBs to within a few arcminutes. With the availability of such accurate positions optical and radio telescopes could be pointed in the direction, the source could be identified and the afterglow of the burst at these longer wavelengths could be observed continuously, upto several weeks after the burst. Detailed measurements in the γ-ray window were made with the BATSE instrument aboard the *Compton Gamma Ray Observatory (CGRO)* - we will review these before describing the measurements of the afterglow and some of the theoretical models that help us understand the phenomena that underline the GRBs which undoubtedly are at cosmological distances and constitute the most energetic events in the Universe.

M. M. Shapiro et al. (eds.), Astrophysical Sources of High Energy Particles and Radiation, 191–201.
© 2001 *Kluwer Academic Publishers. Printed in the Netherlands.*

2. Observations of Gamma Ray Bursts

There have been successive generations of experiments aboard satellites which have provided data on GRBs: Russian - French PHEBUS on GRANAT, WATCH for locating the bursts aboard GRANAT and on the European EUREKA, γ-ray detectors on Ulysses, and most importantly the four instruments aboard the CGRO.

2.1. TIME-STRUCTURE OF GRBS

The temporal structure of GRBs is most diverse, defying detailed classification, with total duration varying from about .01s to about 10^3s and internal structures with substantial changes in intensity occurring over timescales of milliseconds. There are events in the BATSE catalogue in which 90% of the counts in the band 50–300 keV were accumulated within ~ 5 ms, with internal variability as short as ~ 0.2 ms. Before the main burst occurs, in about few percent of the cases, there is a weak precursor. There are cases in which softer X-ray emissions continue after the burst and in some cases thermal X-ray emission both before and after the burst has been seen, e.g. by the *GINGA* satellite. There are also instances where the burst activity in the 50–300 keV based was associated with γ-emission at GeV energies lasting for about 1.5 hours. Examples of the temporal profiles of some bursts showing their diversity may be seen in the review by Fishman et al. (1994).

These profiles indicate that some bursts have a rapid rise and a slower exponential fall off, some have multiple peaks and some have a chaotic spiky behaviour. Even though a very clear bifurcation is not possible it has been found useful to divide the bursts into short bursts and long bursts depending upon whether the burst duration is less or more than 2s about 75% of the observed bursts are long bursts.

The integrated energy fluence in the bursts is distributed widely from $\sim 10^{-5}$ to $\sim 10^{-7}$ erg cm^{-2}. Note that if these are at cosmological distances $\sim 10^{28}$cm then the total energy emitted in γ-rays is very large ranging from $\sim 10^{49}$ to 10^{53} ergs all emitted in a short time; This makes GRBs the most intense and energetic sources of radiation in the Universe.

2.2. SPECTRA

The observed spectra of GRBs are invariably non-thermal in the 100–300 keV range with a flattening at lower energies so that only very few bursts are seen at ~ 10 keV in prompt emission. Some examples of the γ-ray spectra of the bursts are given by Share et al. (1994) and Schaefer et al. (1994). When one plots the product $E^2 F(E)$ where $F(E)$ is the number of γ-rays at E in the interval (E, E+dE) seen through out the burst, it shows a broad peak at ~ 500 keV indicating that the dominant part of the energy release occurs around this energy. The low energy part of the spectrum between 50 keV and 300 keV can be fit with a power-law of the form

$$F(E)dE = E^\alpha dE \tag{1}$$

with observed values of α lying in the range -3.5 to +0.5 with a broad peak around -1.8 to -2. The flat spectra with slopes $-1/2 < \alpha < 1/3$ at the lowest energies suggests that the radiation could be due to the synchrotron process by relativistic electrons accelerated in a shock, such a power-law steepens at high energy and one can fit a new power-law at high energies or following Band et al. (1993)

$$
\begin{aligned}
F(E) &= N_0 E^\alpha e^{-E/E_0} \quad \text{for } E \leq H \text{ and} \\
F(E) &= N_0 \{(\alpha - \beta)E_0\}^{(\alpha-\beta)} e^{(\beta-\alpha)} E^\beta \quad \text{for } E \geq H,
\end{aligned} \tag{2}
$$

with $H = (\alpha - \beta)E_0$ In this form the luminosity peaks at $E_p = \frac{(\alpha+2)}{(\alpha-\beta)} H$ and the distribution in H has a broad peak between 80 keV and 800 keV (Band et al. 1993, Cohen et al. 1998).

Observations place strict upper bounds on (L_{opt}/L_γ) for the bursts. Even at this time it is interesting to note that these non thermal γ-ray spectra show that the source is optically thin and the x-ray emissions can not be attributed to cooling of a plasma heated by γ-rays. However, it is also noted that in many bursts the spectrum is harder during the initial increase in a spike than during the exponentially decreasing phase. Actually the peak in the hardness occurs before the peak luminosity. When there are multiple peaks then the ones that occur later are softer. No spectral lines in the γ-ray region have been observed with certainty even though there were many searches for cyclotron lines associated with strong magnetic fields $> 10^{12}$G, such as those found around neutron stars.

2.3. DISTRIBUTION OF GRBS ON THE SKY

With the accumulation of data over the years the angular distribution of the arrival directions of GRBs is seen to be isotropic with the statistical uncertainties better than $\langle cos\phi \rangle = 0.017\pm0.018$; $\langle sin^2 b - 1/3 \rangle = -0.003\pm 0.009$ As the observations continue the isotropy of the arrival directions is getting established with increasing precision.

2.4. DISTRIBUTION IN SPACE

As we will discuss in detail in the next section the GRBs are now established to be mostly originating in galaxies at redshifts $z \sim 1$. A similar conclusion had been reached earlier by studying the log N - log F plot of the observed events. It would perhaps be best to postpone the discussion of the luminosity function of GRBs until it can directly be derived from a large set of GRBs with measured redshifts. The future mission 'Swift' is expected to provide about 100 redshifts over 3 years of observations.

3. Identification of Optical Counterparts

The accuracy with which γ-ray detectors have been able to locate the GRBs in the sky had been rather poor $\sim 20^o$, except in a few cases where the burst had been detected with good timing. Even this latter method could provide the coordinates of the burst only after considerable time needed for collation of data and analysis. Because of this optical identification had become an onerous search without much success.

These efforts were indeed revolutionized by observations with the Wide Field Camera (WFC) and Narrow Field Instruments (NFI) on BeppoSAX while triggering on the low energy (2-30 keV) part of the GRB spectrum could localize a large number of bursts to within $\sim 3'$ and $50''$ respectively. Apart from the higher angular accuracy with which the source could be located the coordinates be made available world-wide within couple of hours after the burst. Thus to-day we have more than twenty GRBs whose afterglows have been detected and their position knows to better than a second of arc.

With such an accurate determination of the position of the GRBs it has been possible to identify the host galaxies. These are dim galaxies with magnitudes of ~ 24 or larger. It is also clear that the bursts do not come from their core regions, so that early models suggested in analogy with AGNs are ruled out. Even though there are cases where unambiguous identification of the host galaxy has not been possible, the so called "no-host" problem has disappeared; all that one can say is that they do not appear to be associated with bright galaxies.

The redshifts of a large number of the host galaxies and/or intervening material have been observed (see Greiner's home page at http://www.aip.de/People/JGreiner/grbgen.html for details). Several of these host galaxies show star formation rate of $\sim 10 M_\odot/yr$ (Sokolov 2000, Djorgovski 1998). This along with the fact that the typical z is about 1 has led to the suggestion that GRB occurrence is related to the star-forming rates of galaxies.

The optical identifications have clearly shown two significant properties of GRBs that they occur in faint galaxies and that they radiate prodigious amounts of energy ranging from $\sim 10^{48}$ to 3×10^{54} ergs in gamma-rays within a very short time making them the most energetic, intense and hottest sources of radiation in the universe. The customary caveat that these energies are estimated assuming that the emission of radiation is isotropic and beaming will reduce the energy requirement considerably but still does preserve the validity of their exceptional nature. Also the debate that ensued during the early years of the study of GRBs has been now closed in favour of their cosmological origins.

4. Observations of Afterglows

As noted earlier the accurate angular positions made available by the WFC & NFI aboard the BapposSAX satellite allowed the rapid pointing of telescopes in the direction of the event and to study it at optical and radio bands. This has been an international collaborative effort, especially in the optical band since the observations can be carried out only at night. In general from a particular location of an optical telescope only about 10% of the γ-ray bursts will be accessible for observations. Optical observations of afterglows of GRBs are now available with in about 8 hours after the γ-ray trigger. Such observations are carried out in various spectral bands like U, B, V, R, I etc. These observations are continued for weeks or even months, night after night until the glow fades away below detectability.

4.1. TEMPORAL STRUCTURE

The apparent magnitude of GRB000301C in various spectral bands as a function of time is given by Garnavich & Stanek (2000). Note that the intensities decrease as a power-law which after several days steepens to a new power-law. The parameters of this evolution are determined by suitably combining all the data as shown in Fig. 1. taken from Bhargavi & Cowsik (2000) The initial decline in the R-band intensities has a slope $\alpha_1 = -0.7 \pm 0.07$ which steepens after about six days to $\alpha_2 = -2.44 \pm 0.29$. In this particular instance, this change does not occur smoothly but there is an indication of a bimodal distribution. But what is truly remarkable is that the time evolution even across the break is achromatic at least upto about 2 weeks after the burst, suggesting the possibility that gravitational lensing may be a operative. Also Sagar et al. (2000) have suggested that apart from the secular decrease in the intensities there could be fluctuation of the intensities with time constant of a fraction of a day.

Not all bursts show a clear cut steepening of the secular decrease in the afterglow intensities - for example GRB990123 shows no break even upto 100 days after the burst ($t^{-1.12} \pm .03$; Galama et al. 1999). And there are bursts which appear to flatten at late times; this is attributed to the luminosity of a faint host galaxy whose fractional contribution increases as the burst progressively fades away.

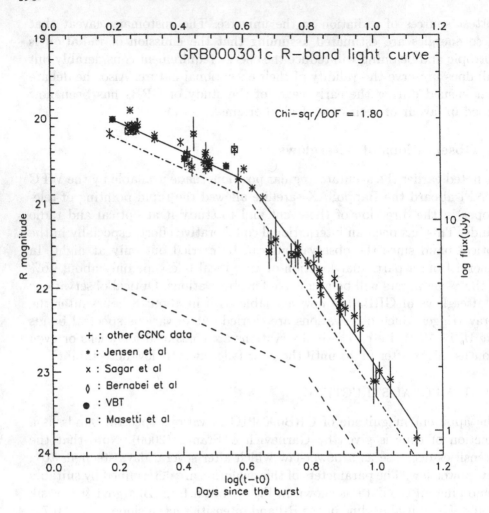

Figure 1. GRB000301c R-band light curve: The dotted and dashed lines represent the major and minor burst which add up to give the light curve shown as a solid line. The data points from VBT are marked by 'filled circles'. (Bhargavi & Cowsik 2000)

4.2. SPECTRAL DISTRIBUTION OF ENERGY FLUX

The observed distribution of energy flux is corrected for reddening and for the sensitivities of different detectors at different frequencies and the spectrum is reconstructed. The amount of energy flux seen in the afterglow per unit frequency interval can generally be fit with power-laws of the form $F(\nu) = F_0 \nu^\beta$ $m^{-2} s^{-1} Hz^{-1}$ with β typically ~ -0.7 in the band from $\sim 10^{12} - 10^{18}$ Hz (2-10 keV) is from THz to the soft X-ray band. An example of such a spectrum is shown in Fig. 2 for GRB 990123 (Galama 1999).

This spectrum is reminiscent of the synchrotron radiation from a non-thermal population of electrons. The spectrum has a broad peak at $\sim 10^{12}$ Hz and turns over progressively to a positive slope at lower frequencies, either because of a low energy cut off in the electron-spectrum or due to synchrotron self-absorption.

5. A generic model

The theoretical understanding of the GRB phenomenon and of the associated afterglows is rapidly converging towards a generic model, but the details are as complex as the observations themselves or worse. We can not say at this time that we can build a plausible model which can explain all the observed features of even a single GRB. On the other hand there is a general consensus as to the broad features of the model - features that were originally suggested for example by Shemi and Piran (1990), Paczyński and Rhoads (1993), Katz (1994a, 1994b), Mészáros and Rees (1997), Sari, Piran and Narayan (1998) and many others. It is impossible to cover all these ideas in this brief lecture and we may refer to a comprehensive review of these ideas by Piran (1999) and supplement it with an evergrowing literature in the press. Here I will only describe the broad features of the generic model which suggested itself by looking at what Ruderman (1975) and Schmidt (1978) called "the compactness problem".

5.1. THE COMPACTNESS PROBLEM

Let F be the energy flux of γ-rays and Δt be the time over which this charges substantially, neglecting, for now, redshift effects, the Luminosity L of the source is given by

$$L = 4\pi D^2 F \qquad (3)$$

If we now assume the size of the source to be the $r_s \sim c\Delta t$ then the energy density inside the source turns out to be

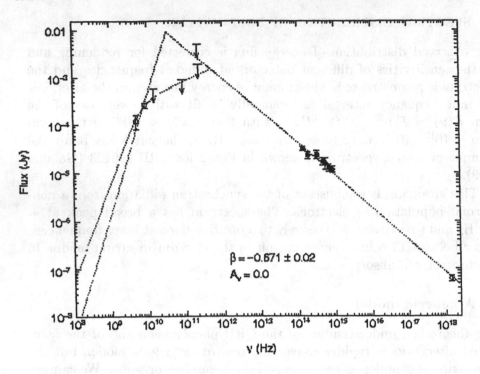

Figure 2. The spectral flux distribution of the afterglow at 1999 January 24.65 UT. (Galama 1999).

$$\rho \approx L(\pi c^3 \Delta t^2)^{-1} \approx 4FD^2/c^3 \Delta t^2$$
$$\approx 10^{22} D_{28}^2 \Delta t_{-2}^2 \; erg \; cm^{-3} \qquad (4)$$

where as usual D_{28} is the distance to the source in units of 10^{28} cm and Δt_2 is in unit of 10^{-2}s. Now the process $\gamma + \gamma \to e^+ + e^-$ will take place inside the source for pairs of gamma rays whose energies satisfy the equation

$$\sqrt{2E_1 E_2(1 - cos\theta)} \quad > \quad 2m_e c^2 \qquad (5)$$

defining the threshold for the pair-creation process with a cross-section approximately equal to the Thompson scattering σ_T. If f_p be the fraction of target photons that satisfy the threshold criterion for a chosen photon its optical depth for pair creation will be

$$\tau \approx f_p \frac{\rho}{\overline{E_\gamma}} c \Delta t \sigma_T \approx 10^{12} \qquad (6)$$

for $\overline{E_\gamma} \sim 1$ MeV, $f_p \sim 1$, $\Delta t \sim 10^2$ s and $D \sim 10^{28}$ cm. With such a large optical depth enough density of $e^+ e^-$ particles will be created in the source to make the source opaque to all photons and consequently thermalising them. This is contrary to the observations of γ-rays which verify that they escape from the source with nonthermal spectra.

The resolution of this paradox occurs when we assume that the source is moving towards the observer with a Lorentz factor Γ which allows an increase in the source size by a factor Γ^2 and a decrease in the pair-fraction f_p by a factor $\Gamma^{-2(\beta+1)} \sim \Gamma^{-3.4}$ so that $\tau \sim \Gamma^{-7.4}$. With Γ of ~ 100 the source becomes optically thin to γ-rays as required. These considerations have given rise to the idea that the sources are relativistically expanding "fire balls".

5.2. THE MODEL

Imagine a large amount of energy $\sim 10^{54}$erg released suddenly inside a small volume of dimensions of the order of 10^7cm embedded in a region of relatively low density ~ 1 atom/cm^3. This leads to the generation of a fireball which is expanding relativistically with Lorentz factor exceeding $\sim 10^2$. There will be a forward shock which will propagate into the external medium and there could also be a reverse shock that moves into the ejected material. The gamma ray emission with all its complexity and spiky charac- ter is attributed to the dissipation of the energy of internal shocks. In these shocks, electrons reach ultrarelativistic energies and radiate γ-ray through synchrotron emission in magnetic fields whose strength approaches near equipartition. All this occurs before the initial explosion has slowed down considerably by accumulating too much material from the external medium by snow-ploughing through it. The afterglow consisting of the slowly fading X-ray, optical and radio components occurs where this initial explosion has slowed down but is still relativistic. Thus according to the generic model the bulk kinetic energy of systematic motion of the expansion of the fireball gets converted into internal energy which is radiated as γ-rays and as the afterglow in collision with the external medium. It was recognized quite early in the study of gamma ray bursts that internal shocks and external shocks play crucial roles in the discipation of the internal energy into the observed radiation and especially in a generating power-law spectra. The steepening of the light curve of the afterglow several days subsequent to the burst is interpreted as due to the flaring up of the jet which was responsible for the burst. The talk by Piran at this school will give the details of the theoretical aspects of such a model.

There are two basic ideas regarding how the explosion is energized:

a. It could be due to the merging of two compact objects : a pair of neutron stars (leading to the formation of a black hole) or a neutron star - black hole pair.

b. The collapse of single object into a black hole - either a rapidly rotating massive star or that of a massive accretion disc on to a central black hole.

In the case (a) the neutron stars are expected to have a large scale height owing to the kinematics of supernova explosion in binaries and in the case (b) the scale height should be small with a closer association with star-forming regions. A recent analysis by Bloom et al. (2000) supports the later scenario.

6. Conclusions

Gamma Ray Bursts are the most energetic and relativistic astronomical phenomena in the universe. A qualitative understanding of the GRBs has been achieved in terms of fireball model where relativistic hydrodynamics with shocks play a crucial role in converting the energy of the explosion into the observed radiation. They are observable upto large redshifts. This feature coupled with their rapid variability makes them ideal candidates for study as gravitational lenses and thus derive cosmological parameters. The study of afterglows will be greatly facilitated by the Indian Astronomical Observatory which has just been set-up with a 2-m Optical/IR telescope at Hanle (longitude= $32°46'46''$, Latitude= $78°57'51''$E) right in the middle of the wide lacuna in such facilities spanning nearly $180°$ from Siding Springs in Australia to La Palma in the Canary Islands.

Acknowledgement

I acknowledge the help of Ms S.G.Bhargavi in preparation of this manuscript.

References

Band, D. L., 1993, ApJ, 413, 281,
Bhargavi, S. G. & Cowsik, R., 2000, ApJL in Press; (astroph/0010308)
Bloom, J. S., Kulkarni, S. R. & Djorgovsky, S. G., 2000, (astroph/0010176)
Cohen, E., Narayan, R. & Piran, T., 1998, ApJ, 500, 888; (astroph/9710064)
Djorgovski, S. G., 1998, ApJ, 508, L17
Fishman, G. J., et al. 1994, ApJS, 92, 229
Fishman, G. J. & Meegan, C. A. 1995, Ann. Rev. Astr. Ap., 33, 415
Galama, T. J., 1999, Nature, 398, 394
Garnavich, P. M., 2000, (astroph/0008049)
Katz, J. I., 1994a, ApJ, 422, 248
Katz, J. I., 1994b, ApJ, 432, L107
Mészáros, P. & Rees, M. J., 1997, ApJ, 476, 232
Paczyński, B. & Rhoads, J. E., 1993, ApJ, 418, L5
Piran, T., 1999, Phy. Rep., 314, 575 ;(astroph/9810256)
Rhoads, J. E. & Fruchter, A. S., 2000, submitted to ApJ ;(astroph/0004057)
Ruderman, M., 1975, Ann. N. Y. Acad. Sci., 262, 164
Sagar, R., et al. 2000, BASI, 28, 499 ;(astroph/0004223)
Sari, R., Piran, T., Narayan, R., 1998, ApJ, 497, L17
Schaefer, B. et al. 1994, *Proc. GRB w/s, 1993, AL, AIP-307 (ed) Fishman, G. J. et al*,
    p-280
Schmidt, W. K. H., 1978, Nature, 271, 525
Share, G. et al. 1994, *Proc. GRB w/s, 1993, AL, AIP-307 (ed) Fishman, G. J. et al*,
    p-283
Shemi, A., Piran, T., 1990, ApJ, 365, L55
Sokolov, V. V. et al. 2000, (astroph/0001357)

SEARCHES FOR TEV γ-RAY EMISSION WITH THE HEGRA CHERENKOV TELESCOPES

For the HEGRA collaboration:

N. Götting

*Universität Hamburg, II. Institut für Experimentalphysik, Luruper Chaussee 149,
D-22761 Hamburg*

Niels.Goetting@desy.de

F. Lucarelli

Universidad Complutense, Facultad de Ciencias Físicas, E-28040 Madrid

Fabrizio.Lucarelli@mpi-hd.mpg.de

M. Tluczykont

*Universität Hamburg, II. Institut für Experimentalphysik, Luruper Chaussee 149,
D-22761 Hamburg*

Martin.Tluczykont@desy.de

Abstract

Searches for Galactic and extragalactic sources of TeV γ-rays are described using the HEGRA stereoscopic system of imaging atmospheric Cherenkov telescopes with its good angular resolution and γ-hadron separation capability. We present results from recent observations of the interaction region of the Monoceros supernova remnant and the Rosette nebula, from the young open star clusters Berkeley 87 and IC 1805, and the giant radiogalaxies M 87, NGC 1275 and Cygnus A. Upper flux limits with energy thresholds between 0.7 and 1.2 TeV are obtained in the range of 0.5 to 2.5 $\times$ 10^{-12}ph.cm^{-2}s^{-1} for the individual objects.

203

M. M. Shapiro et al. (eds.), Astrophysical Sources of High Energy Particles and Radiation, 203–217.

1. Introduction

Imaging atmospheric Cherenkov telescopes (IACTs) are powerful tools to observe very high energy (VHE) γ-radiation from Galactic and extragalactic sources [1]. The HEGRA Cherenkov telescopes operated in stereoscopic observation mode on the Canary island of La Palma have been developed for the investigation of sources of γ-rays in the TeV energy range with special emphasis on a good energy and direction reconstruction (chapter 2).

As the question for the origin of cosmic rays is still open, effort has especially been put on the observation of supernova remnants (SNRs). The SNR Cas-A was just recently detected at the $5\,\sigma$ level with the HEGRA IACT system [2]. Interaction regions between SNRs and dense molecular clouds are also possible sites of TeV γ-ray emission. Observations of the interaction region of the Monoceros SNR and the Rosette nebula as a good candidate of this type have been carried out with the HEGRA Cherenkov telescopes (chapter 3).

Some models also motivate a search for other Galactic sites of shock acceleration. Young open star clusters could also be promising source candidates for the acceleration of cosmic rays. HEGRA IACT system observations of the young open star cluster Berkeley 87 allow to constrain the parameter space allowed for the models as discussed in chapter 4.

All extragalactic sources of TeV γ-rays observed so far are Blazars. However, giant radiogalaxies also show matter ejection in jets, but these jets are not aligned with the observer's line of sight. Several γ-ray production regions belonging to these objects are conceivable. The three giant radiogalaxies M 87, NGC 1275 and Cygnus A have been extensively observed with the HEGRA IACT system. In chapter 5 we present the results of the search for TeV γ-radiation from these objects.

2. The HEGRA Stereoscopic System of Cherenkov Telescopes

The HEGRA[1] collaboration operates six imaging atmospheric Cherenkov telescopes located at the Observatorio del Roque de los Muchachos on the Canary island of La Palma ($28.8°$ N, $17.9°$ W) at a height of $2200\,\mathrm{m}$ above sea level (approx. $800\,\mathrm{g/cm^2}$ overburden). The telescope CT 1 works as a stand-alone detector [3], while the other five telescopes build the stereoscopic IACT system [4].

---

[1] HEGRA is an acronym for "High Energy Gamma Ray Astronomy"

Figure 1. View of one of the five identical Cherenkov telescopes of the HEGRA IACT system (left). The IACT camera consisting of 271 photomultipliers arranged on a hexagonal grid (right).

These telescopes are arranged on the corners of a square area with one telescope in the center. The distance of the outer IACTs to the central one is about 80 m. Each system telescope has a tesselated mirror area of 8.5 m² and a camera consisting of 271 photomultipliers (see Figure 1). The angular size of the pixels is 0.25° leading to a geometrical field of view of approximately 4.3° in diameter.

The HEGRA IACT system operates with an energy threshold of 500 GeV for photons of vertical incidence. A flux sensitivity $\Phi_\gamma(E > 1\,\mathrm{TeV}) \geq 10^{-12}\,\mathrm{ph.\,cm^{-2}\,s^{-1}}$ is provided for an excess of $5\,\sigma$ within 20 hours of observation time. The stereoscopic reconstruction method leads to an energy resolution $\Delta E/E \leq 20\,\%$, an angular resolution $\Delta\Theta/\Theta \leq 0.1°$ and a sufficiently small error $\Delta x_{\mathrm{Core}} \leq 20\,\mathrm{m}$ on the shower core position at the observation level [5].

To search for TeV γ-rays a good data quality is required leading to specific conditions on the weather situation and the detector performance during data taking. Only events with *stereoscopic air shower imaging* are recorded, i. e. events with at least 2 coincident IACT triggers. As the image shape depends on the primary shower energy and the distance from the shower core to the telescope (impact parameter) the standard image parameter *width* is scaled according to the *width* parameter predicted for γ-showers under the same conditions using Monte Carlo simulations. The image parameter *mean scaled width* $(mscw)$ $\langle \tilde{w} \rangle$ is then calculated to be the mean value of the scaled *width* parameters of the single telescopes. A cut on *mscw* provides a separation between γ- and hadron induced showers with a cosmic ray rejection of up to a factor of 100.

3. Search for extended TeV γ-ray Emission from the Monoceros Supernova Remnant - Rosette Nebula Interaction Region

The TeV γ-rays produced in the regions where the supernova remnants (SNRs) are interacting with dense molecular clouds are expected to be detectable using currently operating imaging atmospheric Cherenkov telescopes [6]. The proton/nuclei component of the cosmic rays (CR) accelerated in SNRs through the diffuse shock acceleration may produce π^0-mesons and consequently decay-γ-rays in collisions of the CRs with the dense medium. Such a scenario may well be true for the interaction region of Monoceros SNR with the rather dense Rosette nebula.

We have mapped an extended sky region of $2° \times 2°$ associated with the Monoceros SNR/Rosette nebula, centered towards the hard spectrum X-ray point source SAX J0635+533 recently detected by the Italian-Dutch satellite BeppoSAX [7]. EGRET detected from this region an extended γ-ray emission (3EG J0634+0521) in the energy range from 100 MeV up to 10 GeV at 7σ level of confidence [8]. The X-ray point source SAX J0635+533 is found within the 95% probability circle of the EGRET detection.

The Monoceros SNR lies at a distance of 1.6 kpc [9] and has a radial extension of 50-60 pc. It is about $3 - 15 \cdot 10^4$ yr old with the shock expansion in the Sedov phase. The approximate nucleon density of the Rosette nebula is of 40 cm^{-3}. The optical and radio properties of the Monoceros SNR/Rosette nebula interaction region have been studied in great detail by [9] and [10].

The Whipple group reported an upper limit from the Monoceros SNR/Rosette nebula interaction region of 1.41×10^{-11} ph. cm^{-2} s^{-1} (approx. 30% of the Crab flux) above 500 GeV after 13 hours of observations [11]. This upper limit constrains the current models of γ-ray emission; however, it is still consistent with a possibly rather steep ($\alpha \simeq 2.4$) acceleration spectrum of CRs in SNRs.

3.1. Data Analysis of the Monoceros SNR - Rosette Nebula Observations

The Monoceros SNR/Rosette nebula region was observed with the HEGRA IACT system mostly in ON-source mode for a total of about 43 hours (36.5 hours after data cleaning), accompanied by a limited

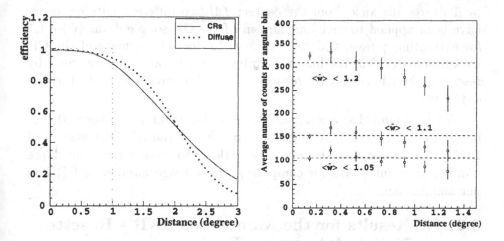

Figure 2. *Left panel*: Detection efficiency versus distance to the center of the field of view (FoV) using Monte Carlo simulations for the CR and diffuse γ-rays. *Right panel*: The CR data rate versus distance from the FoV after application of three different cuts on *mscw* ($\langle \tilde{w} \rangle <$ 1.05, 1.1, 1.2).

number of OFF runs (3.5 hours) [12]. For ON observations the center of the field of view (FoV) of the telescopes was adjusted to the BeppoSAX point source.

Sensitivity over the field of view: In order to search for unidentified or extended sources of TeV γ-rays, the sensitivity over the field of view is an important issue. Using Monte Carlo simulated air showers for cosmic rays and diffuse γ-rays we have studied the detection efficiency as a function of distance from the center of the FoV. The results of the analysis allow to conclude that the detection efficiency is almost constant over the area limited by the distance of 1 degree from the center of the FoV after applying the standard cut on *mean scaled width* (hereafter *mscw*), $\langle \tilde{w} \rangle <$ 1.1 (see Figure 2, left panel). In addition, we have plotted the number of recorded CR events versus the distance from the center of the FoV after applying the cuts $\langle \tilde{w} \rangle \leq$ 1.05, 1.1, 1.2 (see Figure 2, right panel). One can see that the number of CR counts is almost constant within 1 degree from the center of the FoV.

Models of CR background: Three different methods of CR background estimate have been applied here in order to predict the contamination of the CRs per each angular bin over the FoV: *(1)* the average number of background events per search bin was computed for the ON-source region after applying the cut on *mscw* over the FoV restricted by 1 degree distance from the center; *(2)* two different cuts on *mscw* have been applied to each angular bin of the ON-source data: $\langle \tilde{w} \rangle < 1.1$ for extracting γ-rays, and $1.4 < \langle \tilde{w} \rangle < 1.6$ for the estimate of the CR background [13]; *(3)* the limited number of OFF data runs was used by rescaling the CR rates with respect to the observation time for the ON data runs.

We have found that a reasonable size of the search bin is about $0.2° \times 0.2°$. It is larger than the angular resolution of the HEGRA system of IACTs ($0.1°$) and, at the same time, allows to have a relatively large number of angular bins for computing the average number of CR hits per angular bin.

3.2.   Results for the Monoceros SNR - Rosette Nebula Interaction Region

The two-dimensional map of the reconstructed events for a $2° \times 2°$ region of the FoV is shown in Figure 3, *upper panel*. One can see from this Figure that a number of search bins show an excess. Interestingly, most of these bins are concentrated close to the interaction region of the Monoceros SNR with the Rosette nebula.

Based on the different models of the background estimate mentioned above, we have calculated the statistical significance of the excess events using the maximum likelihood method of Li & Ma [14]. The calculated significances are summarized in Table 1.

The three models of the CR background estimate listed above give comparable results. Note that model 2 gives a higher relative content of the CR events as compared with the other two. Nevertheless, possible systematic effects which may distort the shape of the CR distribution at different distances from the center of the FoV, could be present here; we did not investigate in detail these effects and further studies are needed for a definite conclusion on that issue. To estimate possible systematic effects caused by the different night sky background, we have compared the CR rates for angular bins of different size over the FoV using both, data of Crab nebula observations and data of a few pulsars observed with the HEGRA IACT system at relatively small angular distance from Monoceros SNR/Rosette nebula interaction region. The results are consistent within the estimated statistical errors.

The two-dimensional plot of the significance is shown in Figure 3, *lower panel*. This picture is produced after applying a standard polynomial smoothing over the neighboring bins. One can see that the bins with significances above 2.5 are located within the EGRET 95% probability circle and are associated with the interaction region of the Monoceros

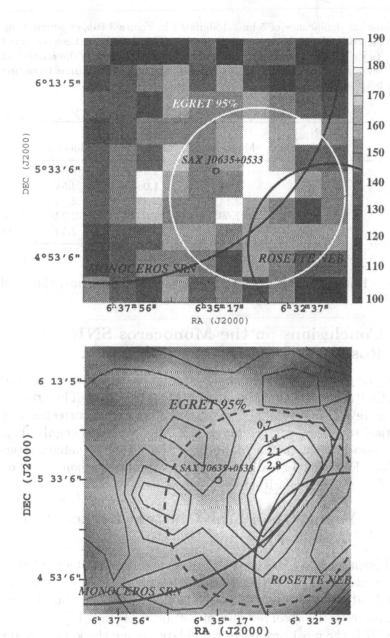

Figure 3.   *Upper panel:* Number of reconstructed events per $0.2° \times 0.2°$ bin over the field of view, after applying the cut $\langle \tilde{w} \rangle < 1.1$. *Lower panel:* Filled contour plot of the reconstructed events superimposed with the contour levels of the calculated significance. The increment of the significance contours is 0.7.

Table 1. Statistical significance of 5 bins designated in Figure 3 (*upper panel*) using the different methods of CR background estimate. The data have been analyzed using the cut $\langle \tilde{w} \rangle < 1.1$ on the mean scaled width parameter. $N_{(i,j)}$ is the number of reconstructed events for the angular bin in the i-th row and j-th column (counting from bottom left).

i	j	$N_{(i,j)}$	Significance [σ]		
			Model 1	Model 2	Model 3
4	6	190	3.34	2.39	2.90
5	7	185	2.91	1.08	2.54
5	8	188	3.14	2.33	2.75
6	7	182	2.69	1.65	2.32
7	7	185	2.92	2.95	2.54

SNR and the Rosette nebula, which can be used as an *a priori* choice of a search window for an extended γ-ray source.

3.3. Conclusions on the Monoceros SNR - Rosette Nebula Interaction Region

The Monoceros SNR/Rosette nebula interaction region was observed with the HEGRA IACT system for 36 hours in 1999/2000. The preliminary analysis reveals in several bins of $0.2° \times 0.2°$ an excess corresponding to a statistical significance in the range of 2.5 to 3σ. These angular bins are closely associated with the Monoceros SNR/Rosette nebula interaction region. Further observations will provide a tighter constraint on possible TeV γ-ray emission from this region.

4. The Young Open Star Clusters Berkeley 87 and IC 1805

The still open question for the origin of cosmic rays has motivated to search for acceleration sites other than shell type super nova remnants [15]. Several authors have argued that Galactic young open star clusters could contribute to the observed flux of cosmic rays [16], [17].

Berkeley 87 is the most promising candidate among the known young open star clusters. At a distance of 0.9 kpc Berkeley 87 is probably part of the Cygnus star forming region ON2. It contains approximately 100 young massive stars of the OB spectral class with strong stellar winds embedded in a very dense interstellar medium and molecular clouds. The most prominent of these stars is the Wolf-Rayet star WR–142, located at the center of the cluster. With a mass of approx. 60 $M_\odot$ and a mass loss rate of $1.68 \cdot 10^{-5} M_\odot$ yr^{-1} [18], WR–142 produces stellar winds of approximately 5200 km/s propagating into the ambient interstellar medium with an estimated Mach number of 26 and thus producing a system of shock fronts. The EGRET source 2EG J2019+3719 has been associated with Berkeley 87 [17]. The observed γ-ray flux above 100 MeV could be explained by a cosmic ray density in the open cluster being a factor 100 higher than the average Galactic density [19]. Extrapolations of the hard EGRET spectrum ($dN/dE \sim E^{-1.91}$) to higher energies are shown in Figure 5 [19]. The predicted fluxes are in the detection range of the HEGRA Cherenkov telescopes.

The young open cluster IC 1805 shows very similar features to Berkeley 87. It is the central cluster of the star association CAS OB6, embedded in the compact H II region W4. It is located at 2.4 kpc distance to the sun and contains several members with hypersonic winds among which the two peculiar stars HD 15570 and HD 15578 both have estimated wind velocities of 3800 km/s [26]. The EGRET γ-ray sources 3EG J0241+6103, 3EG J0229+6151 [27] and the COS B source 2CG 135+01 [28] were detected in the neighbourhood of IC 1805 as well as the hard X-ray source 4U 0241+61 [29].

4.1. Data Analysis and Results for the Young Open Star Clusters

For the analyses presented here, the cut on *mscw* was chosen as $\langle \tilde{w} \rangle < 1.2$. Taking into account the angular resolution of 0.1°, the optimal signal region for a pointlike source is $0 < \Delta\Theta^2 < 0.05 \deg^2$ (where $\Delta\Theta$ is the angle between the reconstructed event and the source direction). The data were taken in the so called *wobble* mode [22], allowing simultaneous ON- and OFF-source observations.

Crab nebula data samples from the periods October 1999 and October/November 2000 were used for upper limit calculations [23] of the Berkeley 87 and IC 1805 data respectively. The γ- and hadron-efficiencies are $\epsilon_\gamma = 0.68$ and $\epsilon_{hadron} = 0.12$ for October 1999 and $\epsilon_\gamma = 0.71$ and $\epsilon_{hadron} = 0.10$ for October/November 2000.

Berkeley 87: In total 6.4 h observation time on Berkeley 87 have been accumulated in August 1999. 5.6 h of the observation were performed with the complete 5 telescope system, while 0.8 h were performed with a 4 telescope system due to technical reasons. The average zenith angle of 15° translates into an energy threshold of the data sample of 700 GeV. The observations were centered on the star WR–142. For the given angular resolution of the HEGRA IACT system the object can be considered as pointlike.

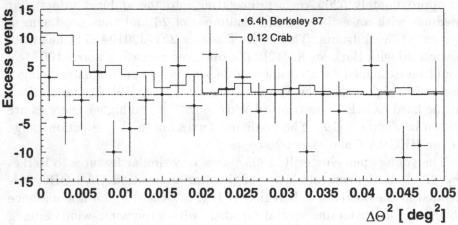

Figure 4. Number of reconstructed excess events vs. angular distance $\Delta\Theta^2$ for Berkeley 87 (dots with statistical errors). Also shown is a histogram of a 6.4h Crab data sample scaled down to the value of the upper limit (0.12 Crab units) obtained for Berkeley 87. The data were taken in the so called *wobble mode*, allowing simultaneous ON- and OFF-source observations.

In Figure 4 the difference between the number of reconstructed γ-ray events in the ON-source and the OFF-source regions is plotted versus the squared angular distance $\Delta\Theta^2$. The 90% confidence level upper limit on the integral flux has been calculated to be: $\Phi_{90\%}(E > 0.7\,\mathrm{TeV}) = 0.12$ in units of the Crab flux [5] corresponding to $\Phi_{90\%}(E > 0.7\,\mathrm{TeV}) = 2.5 \times 10^{-12}\,\mathrm{ph.\,cm^{-2}s^{-1}}$, assuming a source energy spectrum $\Phi_\gamma \sim E^{-2}$. This HEGRA upper limit is indicated in Figure 5.

IC 1805: The young open star cluster IC 1805 has been observed for a total of 10 h in October/November 2000. The data are very recent and the results of this analysis should therefore be considered as preliminary. The average zenith angle of about 34° translates into an energy threshold of 1.15 TeV for the IC 1805 data sample.

No evidence for emission of TeV γ-rays has been found. Assuming a point source like emission from IC 1805, an upper limit $\Phi_{90\%}$ on the integral γ-ray flux has been calculated to $\Phi_{90\%}(E > 1.15\,\mathrm{TeV}) = 0.11$ in units of the Crab flux, corresponding to $\Phi_{90\%}(E > 1.15\,\mathrm{TeV}) = 2 \times 10^{-12}\,\mathrm{ph.\,cm^{-2}s^{-1}}$ in absolute flux units, assuming a source energy spectrum $\Phi_\gamma \sim E^{-2}$.

Conclusion: The derived upper limits on the integral flux from Berkeley 87 and IC 1805 constrain the possible contribution of young open clusters to the observed flux of charged Galactic cosmic rays. In

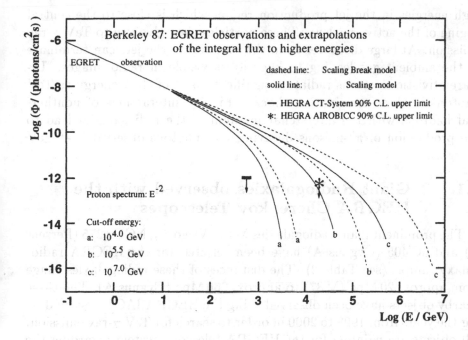

Figure 5. The EGRET γ-ray spectrum (Φ) and predicted spectra, extrapolated from the EGRET observations (E>100 MeV) to higher energies, assuming different input proton spectra and γ-ray production from π^0-decay [19]. The upper limit calculated in this work is indicated as a bar (preliminary). An earlier upper limit determined with the HEGRA AIROBICC detector is also shown (asterisk) [20].

the case of Berkeley 87 the upper limit rules out the predictions by Giovanelli et al. [19]. These results make a significant contribution of young open clusters to the acceleration of charged Galactic cosmic rays seem unlikely.

5.   Search for TeV γ-Ray Emission from Giant Radiogalaxies

Giant radiogalaxies contain huge amounts of mass, e. g. approximately $10^{13} M_\odot$ in the inner 100 kpc of M 87. Distinct large scale jets are developed, most probably by the central engine which is supposed to be a supermassive black hole. In contrast to Blazars, the jet of a radiogalaxy and the observer's line of sight are not aligned. A large amount of nonthermal particles is expected to be confined within giant radiogalaxies.

Due to these properties the emission of TeV γ-radiation might be possible in different production processes and at different regions of giant radiogalaxies, respectively: Charged particles are accelerated to very high energies in the jet production region which is close to the central engine of the active nucleus. These particles then give rise to TeV γ-ray emission. At large distances from the core region the jets can terminate in the ambient gas leading to huge lobes visible in radio images. Ultrarelativistic electrons radiate a significant part of their energy as TeV photons in these termination shocks. Finally, interactions of nonthermal hadrons with the interstellar medium of the radiogalaxies lead to the production of π^0-mesons decaying into photons observable at TeV energies.

5.1. Giant Radiogalaxies observed with the HEGRA Cherenkov Telescopes

The prominent giant radiogalaxies M 87 (Virgo A), NGC 1275 (Perseus A) and 3C 405 (Cygnus A) have been selected for the HEGRA radiogalaxy sample (see Table 2). The distances of these radiogalaxies range from approx. 20 Mpc (M 87) to approx. 260 Mpc (Cygnus A). The three nearby objects have been observed using the HEGRA IACT system during the years from 1998 to 2000 in order to search for TeV γ-ray emission. All objects are *pointlike* for the HEGRA telescope system regarding the instrument's angular resolution. The mean zenith angles of $\langle\vartheta\rangle \approx 15°$ lead to an energy threshold for these observations of $E_{\mathrm{thr}} \approx 700\,\mathrm{GeV}$. Observations of the Crab nebula have been performed with similar zenith angle distributions during the same observation seasons. As the Crab nebula is the "standard candle" for TeV γ-ray astronomy in the northern hemisphere these data sets allow to calculate upper limits on the TeV γ-ray flux. In addition, the Crab data provide a control of system-

Table 2. Properties and preliminary results on the observations of the HEGRA radiogalaxy sample and the Crab nebula. The calculation of upper limits on the TeV γ-ray flux is described in the text.

Object Name	Redshift z	Time (h)	E_{thr} (GeV)	excess events $N_{\mathrm{ON}} - N_{\mathrm{OFF}}$			$\Phi_{90\%}^{a}$ (Crab)^{b}	$\Phi_{90\%}$ (abs.)^{c}
M 87	0.0043	44.1	720	28	±	57	0.030	0.92
NGC 1275	0.0176	44.4	690	-80	±	55	0.015	0.49
Cygnus A	0.0561	11.1	700	-37	±	28	0.033	1.05
Crab		81.4	680	6929	±	112	-	-

^{a}Upper limit on the TeV γ-ray flux Φ_γ at 90 % confidence level
^{b}Flux limits in units of the flux of the Crab nebula
^{c}Flux limits in units of $10^{-12}\,\mathrm{ph.cm}^{-2}\,\mathrm{s}^{-1}$

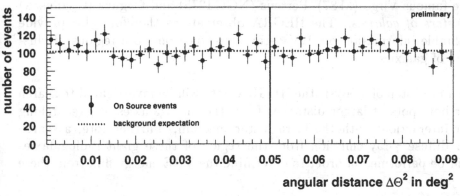

Figure 6.   Number of reconstructed events from the M 87 data set (preliminary) as a function of the squared angular distance $\Delta\Theta^2$ after applying the *mean scaled width* cut $0.5 < \langle \tilde{w} \rangle < 1.2$. The horizontal line indicates the expected number of background events while the vertical line shows the cut $\Delta\Theta^2 < 0.05 \deg^2$ applied in the search for point sources. The statistical uncertainties are indicated by the error bars.

atic effects and give the possibility of independently checking the data quality.

5.2.   Results for the Giant Radiogalaxies

The result of the M 87 observations is shown as an example in Figure 6. The event distribution from the ON-source region is plotted versus the squared angular distance $\Delta\Theta^2$ from the M 87 position. The expected mean number of background events has been derived from an OFF-source region. Data from the object and the OFF-source region are obtained simultaneously by operating the telescopes in the *wobble* mode. Regarding the HEGRA IACT system's angular resolution a direction cut is set to $\Delta\Theta^2 \leq 0.05 \deg^2$. For the γ-hadron separation the image shape cut on *mscw* is applied as $0.5 < \langle \tilde{w} \rangle < 1.2$. There is no indication of TeV γ-ray emission in the M 87 data.

The analysis of the HEGRA IACT system data has not resulted in any evidence for TeV γ-ray emission from the sample of giant radio-galaxies (see Table 2). Upper limits on the TeV γ-ray flux were calculated using contemporaneous observations of the Crab nebula following the procedure described by Helene [23]. A conversion from Crab units to absolute flux units can be obtained assuming a Crab-like spectrum around 1 TeV [30]:

$$\Phi_\gamma(E > 1\,\mathrm{TeV}) = 1.8 \cdot 10^{-11} (E/\mathrm{TeV})^{-1.59}\,\mathrm{cm}^{-2}\mathrm{s}^{-1}$$

The three observed giant radiogalaxies are also the central objects of the famous Virgo (M 87), Perseus (NGC 1275) and Cygnus (Cygnus A) *clusters of galaxies*. The HEGRA observations therefore also provide valuable information on TeV γ-ray emission from core regions of clusters of galaxies.

In a next step of analysis the HEGRA data will be investigated to search for hot spots at larger distances from the radiogalaxies, e. g. searching for interactions with the intracluster medium. Furthermore, a search for diffuse γ-ray emission from the region of these giant radiogalaxies will be performed in order to determine the VHE energy flux from these objects.

6. Summary

The recent investigation of Galactic and extragalactic objects with the HEGRA stereoscopic IACT system has not revealed evidence for TeV γ-ray emission. The interaction region of the Monoceros supernova remnant and the Rosette nebula shows an excess of more than $2\,\sigma$ in several search bins of the size $0.2° \times 0.2°$. Upper flux limits on TeV γ-radiation from the young open star clusters Berkeley 87 and IC 1805 could be obtained to be $\Phi_{90\%}(E > 0.70\,\mathrm{TeV}) = 2.5 \times 10^{-12}$ ph. cm^{-2}s^{-1} and $\Phi_{90\%}(E > 1.15\,\mathrm{TeV}) = 2.0 \times 10^{-12}$ph. cm^{-2}s^{-1}, respectively, assuming a soure energy spectrum with a spectral index of 2. The investigation of the giant radiogalaxies M 87, NGC 1275 and Cygnus A allows to calculate upper limits $\Phi_{90\%}(E > 0.7\,\mathrm{TeV})$ on the TeV flux between 0.49 and 1.05×10^{-12}ph. cm^{-2}s^{-1} under the assumption of a Crab-like spectrum.

Acknowledgements

The support of the German Ministry for Research and Technology BMBF and of the Spanish Research Council CYCIT is gratefully acknowledged. We thank the Instituto de Astrofísica de Canarias (IAC) for the use of the HEGRA site at the Observatorio del Roque de los Muchachos (ORM) and for the excellent working conditions on La Palma. We gratefully acknowledge the technical support staff of the Heidelberg, Kiel, Munich, and Yerevan Institutes.

References

[1] Aharonian, F. A., et al. (1997). *Astropart. Phys.* **6**, 369.

[2] Pühlhofer, G., for the HEGRA collaboration,
to appear in the Proc. of the Heidelberg Gamma Ray Symposium, 2000.

[3] Mirzoyan R., et al. (1994). *Nucl. Instr. Meth. A* **351**, 513-526.

[4] Daum A., et al. (1997). *Astropart. Phys.* **8**, 1-11.

[5] Konopelko A., et al. (1999). *Astropart. Phys.* **10**, 275-289.

[6] Aharonian, F., Drury, L. O., Völk, H. J. (1994). *A&A* **285**, 645-647.

[7] Kaaret, P., et al. (1999). *ApJ* **523**, 197-202.

[8] Jaffe, T., et al. (1997). *ApJ Letters* **484**, L129-L131.

[9] Davies, R. D., et al. (1978). *A&AS* **31**, 271-284.

[10] Graham D. A., et al. (1982). *A&A* **109**, 145-154.

[11] Lessard, R. W.,
to appear in the Proc. of the Heidelberg Gamma Ray Symposium, 2000.

[12] Lucarelli, F., et al., for the HEGRA Collaboration,
to appear in the Proc. of the Heidelberg Gamma Ray Symposium, 2000.

[13] Rowell, G. P., for the HEGRA collaboration,
to appear in the Proc. of the Heidelberg Gamma Ray Symposium, 2000.

[14] Ti-pei Li, Yu-qian Ma (1983). *ApJ* **272**, 317-324.

[15] Drury L. O'C., Aharonian F.A., and Völk H.J. (1994). *Astron. Astrophys.* **287**, 959.

[16] Manchanda R.K., Polcaro V.F., and Norci L. (1996). *Astron. Astrophys.* **305**, 457.

[17] Polcaro V.F. et al. (1991). *Astron. Astrophys.* **252**, 590.

[18] Barlow, M.J. (1990). *Proc. IAU Symp.* **No. 143**, Sanur (Bali, Indonesia), June 18-22.

[19] Giovanelli F., Bednarek W., and Karakula S. (1996). *J. Phys. G Part. Phys.* **22**, 1223.

[20] Prahl, J. (1999). *PhD Thesis, dissertation.de*, Hamburg.

[21] Daum, A., et al. (1997). *ApJ* **322**, 706.

[22] Aharonian F. A., et al. (1997). *A & A* **327**, L5-L8.

[23] Helene, O. (1983). *Nucl. Inst. Meth. A* **212**, 319.

[24] Konopelko, A. K., Pühlhofer, G. (1999), *Proc. of the 26th ICRC* **Vol. 3**, 444.

[25] Tluczykont, M., for the HEGRA collaboration,
to appear in the Proc. of the Heidelberg Gamma Ray Symposium, 2000.

[26] Burki G., et al. (1981). *Astron. Astrophys. L* **79**, 13, L69-L73.

[27] Hartmann R.C., et al. (1999). *Astrophys. J., Suppl. Ser.* **123**, 79-202.

[28] Swanenburg B.N., et al. (1981). *Astrophys. J.* **243**, L69-L73.

[29] Forman W., et al. (1978). *Astrophys. J., Suppl. Ser.* **38**, 357-412.

[30] Aharonian F. A., et al. (2000). *ApJ* **539**, 317-324.

VARIABLE GAMMA RAY SOURCES, 1: INTERACTIONS OF FLARE ENERGETIC PARTICLES WITH SOLAR AND STELLAR WINDS

Lev I. DORMAN

Israel Cosmic Ray Center and Emilio Segre' Observatory, affiliated to Tel Aviv University, Technion, Israel Space Agency, and IZMIRAN Russian Ac. of Sci.

ICRC&ESO, P.O.B. 2217, Qazrin 12900, ISRAEL

1. Introduction

The generation of gamma rays (GR) by interaction of flare energetic particles (FEP) with stellar wind matter shortly was considered in [1,2]. Here we will give a development of this research with much more details. As example, we will consider in the first the situation with GR generation in the interplanetary space by solar FEP in periods of great events, determined mainly by 3 factors:

1st- by space-time distribution of solar FEP in the Heliosphere, their energetic spectrum and chemical composition (see review in [3-10]; for this distribution can be important nonlinear collective effects (especially for great events) of FEP pressure and kinetic stream instability [11-14].

2nd- by the solar wind matter distribution in space and its change during solar activity cycle; for this distribution will be important also pressure and kinetic stream instability of galactic cosmic rays (CR) as well as of solar FEP (especially in periods of very great events) [12-14, 15].

3rd- by properties of solar FEP interaction with solar wind matter accompanied with GR generation through decay of neutral pions [16-18].

After consideration of these 3 factors we will calculate expected GR emissivity space-time distribution, and expected fluxes of GR for measurements on the Earth's orbit in dependence of time after the moment of FEP generation for different directions of GR observations. We calculate expected fluxes also for different distances from the Sun inside the Heliosphere and outside. We expect that the same 3 factors will be important for GR generation by stellar FEP in stellar winds, but for some types of stars the total energy in FEP is several orders higher than in solar flares and lost matter speed is several orders higher than from the Sun [19-22].

Observations of GR generated in interactions of solar FEP with solar wind matter can give for the periods of great events valuable information on solar wind matter 3d-distribution as well as on properties of solar FEP and its propagation parameters. Especially important will be observations of GR generated in interactions of stellar FEP with stellar wind matter. It will be shown down that in this case can be obtained

M. M. Shapiro et al. (eds.), Astrophysical Sources of High Energy Particles and Radiation, 219–230.
© 2001 *Kluwer Academic Publishers. Printed in the Netherlands.*

information on total energy and energetic spectrum in stellar FEP, on mode of its propagation as well as information on stellar wind matter distribution.

2. The 1st Factor: Solar FEP Space-Time Distribution

2.1. MODEL OF SOLAR FEP PROPAGATION

The problem of solar FEP generation and propagation through the solar corona and in the interplanetary space as well as its energetic spectrum and chemical and isotopic composition was reviewed in [3-10]. In the first approximation according to numeral data of observations of many events for about 5 solar cycles the time change of solar FEP and energy spectrum change can be described by the solution of isotropic diffusion (characterized by the diffusion coefficient $D_i(E_k)$) from some pointing instantaneous source $Q_i(E_k, \mathbf{r}, t) = N_{oi}\delta(\mathbf{r})\delta(t)$ of solar FEP of type i (protons, α – particles and heavier particles, electrons) by

$$N_i(E_k, \mathbf{r}, t) = N_{oi}(E_k)\left[2\pi^{1/2}(D_i(E_k)t)^{3/2}\right]^{-1} \times \exp\left(-r^2/(4D_i(E_k)t)\right), \quad (1)$$

where $N_{oi}(E_k)$ is the energetic spectrum of total number of solar FEP in the source. At the distance $r = r_1$ the maximum of solar FEP density

$$N_{i\max}(r_1, E_k)/N_{oi}(E_k) = 2^{1/2}3^{3/2}\pi^{-1/2}\exp(-3/2)r_1^{-3} = 0.925\, r_1^{-3} \quad (2)$$

will be reach according to Eq. (1) at the moment

$$t_{\max}(r_1, E_k) = r_1^2/6D(E_k), \quad (3)$$

and the space distribution of solar FEP density at this moment will be

$$N_i(r, E_k, t_{\max})/N_{oi}(E_k) = (54/\pi)^{1/2}r_1^{-3}\exp\left(-3r^2/2r_1^2\right) = 4.146\, r_1^{-3}\exp\left(-3r^2/2r_1^2\right). (4)$$

2.2. ENERGY SPECTRUM OF SOLAR FEP IN THE SOURCE

According to numeral experimental data, the energetic spectrum of generated solar energetic particles in the source approximately can be described as (see review in [9]):

$$N_{oi}(E_k) \approx N_{oi}(E_k/E_{k\max})^{-\gamma}, \quad (5)$$

where γ increases with increasing of energy from about $0 \div 1$ at $E_k \leq 1$ $GeV/nucleon$ to about $6 \div 7$ at $E_k \approx 10 + 15$ $GeV/nucleon$. Parameters N_{oi} and γ are changing sufficiently from one event to other: for example, for the greatest observed event of February 23, 1956 $N_{oi} \approx 10^{34} + 10^{35}$, in the event of November 15, 1960 $N_{oi} \approx 3 \times 10^{32}$, in the event of July 18, 1961 $N_{oi} \approx 4 \times 10^{31}$, in the event of May 23, 1967 $N_{oi} \approx 10^{31}$. For the greatest observed event of February 23, 1956 parameter γ had values ≈ 1.2 at $E_k \approx 0.3$ $GeV/nucleon$, $\gamma \approx 2.2$ at $E_k \approx 1$ $GeV/nucleon$, $\gamma \approx 4$ at $E_k \approx 5 + 7$ $GeV/nucleon$, and $\gamma \approx 6 \div 7$ at $E_k \approx 10 + 15$ $GeV/nucleon$; this change of

γ is typical for many great solar energetic particle events (see in [3-5] about event of February 23, 1956, and review about many events in [4-10]). Approximately the behavior of value γ in Eq. (5) can be described as

$$\gamma = \gamma_o + \ln(E_k/E_{ko}),$$ (6)

where parameters γ_o and E_{ko} are different for individual events, but typically they are in intervals $2 \le \gamma_o \le 5$ and $2 \le E_{ko} \le 10$ *GeV/nucleon*. The position of maximum in Eq. (5) with taking into account Eq. (6) is determined by

$$E_{k\,max} = E_{ko}\,exp(-\gamma_o), \quad N_{oi}(E_{k\,max}) = N_{oi}.$$ (7)

2.3. THE TOTAL ENERGY OF SOLAR FEP

The total energy contained in FEP will be according to Eq. (5)-(7):

$$E_{tot} = N_{oi}\int_0^\infty E_k(E_k/E_{k\,max})^{-\gamma_o - \ln(E_k/E_{ko})}\,d(E_k/E_{k\,max}) = bN_{oi}E_{k\,max},$$ (8)

where

$$b = \int_0^\infty x^{1-\ln x}\,dx = 4.82.$$ (9)

For great solar FEP events $E_{tot} \approx 10^{31} + 10^{32}$ *erg* and more [3-10], for great stellar FEP events $E_{tot} \approx 10^{35} + 10^{37}$ *erg* [19-22].

2.4. DEPENDENCE OF FEP TRANSPORT PATH FROM PARTICLE ENERGY

In Eq. (1)

$$D_i(E_k) = \Lambda_i(E_k)V(E_k)/3$$ (10)

is the diffusion coefficient, $\Lambda_i(E_k)$ is the transport path for particle scattering in the interplanetary space, $V(E_k)$ is the particle velocity in dependence of kinetic energy per nucleon E_k:

$$V(E_k) = c\left(1 - \left(1 + E_k/m_nc^2\right)^{-2}\right)^{1/2},$$ (11)

where m_nc^2 is the rest energy of nucleon. According to numeral experimental data and theoretical investigations $\Lambda_i(E_k)$ have a bride minimum in the region 0.1-0.5 *GeV/nucleon* and increases with energy decreasing lower than this region as about $\propto E_k^{-1}$ (caused by "tunnel" effect for particles with curvature radius in the interplanetary magnetic field (IMF) smaller than smallest scale of hydro-magnetic turbulence, see in [23]) as well as with energy increasing over this interval as $\propto E_k^\beta$, where β depends from the spectrum of turbulence and usually increases from 0 up to about 1 for high energy particles of few *GeV/nucleon* and then up to about 2 for very high energy particles with curvature radius in IMF bigger than biggest scale of magnetic inhomogeneities in IMF (according to investigations of galactic cosmic ray modulation in the Heliosphere it will be at $E_k \geq 20 + 30$ $GeV/nucleon$). For calculations of expected space-time distribution of gamma ray emissivity we try to describe this dependence for the most part of spectrum what is important for gamma-ray emission (from about 0.01 $GeV/nucleon$ up to about 20 $GeV/nucleon$) approximately as

$$\Lambda_i(E_k) \approx \Lambda_i(W,r,t) \left(\frac{E_1}{E_k} + \frac{E_k}{E_2} + \left(\frac{E_k}{E_3} \right)^2 \right). \qquad (12)$$

To determine parameters E_1, E_2, E_3 we use observations of solar cosmic ray events as well as observations of galactic cosmic ray modulation in the interplanetary space. The time-dependencies of galactic cosmic ray primary fluxes for effective rigidities $R = 2, 5,$ 10 and 25 GV were found in [24] on the basis of ground measurements of muon and neutron components as well as measurements in stratosphere on balloons and in space on satellites and space-crafts. The residual modulation (relative to the flux out of the Heliosphere) for $R \approx 10\,GV$ in the minimum and maximum of solar activity was determined as 6 and 24 % (what is in good agreement with results on CR-SA hysteresis effects [25-30]). According to convection-diffusion model of cosmic ray cycle modulation [31, 32, 28] the slope of the residual spectrum $\Delta D(R)/D_0(R) \propto R^{-\beta}$ reflects the dependence $\Lambda(R) \propto \left(\Delta D(R)/D_0(R)\right)^{-1} \propto R^{\beta}$. In [24] the spectral index β was determined as $\beta \approx 0.4$ at $2 - 5\,GV$, $\beta \approx 1.1$ at $5 - 10\,GV$, and $\beta \approx 1.6$ at $10 - 25\,GV$. Eq. (12) will be in agreement with these results and with data on FEP events in smaller energy region if we choose $E_1 \approx 0.05$ $GeV/nucleon$, $E_2 \approx 2$ $GeV/nucleon$, and $E_3 \approx 5$ $GeV/nucleon$.

2.5. DEPENDENCE OF TRANSPORT PATH FROM SOLAR ACTIVITY LEVEL

The dependence of the transport path from the level of solar activity is characterized by $\Lambda_i(W)$. This parameter can be determined from investigations of galactic cosmic ray modulation in the interplanetary space on the basis of observations by neutron monitors and muon telescopes for several solar cycles. According to [25-30] $\Lambda_i(W) \propto W^{-1/3}$ for the period of high solar activity and $\Lambda_i(W) \propto W^{-1}$ for the period of low solar activity. According to [29-30], the hysteresis phenomenon in the connection of long-term cosmic ray intensity variation with solar activity cycle can be explained well by the analytical approximation of this dependence, taking into account the time lag of processes in the interplanetary space relative to caused processes on the Sun:

$$\Lambda_i(W,r,t) = \Lambda_i(W_{\max}) \times \left(W\left(t - \frac{r}{u}\right) \middle/ W_{\max} \right)^{-\frac{1}{3} - \frac{2}{3}\left(1 - W\left(t - \frac{r}{u}\right) \middle/ W_{\max}\right)}, \qquad (13)$$

where $W_{\max}$ is the sunspot number in maximum of solar activity and $\Lambda_i(W_{\max}) \approx 10^{12} cm$

3. 2nd-Factor: Space-Time Distribution of Solar Wind Matter

The detail information on the 2nd-factor for distances smaller than 5 AU from the Sun was obtained in the last years by the mission of Ulysses. Important information for bigger distances (up to 60-70 AU) was obtained from missions Pioneer 10, 11, Voyager 3, 4 , but only not far from the ecliptic plane. If we assume for the first approximation the model of Parker [31] of radial solar wind expanding into the interplanetary space which is in good according with all available data of direct measurements in the Heliosphere, then the behavior of matter density of solar wind will be described by the relation

$$n(r,\theta) = n_1(\theta) u_1(\theta) r_1^2 / (r^2 u(r,\theta)), \qquad (14)$$

where $n_1(\theta)$ and $u_1(\theta)$ are the matter density and solar wind speed at the helio-latitude θ on the distance $r = r_1$ from the Sun $(r_1 = 1 AU)$. The dependence $u(r,\theta)$ is determined by the interaction of solar wind with galactic CR and anomaly component of CR, with interstellar matter and interstellar magnetic field, by interaction with neutral atoms penetrating from interstellar space inside the Heliosphere, by the nonlinear processes caused by these interactions (see [14, 15]). According to the recent calculations of Roux and Fichtner [15] the change of solar wind velocity can be described approximately as

$$u(r) \sim u_1(1 - b(r/r_o)), \qquad (15)$$

where the distance to the terminal shock wave $r_o \approx 74\,AU$ and parameter $b \approx 0.13 \div 0.45$ in dependence of subshock compression ratio (from 3.5 to 1.5) and from injection efficiency of pickup protons (from 0 to 0.9).

4. 3-rd Factor: Gamma Ray Generation by FEP in the Heliosphere

4.1. GENERATION OF NEUTRAL PIONS

According to [16-18] the neutral pion generation caused by nuclear interactions of energetic protons with hydrogen atoms through reaction $p + p \rightarrow \pi^o +$ anything will be determined by

$$F_{pH}^{\pi}(E_{\pi},r,\theta,t) = 4\pi n(r,\theta,t) \int_{E_{k\min}(E_{\pi})}^{\infty} dE_k N_p(E_k,r,t)(\varsigma \sigma_{\pi}(E_k))(dN(E_k,E_{\pi})/dE_{\pi}), \quad (16)$$

where $n(r,\theta,t)$ is determined by Eq. (14), $E_{k\,min}\langle E_\pi\rangle$ is the threshold energy for pion generation, $N_p(E_k,r,t)$ is determined by Eq. (1), $\langle\varsigma\sigma_\pi(E_k)\rangle$ is the inclusive cross section for reactions $p+p\to\pi^o+\text{anything}$, and $\int\limits_0^\infty(dN(E_k,E_\pi)/dE_\pi)dE_\pi=1$.

4.2. SPACE-TIME DISTRIBUTION OF GAMMA RAY EMISSIVITY

Gamma ray emissivity caused to nuclear interaction of FEP protons with solar wind matter will be determined according to [16-18] by

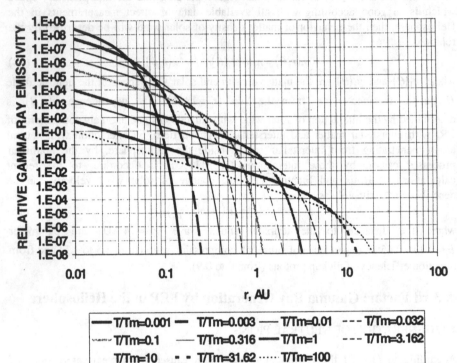

Figure 1. Expected space distribution of gamma ray emissivity for different time T after FEP generation in units of time maximum Tm on 1 AU, determined by Eq. (3). The curves are from T/Tm=0.001 up to T/Tm=100.

$$F_{pH}^\gamma\left(E_\gamma,r,\theta,t\right)=2\int\limits_{E_{\pi\,min}\left(E_\gamma\right)}^\infty dE_\pi\left(E_\pi^2-m_\pi^2c^4\right)^{-1/2}F_{pH}^\pi\left(E_\pi,r,\theta,t\right),, \qquad (17)$$

where $E_{\pi\,min}\left(E_\gamma\right)=E_\gamma+m_\pi^2c^4/4E_\gamma$. Let us introduce Eq. (1) in (16) and (17) by taking into account Eq. (14):

$$F_{pH}^{\gamma}\left(E_{\gamma},r,\theta,t\right)=B(r,\theta,t)\int_{E_{\pi min}\left(E_{\gamma}\right)}^{\infty}\left(E_{\pi}^2-m_{\pi}^2c^4\right)^{-1/2}dE_{\pi}\times$$

$$\int_{E_{k min}\left(E_{\pi}\right)}^{\infty}N_{op}\left(E_k\right)\left(\varsigma\sigma_{\pi}\left(E_k\right)\right)\left(t/t_1\right)^{-3/2}exp\left(-3r^2t_1/2r_1^2t\right)dE_k,$$

$$(18)$$

Where $B(r,\theta,t)=3^{3/2}2^{7/2}\pi^{1/2}r_1^2n_1(\theta,t)u_1(\theta,t)/r^2u(r,\theta,t)$ and $t_1=r_1^2/6D_p(E_k)$ is the time in which the density of FEP, at a distance of 1 AU, reaches the maximum value. The space distribution of gamma ray emissivity for different t/t_1 will be determined mainly by function $r^{-2}(t/t_1)^{-3/2}exp\left(-3r^2t_1/2r_1^2t\right)$, where t_1 corresponds to some effective value of E_k in dependence of E_{γ}, according to Eq. (16) and (17). The biggest gamma ray emission is expected in the inner region $r\leq r_i=r_1(2t/3t_1)^{1/2}$ where the level of emission $\propto r^{-2}(t/t_1)^{-3/2}$. Out of this region gamma ray emissivity decreases very quickly with r as $\propto r^{-2}exp\left(-(r/r_i)^2\right)$. For an event with total energy 10^{32} ergs at $t=t_1=10^3$ sec, $r_i=10^{13}$ cm, $n_1(\theta,t)\approx5\,cm^{-3}$ $D_p(E_k)\approx4\times10^{22}\,cm^2/sec$, we obtain $F_{pp}^{\gamma}(E_{\gamma}>0.1GeV,r)\approx10^8\,r^{-2}\,ph.cm^{-3}s^{-1}$ (on the distance 5 solar radius it gives 10^{-15} $ph.cm^{-3}s^{-1}$). Eq. (18) describes the space-time variations of gamma ray emissivity distribution from interaction of solar energetic protons with solar wind matter (see some examples in Figure 1).

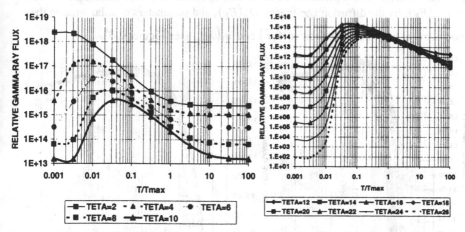

Figure 2. Expected GR fluxes for directions 2° to 26° from the Sun in dependence of T/Tmax, where Tmax was determined by Eq. (3).

5. Expected Angle Distribution and Time Variations of Gamma Ray Fluxes for Observations Inside the Heliosphere

Let us assume that the observer is inside the Heliosphere, on the distance $r_{obs} \le r_o$ from the Sun and helio-latitude θ_{obs} (here r_o is the radius of Heliosphere). The sight line of observation we can determine by the angle θ_{sl}, computed from the equatorial plane from direction to the Sun to the North. In this case the expected angle distribution and time variations of gamma ray fluxes will be

$$\Phi_{pH}^{\gamma}\left(E_{\gamma}, r_{obs}, \theta_{sl}, t\right) = \int_{0}^{L_{\max}(\theta_{sl})} F_{pH}^{\gamma}\left(E_{\gamma}, L(r_{obs}, \theta_{sl}), t\right) dL. \tag{19}$$

In Eq. (19) GR emissivity

$$F_{pH}^{\gamma}\left(E_{\gamma}, L(r_{obs}, \theta_{sl}), t\right) = F_{pH}^{\gamma}\left(E_{\gamma}, r, \theta, t\right) \tag{20}$$

is determined by Eq. (18) taking into account that

$$r = \left(r_{obs}^2 + L^2 + 2r_{obs}L\Delta\theta\right)^{1/2}, \quad \theta = \theta_{obs} + \arccos\left(\frac{r_{obs}^2 + r_{obs}L\Delta\theta}{r_{obs}\left(r_{obs}^2 + L^2 + 2r_{obs}L\Delta\theta\right)^{1/2}}\right). \tag{21}$$

where $\Delta\theta = \theta_{sl} - \theta_{obs}$. In Eq. (19)

$$L_{\max} = \frac{r_o}{\sin\Delta\theta}\sin\left[\Delta\theta - \arcsin\left(\frac{r_{obs}}{r_o}\sin\Delta\theta\right)\right]. \tag{22}$$

According to (18) – (22) the expected angle distribution and time variations of gamma ray fluxes for local observer ($r_{obs} \le r_o$) from interaction of solar energetic protons with solar wind matter will be determined by the energy spectrum of proton generation on the Sun $N_{op}(E_k)$, by the diffusion coefficient $D_p(E_k)$, and parameters of solar wind in the period of event near the Earth orbit $n_1(\theta, \tilde{t})$ and $u_1(\theta, \tilde{t})$ (see Figures 2-3).

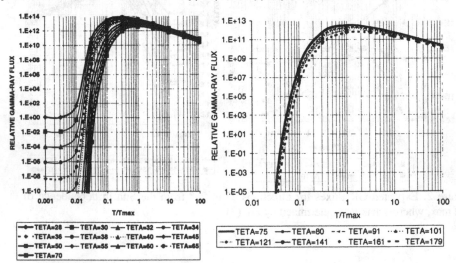

Figure 3. The same as in Figure 2, but for directions 28° to 179° from the Sun.

In the case of spherical symmetry we obtain

$$\Phi_{pH}^{\gamma}\left(E_{\gamma}, r_{obs}, \varphi, t\right) \approx F_{pH}^{\gamma}\left(E_{\gamma}, r = r_{obs}\sin\varphi, t\right)\left(\theta_{max} - \theta_{min}\right)r_{obs}\sin\varphi \,, \quad (23)$$

where φ is the angle between direction on the star and direction of observation, $\theta_{max} = arccos\left(r_{obs}\sin\varphi/r_i\right)$; $\theta_{min} = -arccos\left(r_{obs}\sin\varphi/r_i\right)$ if $r_{obs} > r_i$ and $\theta_{min} = \varphi - \pi/2$, if $r_{obs} \le r_i$. For the great solar event with the total energy in FEP 10^{32} ergs for $r_{obs} = 1AU$, Eq. (23) gives

$$\Phi_{pH}^{\gamma}\left(E_{\gamma} > 0.1GeV, r_{obs}, \varphi, t\right) \approx \frac{6.7 \times 10^{-6}}{\sin\varphi}\left(\frac{t}{t_1}\right)^{-\frac{3}{2}} exp\left(-\frac{3t_1\sin^2\varphi}{2t}\right)ph.cm^{-2}s^{-1} \quad (24)$$

6. GAMMA RAYS FROM INTERACTION OF FLARE ENERGETIC PARTICLES WITH STELLAR WIND MATTER

Let us suppose that some observer is on the distance $r_{obs} \gg r_o$, where r_o is radius of stellar-sphere. In this case

$$\Phi_{pH}^{\gamma}\left(E_{\gamma}, r_{obs}, t\right) = 2\pi r_{obs}^{-2}\int_{-\pi/2}^{\pi/2}\cos\theta d\theta\int_{0}^{r_o} r^2 dr F_{pH}^{\gamma}\left(E_{\gamma}, r, \theta, t\right) \,, \quad (25)$$

where $F_{pH}^{\gamma}\left(E_{\gamma}, r, \theta, t\right)$ was determined by Eq. (18). For spherical-symmetrical modes of FEP propagation and stellar wind matter distribution, we obtain

$$\Phi_{pH}^{\gamma}\left(E_{\gamma}, r_{obs}, t\right) = 4\pi r_{obs}^{-2} F_{pH}^{\gamma}\left(E_{\gamma}, r_1, t_1\right)\left(t/t_1\right)^{-1}r_1^3\Phi\left(\frac{r_o}{r_1}(3t_1/t)^{1/2}\right)ph.cm^{-2}s^{-1} \,, \quad (26)$$

where $\Phi(x)$ is the probability function. For a flare star with total energy in FEP event 10^{36} ergs and $n_1(\theta,t) \approx 500\,cm^{-3}$ the expected emissivity at $t = t_1 = 10^3$ sec will be $F_{pp}^{\gamma}\left(E_{\gamma} > 0.1GeV, r\right) \approx 10^{14}\,r^{-2}\,ph.cm^{-3}.sec^{-1}$. For this case Eq. (26) gives

$$\Phi_{pH}^{\gamma}\left(E_{\gamma} > 0.1GeV, r_{obs}, t\right) = 2 \times 10^{28}\,r_{obs}^{-2}\left(t/t_1\right)^{-1}\Phi\left(\frac{r_o}{r_1}(3t_1/t)^{1/2}\right)ph.cm^{-2}s^{-1} \,. \quad (27)$$

According to (27) for $t_1 = 10^3$ s at $t = 10$ s and 100 s the value $\frac{r_o}{r_1}(3t_1/t)^{1/2} \gg 1$ and $\Phi(x) \approx 1$, that at distance $r_{obs} = 10^{19}\,cm$ (about 3 ps) expected flux $\Phi_{pH}^{\gamma}\left(E_{\gamma} > 0.1GeV, r_{obs}, t\right)$ will be 2×10^{-8} and 2×10^{-9} $ph.cm^{-2}s^{-1}$, correspondingly. Eq. (27) shows that the total flux of gamma-rays from stellar wind generated by FEP interaction with wind matter must fall inverse proportional with time and does not depend from the details of the event. It is important for separation of GR generation in stellar wind from direct generation in stellar flare.

7. DISCUSSION AND CONCLUSIONS

Estimations according to Eq. (24) show that in periods of great solar FEP events with total energy $\approx 10^{32}$ ergs the expected flux of GR with energy > 100 MeV in direction 2° from the Sun at $t/t_1 = 1/3$ reaches $\approx 10^{-3}$ $ph.cm^{-2}.sr^{-1}.sec^{-1}$. It means that according to Figure 2 expected flux in the same direction at $t/t_1 = 1/30$ reaches $\approx 2 \times 10^{-2}$ $ph.cm^{-2}.sr^{-1}.sec^{-1}$. In direction 30° from the Sun (see Figure 3) expected GR fluxes are much smaller: the maximum will be at $t/t_1 = 1/3$ and reaches value only $\approx 10^{-5}$ $ph.cm^{-2}.sr^{-1}.sec^{-1}$. Expected GR fluxes are characterized by great specific time variations, which depend from direction of observations relative to the Sun, total FEP flux from the source, parameters of FEP propagation (summarized in value of t_1), and properties of solar wind (see Figure 1 for expected GR emissivity space-time distribution and Figures 2 and 3 for expected GR fluxes). It is important that present GR telescopes might measure expected GR fluxes in periods of great FEP events. These observations of gamma rays generated in interactions of FEP with solar wind matter can give important information on solar wind 3d-distribution as well as on properties of solar FEP and its propagation parameters. Moreover, the monitoring of GR observations in directions at few degrees from the Sun can give important possibility to predict expected radiation hazard from FEP on the Earth and in space. Important information can be obtained also by GR monitoring of nearest stars with great FEP events (energy spectrum of FEP and parameters of propagation, stellar wind matter space-time distribution).

8. ACKNOWLEDGMENTS

I would like to thank Prof. Maurice Shapiro and Prof. John Wefel for the possibility to attend the school of cosmic ray astrophysics, for very interesting discussions and fruitful comments, and the staff of Ettore Majorana Center for very kind hospitality. This research was supported by Israel Cosmic Ray Center and Emilio Segre' Observatory, affiliated to Tel Aviv University, Technion, and Israel Space Agency.

9. REFERENCES

1. Dorman, L.I. (1996) Cosmic ray nonlinear processes in gamma-ray sources, *Astronomy and Astrophysics*, Suppl. Ser., **120**, No. 4, 427-435.
2. Dorman, L.I. (1997) Angle distribution and time variation of gamma ray flux from solar and stellar winds, 1. Generation by flare energetic particles, in C.D. Dermer, M.S. Strickman, and J.D. Kurfess (eds.), *Proc. 4th Compton Symposium* Williamsburg, pp.1178-1182.
3. Dorman, L.I. (1957) *Cosmic Ray Variations*, Gostekhtheorizdat, Moscow (English translation published in 1958 by Research Division of Ohio Air-Force Base, USA).
4. Dorman, L.I. (1963) *Cosmic Ray Variations and Space Research*, Nauka, Moscow.

5. Dorman L.I. (1963) *Astrophysical and Geophysical Aspects of Cosmic Rays*, in J.G. Wilson and S.A.Wouthuysen (eds.) *Progress of Cosmic Ray and Elementary Particle Physics*, North-Holland Publ. Co., Amsterdam, Vol. 7.

6. Dorman, L.I., and Miroshnichenko, L.I. (1968) *Solar Cosmic Rays*, Fizmatgiz, Moscow (English translation published for the National Aeronautics and Space Administration and National Science Foundation in 1976 (TT 70-57262/ NASA TT F-624), Washington, D.C.)

7. Dorman, L.I. (1974) *Cosmic Rays: Variations and Space Exploration.* North-Holland Publ. Co., Amsterdam.

8. Dorman, L.I. (1978) *Cosmic Rays of Solar Origin,* VINITI, Moscow.

9. Dorman, L.I. and Venkatesan, D. (1993) Solar cosmic rays *Space Sci. Rev.* **64**, 183-362.

10. Stoker, P.H. (1995) Relativistic solar cosmic rays, *Space Sci. Rev.,* **73**, 327

11. Berezinskii, V.S., Bulanov, S.V., Ginzburg, V.L., Dogiel, V.A., and Ptuskin, V.S. (1990) *Cosmic Ray Astrophysics*, Fyzmatgiz, Moscow.

12. Dorman, L.I., Ptuskin, V.S. and Zirakashvili, V.N. (1990) Outer Heliosphere: pulsations, cosmic rays and stream kinetic instability, in S. Grzedzielski and D.E. Page (eds.) *Physics of the Outer Heliosphere*, Pergamon Press, pp. 205-209.

13. Zirakashvili, V.N., Dorman, L.I., Ptuskin, V.S. and Babayan, V.Kh. (1991) Cosmic ray nonlinear modulation in the outer Heliosphere. *Proc.22-th Intern. Cosmic Ray Conf.*, Dublin, Vol.3, pp 585-588.

14. Dorman, L.I. (1995) Cosmic ray nonlinear effects in space plasma, 2. Dynamic Heliosphere, in M.M. Shapiro, R. Silberberg and J.P.Wefel (eds.). *Currents in High Energy Astrophysics* , Kluwer Academic Publishers., Dordrecht/Boston /London, NATO ASI Serie, Vol. 458, pp. 193-20813.

15. Le Roux, J.A. & Fichtner, H., *Ap. J*, **477**, L115 (1997).

16. Stecker, F.W. (1971) *Cosmic Gamma Rays*, Mono Book Co, Baltimore.

17. Dermer, C.D. (1986) *A&A,* **157**, 223

18. Dermer, C.D. (1986) *Ap. J*, **307**, 47

19. Gershberg R.E., Mogilevskij E.I. & Obridko V.N., 1987, Kinem. and Phys. of Celestial Bodies, **3**, No. 5, 3.

20. Gershberg R.E. & Shakhovskaya N.I., 1983, Astrophys. Space Science, **95**, No. 2, 235.

21. Korotin S.A. & Krasnobaev V.I., 1985, Izvesia Krimea Astrophysical Observatory, **73**, 131.

22. Kurochka L.N., 1987, Astronomy J. (Moscow), **64**, 443.

23. Dorman, L.I. (1975) *Experimental and Theoretical Foundations of Cosmic Ray Astrophysics*, Fizmatgiz, Moscow.

24. Belov, A.V., Gushchina, R.T., Dorman, L.I., and Sirotina, I.V. (1990) Rigidity spectrum of cosmic ray modulation, *Proc. 21-th Intern. Cosmic Ray Conf.*, Adelaide, Vol. 6, pp. 52-55.

25. Dorman, I.V., and Dorman, L.I., (1967) Solar wind properties obtained from the study of the 11-year cosmic ray cycle, *J. Geophys. Res.* 72, 1513-1520.

26. Dorman, I.V. & Dorman, L.I. (1967) Propagation of energetic particles through interplanetary space according to the data of 11-year cosmic ray variations. *J. Atmosph. and Terr. Phys.*, **29**, No.4, 429-449.

27. Dorman, I.V. & Dorman, L.I. (1968) Hysteresis phenomena in cosmic rays, properties of solar wind and energetic spectrum of different nuclei in the Galaxy, *Proc. 5-th All-Union Winter School on Cosmophysics*, Apatity, pp.183-196.

28. Dorman, L.I. (1975) *Variations of Galactic Cosmic Rays*, Moscow University Press, Moscow

29. Dorman, L.I., Villoresi, G., Dorman, I.V., Iucci, N., and Parisi, M. (1997) High rigidity CR-SA hysteresis phenomenon and dimension of modulation region in the Heliosphere in dependence of particle rigidity, *Proc. of 25- th Intern. Cosmic Ray Conference,* Durban (South Africa), Vol. 2, pp 69-72.

30. Dorman, L.I., Villoresi, G., Dorman, I.V., Iucci, N., and Parisi, M. (1997) Low rigidity CR-SA hysteresis phenomenon and average dimension of the modulation region and Heliosphere, *Proc. of 25- th Intern. Cosmic Ray Conference,* Durban (South Africa), Vol. 2, pp 73-76.

31. Parker, E.N. (1963) *Interplanetary Dynamically Processes*, Intersci. Publ., New York-London.

32. Dorman, L.I. (1959) To the theory of cosmic ray modulation by solar wind, *Proc. of 6-th Intern. Cosmic Ray Conf.*, Moscow, Vol. 4, pp. 328-334.

VARIABLES GAMMA RAY SOURCES, 2: INTERACTIONS OF GALACTIC COSMIC RAYS WITH SOLAR AND STELLAR WINDS

Lev I. DORMAN

Israel Cosmic Ray Center and Emilio Segre' Observatory, affiliated to Tel Aviv University, Technion, Israel Space Agency, and IZMIRAN Russian Ac. of Sci.

ICRC&ESO, P.O.B. 2217, Qazrin 12900, ISRAEL

Abstract. By data obtained from investigations of hysteresis phenomenon in dependence of galactic cosmic ray intensity from solar activity we determine the change of cosmic ray density distribution in the Heliosphere during solar cycle in dependence of particle energy. On the basis of observation data and investigations of cosmic ray nonlinear processes in the Heliosphere we determine the space-time distribution of solar wind matter. Then we calculate the generation of gamma-rays by decay of neutral pions generated by nuclear interactions of modulated galactic cosmic rays with solar wind matter and determine the expected space-time distribution of gamma-ray emissivity. On the basis of these results we calculate the expected time variation of the angle distribution and spectra of gamma ray fluxes generated by interaction of modulated galactic cosmic rays with solar wind matter for local (inside the Heliosphere) and distant observers (for stellar winds).

1. Introduction

The generation of gamma rays (GR) by interaction of galactic cosmic rays (CR) with solar and stellar winds matter shortly was considered in [1,2]. Here we will give a development of this research with much more details. We will consider the situation with GR generation in the interplanetary space by galactic CR and expected time variations of gamma ray fluxes in dependence of direction of observations and in connection with solar activity (SA) cycle (about 11 years from minimum to minimum of SA) and with solar magnetic cycle (with period about 22 years including odd and even SA cycles from maximum to maximum SA). Space-time distribution of GR emissivity will be determined mainly by 3 factors:

1st- by space-time distribution of galactic CR in the Heliosphere, their energetic spectrum and chemical composition; for this distribution can be important nonlinear collective effects of galactic CR pressure and kinetic stream instability [3-7].

2nd- by the solar wind matter distribution in space and its change during solar activity cycle; for this distribution will be important also pressure and kinetic stream instability of galactic CR [4-7].

3rd- by properties of galactic CR interaction with solar wind matter accompanied with GR generation through decay of neutral pions [8-10].

231

M. M. Shapiro et al. (eds.), Astrophysical Sources of High Energy Particles and Radiation, 231–243.
© 2001 *Kluwer Academic Publishers. Printed in the Netherlands.*

After consideration of these 3 factors we will calculate expected GR emissivity space-time distribution, and expected fluxes of GR for measurements on the Earth's orbit in dependence of the level of solar activity for different directions of GR observations. We will calculate expected GR fluxes also for different distances from the Sun inside the Heliosphere (local observations) and outside (distant observations). We expect that the same 3 factors will be important for GR generation by galactic CR in stellar winds, but for some types of stars the lost matter speed is several orders higher than from the Sun.

Observations of GR generated in interactions of galactic CR with solar wind matter can give valuable information on solar wind matter 3d-distribution as well as on properties of galactic CR global modulation and its propagation parameters. Especially important will be observations of GR generated in interactions of galactic CR with stellar wind matter. It will be shown that in this case can be obtained important information on galactic CR modulation in stellar-sphere as well as information on stellar activity and stellar wind.

2. Factor 1-st: Galactic Cosmic Ray Space-Time Distribution
2.1. CONVECTION-DIFFUSION MODULATION

The problem of galactic CR propagation through the interplanetary space as well as modulation of its intensity and energetic spectrum (1-st factor) was reviewed in [11-15], and with taking into account CR nonlinear processes in [6]. According to this research, the convection-diffusion modulation of energy spectra of proton component of galactic CR can be described in the queasy-stationary approximation of the spherical symmetrical geometry [16, 17] as

$$N_p(r, E_k) = N_p(E_k) exp\left(-\frac{\gamma+2}{3} \int_r^{r_o} \frac{u(r,t)dr}{D_p(r, E_k, t)}\right), \qquad (1)$$

where according to [18]

$$N_p(E_k) = 2.2 \times \left(E_k + m_p c^2\right)^{-2.75} \ proton.sr^{-1}.cm^{-2}.s^{-1}.GeV^{-1} \qquad (2)$$

is the differential energy spectrum of proton component of galactic CR outside of the Heliosphere, $u(r,t)$ is the solar wind velocity, and

$$D_p(r, E_k, t) = \Lambda_p(r, E_k, t) V(E_k)/3 \qquad (3)$$

is the diffusion coefficient, $\Lambda_p(r, E_k, t)$ is the transport path for particle scattering, $V(E_k)$ is the particle velocity in dependence of kinetic energy per nucleon E_k:

$$V(E_k) = c\left(1 - \left(1 + E_k/m_n c^2\right)^{-2}\right)^{1/2}. \qquad (4)$$

2.2. ON THE DEPENDENCE OF TRANSPORT PATH FROM PARTICLE ENERGY

According to numeral experimental data and theoretical investigations $\Lambda_i(E_k)$ have a bride minimum in the region 0.1-0.5 $GeV/nucleon$ and increases with energy decreasing lower than this region as about $\propto E_k^{-1}$ (caused by "tunnel" effect for particles with curvature radius in the interplanetary magnetic field (IMF) smaller than smallest scale of hydromagnetic turbulence, see in [19]) as well as with energy increasing over this interval as $\propto E_k^{\gamma}$, where γ depends from the spectrum of turbulence and usually increases from 0 up to about 1 for high energy particles of few $GeV/nucleon$ and then up to about 2 for very high energy particles with curvature radius in IMF bigger than biggest scale of magnetic inhomogeneities in IMF (according to investigations of galactic cosmic ray modulation in the Heliosphere it will be at $E_k \geq 20 + 30$ $GeV/nucleon$). For calculations of expected space-time distribution of gamma ray emissivity we try to describe this dependence approximately as

$$\Lambda_i(E_k) \approx \Lambda_i(W, r, t)\left(\frac{E_1}{E_k} + \frac{E_k}{E_2} + \left(\frac{E_k}{E_3}\right)^2\right). \tag{5}$$

To determine parameters E_1, E_2, E_3 we use observations of solar cosmic ray events as well as observations of galactic cosmic ray modulation in the interplanetary space. The time-dependencies of galactic cosmic ray primary fluxes for effective rigidities $R = 2, 5,$ 10 and 25 GV were found in [20] on the basis of ground measurements (muon and neutron components) as well as measurements in stratosphere on balloons and in space on satellites and space-crafts. The residual modulation (relative to the flux out of the Heliosphere) for $R \approx 10\,GV$ in the minimum and maximum of solar activity was determined as 6 and 24 % (what is in good agreement with results for hysteresis effect obtained on the basis of neutron monitor data [22-29]). According to convection-diffusion model of cosmic ray cycle modulation, the slope of the residual spectrum $\Delta D(R)/D_0(R) \propto R^{-\gamma}$ reflects the dependence

$$\Lambda(R) \propto \left(\Delta D(R)/D_0(R)\right)^{-1} \propto R^{\gamma}. \tag{6}$$

In [20, 21] the spectral index γ was determined as $\gamma \approx 0.4$ at $2 - 5\,GV$, $\gamma \approx 1.1$ at $5 - 10\,GV$, and $\gamma \approx 1.6$ at $10 - 25\,GV$. Eq. (5) will be in agreement with these results and with data on FEP events in smaller energy region if we choose $E_1 \approx 0.05$, $E_2 \approx 2$, and $E_3 \approx 5$ $GeV/nucleon$.

2.3. TRANSPORT PATH DEPENDENCE FROM SOLAR ACTIVITY LEVEL

The dependence of the transport path from the level of solar activity is characterized by $\Lambda_i(W)$. This parameter can be determined from investigations of galactic cosmic ray modulation in the interplanetary space on the basis of observations by neutron monitors and muon telescopes for several solar cycles. According to [22-28] $\Lambda_i(W) \propto W^{-1/3}$ for the period of high solar activity and $\Lambda_i(W) \propto W^{-1}$ for the period of low solar activity. According to [22-29], the hysteresis phenomenon in the connection of long-term cosmic ray intensity variation with solar activity cycle can be explained well by the analytical approximation of this dependence, taking into account the time lag of processes in the interplanetary space relative to caused processes on the Sun:

$$\Lambda_i(W,r,t)=\Lambda_i(W_{max})\times\left(W\left(t-\frac{r}{u}\right)\Big/W_{max}\right)^{-\frac{1}{3}-\frac{2}{3}\left(1-W\left(t-\frac{r}{u}\right)\Big/W_{max}\right)},\qquad (7)$$

where $W_{max}\approx 200$ and $\Lambda_i(W_{max})\approx 10^{12}cm$.

2.4. APPROXIMATION BY USING OF MODULATION PARAMETER $B(t)$

In the some rough approximation the convection-diffusion global modulation described by Eq. (1) can be determined as

$$N_{gp}(r,E_k)\approx N_{gp}(E_k)\exp\left(-\frac{B(t)}{R^\gamma\beta}\left(1-\frac{r}{r_o}\right)\right),\qquad (8)$$

where r_o is the dimension of modulation region, parameter γ determines the dependence $\Lambda(R)\propto R^\gamma$ (see Section 2.2),

$$R=cp/Ze=\left(E_k^2+2E_km_pc^2\right)^{1/2}\Big/e\qquad (9)$$

is the particle rigidity for protons (in GV), and

$$\beta=V/c=\left(E_k^2+2E_km_pc^2\right)^{1/2}\Big/\left(E_k+m_pc^2\right)\qquad (10)$$

is the particle velocity for protons in units of light speed c, and $B(t)$ is the parameter of modulation. According to [22-30] the parameter $B(t)$ changes with solar activity in the first approximation is inverse proportional to (7). Near the minimum of solar activity $B_{min}\approx(0.3+0.4)GV$. In the maximum of solar activity the modulation became higher and $B_{max}\approx(1.6+2.5)GV$ for different solar cycles in dependence of direction of solar general magnetic field and sign of cosmic ray particles charge (influence of drift effects on the galactic cosmic ray modulation, see Section 2.5, Figure 1). For α – particles in galactic cosmic rays the distribution of the modulated spectrum will be

$$N_{g\alpha}(r,E_k)\approx N_{g\alpha}(E_k)\exp\left(-\frac{B(t)}{R^\gamma\beta}\left(1-\frac{r}{r_o}\right)\right),\qquad (11)$$

where $B(t),R,\gamma$, and β are the same as in Eq. (8) for protons, and $N_{g\alpha}(E_k)$ is the α - particle spectrum outside of the Heliosphere, which according to [18]

$$N_{g\alpha}(E_k)=0.07\left(E_k+m_pc^2\right)^{-2.75}\alpha\text{ - }particle.sr^{-1}.cm^{-2}.s^{-1}(GeV/nucleon)^{-1},\qquad (12)$$

and E_k is the kinetic energy of α - particle per nucleon. For more heavier particles we have equation, similar to (11), but

$$R=cp/Ze=(A/Ze)\left(E_k^2+2E_km_pc^2\right)^{1/2}.\qquad (13)$$

2.5. ROLE OF DRIFT EFFECTS IN GLOBAL MODULATION

Described above is the modulation of galactic CR caused by convection-diffusion processes. To this modulation necessary to add modulation caused by drift effects what can be determined mainly by the value of tilt-angle between neutral current sheet plane and solar equator plane [33]. This modulation changes the sign in periods of solar magnetic field reverses. The amplitude of drift modulation A_{dr} as well as dimension of modulation region r_o where determined in [31] as average for even and odd cycles 19-22 (see Figure 1), and in [32] specially for cycle 22 in dependence of effective particle rigidity R_{ef}. For galactic electrons the modulation in the Heliosphere will be determined also by Eq. (1) or (8), but the drift effects will be opposite in comparison with protons and α-particles.

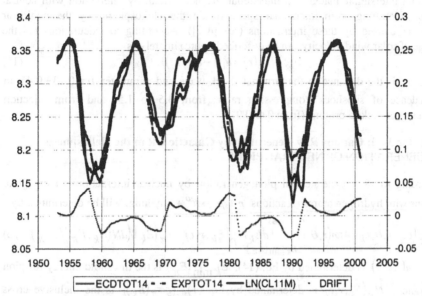

Figure 1. Observed long-term modulation of galactic CR in 1953-2000 according to Climax NM data (effective rigidity $R_{ef} \approx 10\,GV$, curve LN(CL11M)) in comparison with expected (curve EXPTOT14) at $r_o = 14\,\text{av.month} \times u_{av} \approx 108$ AU (for this period $u_{av} \approx 7.7$ AU/av.month). Convection-diffusion modulation (curve ECDTOT14) and drift modulation (curve DRIFT) are also shown. Left Y-scale for natural logarithm of Climax NM counting rate, right Y-scale for drift modulation. Interval between two horizontal lines corresponds to 5% variation.

3. Factor 2-nd: Space-Time Distribution of Solar Wind Matter

The detail information on the 2nd-factor for distances smaller than 5 AU from the Sun was obtained in the last years by the mission of Ulysses. Important information for bigger distances (up to 60-70 AU) was obtained from missions Pioneer 10, 11, Voyager 3, 4, but only not far from the ecliptic plane. Let us assume for the first approximation the model of Parker [16] of radial solar wind expanding into the interplanetary space which is in good agreement with all available data of direct measurements in the Heliosphere. According to this model the behavior of matter density of solar wind will be described by the relation

$$n(r,\theta) = n_1(\theta)u_1(\theta)r_1^2 / \left(r^2 u(r,\theta)\right), \tag{14}$$

where $n_1(\theta)$ and $u_1(\theta)$ are the matter density and solar wind speed at the helio-latitude θ on the distance $r = r_1$ from the Sun $(r_1 = 1AU)$. The dependence $u(r,\theta)$ is determined by the interaction of solar wind with galactic CR and anomaly component of CR, with interstellar matter and interstellar magnetic field, by interaction with neutral atoms penetrating from interstellar space inside the Heliosphere, by the nonlinear processes caused by these interactions (see [4-7]). According to calculations [7] the change of solar wind velocity can be described approximately as

$$u(r) \approx u_1(1 - b(r/r_o)), \tag{15}$$

where r_o is the distance to the terminal shock wave and parameter $b \approx 0.13 + 0.45$ in dependence of subshock compression ratio (from 3.5 to 1.5) and from injection efficiency of pickup protons (from 0 to 0.9).

4. Factor 3-rd: Gamma Ray Generation by Galactic CR in the Heliosphere
4.1. GENERATION OF NEUTRAL PIONS

According to [8-10] the neutral pion generation by nuclear interactions of energetic protons with hydrogen atoms (reactions $p + p \to \pi^o$ + anything) will be determined by

$$F_{pH}^{\pi}(E_{\pi}, r, \theta, t) = 4\pi n(r,\theta,t) \int_{E_{k\,min}(E_{\pi})}^{\infty} dE_k N_p(E_k, r, t)\langle\varsigma\sigma_{\pi}(E_k)\rangle\langle dN(E_k, E_{\pi})/dE_{\pi}\rangle \tag{16}$$

where $n(r,\theta,t)$ is determined by Eq. (14), $E_{k\,min}(E_{\pi})$ is the threshold energy for pion generation, $N_p(E_k, r, t)$ is determined by Eq. (1), $\langle\varsigma\sigma_{\pi}(E_k)\rangle$ is the inclusive cross section for reactions $p + p \to \pi^o$ + anything , and $\int_0^{\infty}\left(dN(E_k, E_{\pi})/dE_{\pi}\right)dE_{\pi} = 1$.

4.2. GAMMA RAY GENERATION BY GALACTIC COSMIC RAY PROTONS

Gamma ray emissivity caused to nuclear interaction of galactic CR protons with solar wind matter will be determined according to [8-10] by

$$F_{pH}^{\gamma}(E_{\gamma}, r, \theta, t) = 2 \int_{E_{\pi\,min}(E_{\gamma})}^{\infty} dE_{\pi}\left(E_{\pi}^2 - m_{\pi}^2 c^4\right)^{-1/2} F_{pH}^{\pi}(E_{\pi}, r, \theta, t), \tag{17}$$

where $E_{\pi\,min}(E_{\gamma}) = E_{\gamma} + m_{\pi}^2 c^4 / 4E_{\gamma}$.

4.3. SPACE-TIME DISTRIBUTION OF GAMMA RAY EMISSIVITY

Let us introduce Eq. (2), (8)-(10) and (15) in (16) and then obtained result in (17):

$$F_{pH}^{\gamma}\left(E_{\gamma},r,\theta,t\right)=2.2\left(\frac{8\pi n_1(\theta,t)n_1^2}{r^2(1-b(r/r_o))}\right)_{E_{\pi}\min(E_{\gamma})}^{\infty}\int dE_{\pi}\left(E_{\pi}^2-m_{\pi}^2c^4\right)^{-1/2}\int_{E_k\min(E_{\pi})}^{\infty}dE_k\times$$

$$\left\langle\varsigma\sigma_{\pi}(E_k)\right\rangle\left(E_k+m_pc^2\right)^{-2.75}\exp\left(-\frac{B(t)\left(E_k+m_pc^2\right)e}{\left(E_k^2+2E_km_pc^2\right)^{(\gamma+1)/2}}\left(1-\frac{r}{r_o}\right)\right).$$ (18)

The expected gamma ray emissivity distribution from interaction of α-particles with solar wind matter will be determined by introducing Eq. (9)-(12), (15) in (16) and (17):

$$F_{\alpha H}^{\gamma}\left(E_{\gamma},r,\theta,t\right)=0.07\left(\frac{8\pi n_1(\theta,t)n_1^2}{r^2(1-b(r/r_o))}\right)_{E_{\pi}\min(E_{\gamma})}^{\infty}\int dE_{\pi}\left(E_{\pi}^2-m_{\pi}^2c^4\right)^{-1/2}\int_{E_k\min(E_{\pi})}^{\infty}dE_k\times$$

$$\left\langle\varsigma\sigma_{\pi}(E_k)\right\rangle\left(E_k+m_pc^2\right)^{-2.75}\exp\left(-\frac{B(t)\left(E_k+m_pc^2\right)(e/2)}{\left(E_k^2+2E_km_pc^2\right)^{(\gamma+1)/2}}\left(1-\frac{r}{r_o}\right)\right).$$ (19)

For the radial extended of solar or stellar wind the space-time distributions of GR emissivity according to Eq. (18) will be mainly determined by the function

$$F_{pH}^{\gamma}\left(E_{\gamma},r,t\right)=F_{pH}^{\gamma}\left(E_{\gamma}\right)\frac{n_1(\theta,t)}{n_o}\left(1-\frac{br}{r_o}\right)^{-1}\left(\frac{n_1}{r}\right)^2\exp\left(-A(E_{\gamma},t)\left(1-\frac{r}{r_o}\right)\right),$$ (20)

where $F_{pH}^{\gamma}(E_{\gamma})$ is the emissivity spectrum from galactic CR protons in the interstellar space (as background emissivity from interstellar matter with density n_o according to [9, 10]) and $n_1(\theta,t)$ is the density of solar or stellar wind on the latitude θ on the distance $n_1=1$ AU from the star. In (20)

$$A(E_{\gamma},t)=B(t)/\left(R^{\gamma}\beta\right)_{ef}(E_{\gamma})$$ (21)

and $\left(R^{\gamma}\beta\right)_{ef}(E_{\gamma})$ is some effective value of $R^{\gamma}\beta$ for particles responsible for GR generation with energy E_{γ}. According to [9, 10] the expected GR emissivity from all particles in galactic CR $F^{\gamma}(E_{\gamma},r,t)$ will increase in about 1.45 times if we take into account also α- particles and heavier particles in galactic CR:

$$F^{\gamma}\left(E_{\gamma},r,t\right)=1.45F_{pH}^{\gamma}\left(E_{\gamma}\right)\frac{n_1(\theta,t)}{n_o}\left(1-\frac{br}{r_o}\right)^{-1}\left(\frac{n_1}{r}\right)^2\exp\left(-A(E_{\gamma},t)\left(1-\frac{r}{r_o}\right)\right).$$ (22)

5. Expected Angle Distribution of Gamma Ray Fluxes From Solar Wind

5.1. GENERAL CASE

Let us assume that the observer is inside the Heliosphere, on the distance $r_{obs}\leq r_o$ from the Sun and helio-latitude θ_{obs}. The sight line of observation we can determine by the angle θ_{sl}, computed from the equatorial plane from anti-Sun direction to the North. In this case the expected angle distribution and time variations of gamma ray fluxes for local observer from interaction of galactic CR with solar wind matter will be

$$\Phi^\gamma\left(E_\gamma,r_{obs},\theta_{sl},t\right)=1.45\times\int_0^{L_{max}(\theta_{sl})}dL\ F_{pH}^\gamma\left(E_\gamma,L(r_{obs},\theta_{sl}),t\right). \tag{23}$$

In Eq. (23) $F_{pH}^\gamma\left(E_\gamma,L(r_{obs},\theta_{sl}),t\right)=F_{pH}^\gamma\left(E_\gamma,r,\theta,t\right)$ determined by Eq. (18), and

$$r=r(L)=\left(r_{obs}^2+L^2+2r_{obs}L\Delta\theta\right)^{1/2},\Delta\theta=\theta_{sl}-\theta_{obs},\ \theta=\theta(L)=$$
$$\theta_{obs}+\arccos\left(\left(r_{obs}^2+r_{obs}L\Delta\theta\right)\Big/\left(r_{obs}\left(r_{obs}^2+L^2+2r_{obs}L\Delta\theta\right)^{1/2}\right)\right), \tag{24}$$

$$L_{max}\left(\theta_{sl}\right)=r_o\sin\left[\Delta\theta-\arcsin\left(\frac{r_{obs}}{r_o}\sin\Delta\theta\right)\right]\Big/\sin\Delta\theta. \tag{25}$$

5.2. SPHERICALLY-SYMMETRICAL CASE

For the spherically-symmetrical problem for observation on the distance r_{obs} in direction determined by the angle φ between direction of observation and direction to the Sun, on the basis of Eq. (23)-(25) we obtain

$$\Phi^\gamma\left(E_\gamma,r_{obs},\varphi,t\right)=1.45F_{pH}^\gamma\left(E_\gamma\right)\left(n_1(t)/n_o\right)G\left(r_{obs},\varphi,b,A(E_\gamma,t)\right), \tag{26}$$

where $A(E_\gamma,t)$ is determined by (21) and

$$G=r_1^2\left(r_{obs}\sin\varphi\right)^{-1}\int_{\theta_{min}}^{\theta_{max}}\left(1-\frac{kr_{obs}\sin\varphi}{r_o\cos\theta}\right)^{-1}\exp\left(-A(E_\gamma,t)\left(1-\frac{kr_{obs}\sin\varphi}{r_o\cos\theta}\right)\right)d\theta, \tag{27}$$

where

$$\theta_{max}=\arccos(r_{obs}\sin\varphi/r_o),\ \theta_{min}=-(\pi/2)+\varphi. \tag{28}$$

We calculate (27) numerically for $r_{obs}=r_1=1\,AU=1.5\times10^{13}\,cm$, b=0.13, 0.30, 0.45, φ=2, 10, 45, 90, and 178° and $A(E_\gamma,t)=0$ (no modulation), and 0.2, 0.4, 0.8, 1.6, 3.2, 6.4 and 12.8. The case $E_\gamma\geq100\,MeV$ corresponds to $R^\gamma\beta\geq2\,GV$ what means $A(E_\gamma,t)\leq0.2$ in minimum of solar activity and $A(E_\gamma,t)\leq1.2$ in maximum of solar activity). The dependence from b is sufficient only for $A(E_\gamma,t)\geq6.4$; in other cases G is about the same for b=0.13, 0.30, 0.45. In Table 1 we show values $G(r_{obs},\varphi,k,A)$, sufficient for cases of generation gamma radiation with $E_\gamma\geq100\,MeV$ ($A\leq1.2$) and for $E_\gamma<100\,MeV$ (corresponds to $A>1.2$).

Table 1. Parameter $G(r_{obs}, \varphi, b, A)$ (in cm) according to Eq. (27) for $r_{obs} = 1$ AU, $b = 0.3$ in dependence of φ (in degrees) and of $A(E_\gamma, t)$ (characterized both effective E_γ and level of solar activity).

φ	A=0	A=0.2	A=0.4	A=0.8	A=1.6	A=3.2
2°	1.3E+15	1.1E+15	9.0E+14	6.0E+14	2.7E+14	5.5E+13
10°	2.6E+14	2.1E+14	1.7E+14	1.2E+14	5.3E+13	1.1E+13
45°	5.0E+13	4.1E+13	3.4E+13	2.3E+13	1.1E+13	2.3E+12
90°	2.4E+13	2.0E+13	1.6E+13	1.1E+13	5.3E+12	1.2E+12
178°	1.5E+13	1.2E+13	1.0E+13	7.2E+12	3.5E+12	8.7E+11

6. Gamma Ray Fluxes from Stellar Winds

In this case $r_{obs} \gg r_0$, and the expected gamma ray fluxes will be

$$\Phi^\gamma(E_\gamma, r_{obs}, t) = 1.45 \times 2\pi \, r_{obs}^{-2} \int_{-\pi/2}^{\pi/2} \cos\theta d\theta \int_0^{r_0} r^2 F_{pH}^\gamma(E_\gamma, r, \theta, t) dr$$

$$\approx 2.9\pi \, r_{obs}^{-2} F_{pH}^\gamma(E_\gamma) \frac{n_1(t)}{n_o} r_1^2 r_o J(b, A(E_\gamma, t))$$

(29)

where

$$J(b, A) = \exp(-A) \times \left[\left(1 + \frac{A}{2} + \frac{A^2}{6} + \ldots\right) + b\left(\frac{1}{2} + \frac{A}{3} + \ldots\right) + b^2\left(\frac{1}{3} + \frac{A}{4} + \ldots\right) + \ldots \right]. \quad (30)$$

Values of $J(b, A)$ for $E_\gamma \geq 100$ MeV are shown in Table 2.

Table 2. Parameter $J(b, A)$ in dependence of b and A.

b	A=0	A=0.1	A=0.2	A=0.3	A=0.4	A=0.6	A=0.8
0.13	1.07	1.02	0.97	0.93	0.88	0.80	0.73
0.30	1.18	1.13	1.08	1.03	0.98	0.89	0.81
0.45	1.29	1.24	1.18	1.13	1.08	0.98	0.90

7. DISCUSSION AND CONCLUSIONS

Obtained results allowed estimate expected distribution of GR emissivity in the Heliosphere or in the some stellar-sphere, to estimate expected fluxes of GR and their time variations for observations of GR from solar wind or from nearest stellar winds. According to Eq. (22) GR emissivity in the interplanetary space will be bigger than in the interstellar space only in the inner part of the Heliosphere at $r < 3$ AU and with decreasing of r GR emissivity will increase about $\propto r^{-2}$. In the main part of Heliosphere GR emissivity will be many times smaller than in the interstellar space. It means that the Heliosphere as well as many stellar-spheres can be considered as holes in the galactic background GR emissivity distribution.

According to Eq. (26) and Table 1 the biggest expected GR flux from solar wind

($n_1 \approx 5 cm^{-3}$) in direction 2° from the Sun near minimum of solar activity will be about the same as from interstellar matter ($n_o \approx 0.1 cm^{-3}$) with dimension $\approx 10^{17}$ cm (on about two order more than the dimension of Heliosphere). This expected GR flux decreases with time several times in the maximum of solar activity and decreases in about two order with increasing angle φ up to the opposite direction from the Sun (see Table 2).

Let us compare this variable GR flux with background GR flux from galactic CR interactions with interstellar matter. In direction perpendicular to the disc plane background GR flux is formatted on the distance about 200 $pc \approx 6 \times 10^{20} cm$, therefore this background GR flux will be about 6×10^3 times more than the biggest expected from solar wind GR flux in direction 2° from the Sun. From this follows that now it is not possible to measure GR from solar wind generated by galactic CR modulated by solar activity cycle. But may be in future with increasing of accuracy of GR telescopes and by using the big variability of this very weak GR source will be possible to investigate this phenomenon and obtain some additional information on solar wind matter distribution and galactic CR modulation in the Heliosphere.

If measurements of GR from solar wind generated by galactic CR will be made outside the Heliosphere, that can be observed following effect: in directions not far from the Sun this GR flux will be about two order bigger than from interstellar matter of the same dimension as Heliosphere, but in measurements at big angles from the Sun will be observed GR flux much smaller than expected from interstellar matter of the same dimension. According to Eq. (29) the ratio of total GR flux from solar wind or from stellar wind Φ^γ_{SW} to the GR flux from the same volume of interstellar medium Φ^γ_{IM} will be

$$\Phi^\gamma_{SW} / \Phi^\gamma_{IM} \approx 1.5 n_1 r_1^2 J(b, A(E_\gamma, t)) / (n_o r_o^2), \qquad (31)$$

what will be change in time according to Table 2 with solar or stellar cycle and depends from E_γ. Table 2 shows that $J(b, A)$ for the Heliosphere increases from 0.73 to 1.29 with decreasing $A(E_\gamma, t)$ from 0.8 to 0 and with increasing b from 0.13 to 0.45, that for the rough estimations we can put $J(b, A) \approx 1$. In this case Eq. (31) for solar wind ($r_1 = 1 AU, r_o \approx 100 AU, n_1 \approx 5 cm^{-3}, n_o \approx 0.1 cm^{-3}$) gives $\approx 7.5 \times 10^{-3}$. It means that Heliosphere can be considered as a deep variable hole in background GR emissivity distribution. The value of ratio (31), stellar wind density n_{1St} on the distance 1 AU from the star and dimension of hole r_{oSt} are determined mainly by the value of mass loosing by star $\dot{M}_{St}$ (for the Sun $\dot{M}_{Sun} \approx -10^{-14} M_{Sun}/year$) and speed of wind u_{1St} on the distance 1 AU from the star:

$$n_{1St} = n_1 u_1 \dot{M}_{St} / (u_{1St} \dot{M}_{Sun}), \ r_{oSt} = r_o (u_{1St} \dot{M}_{St} / u_1 \dot{M}_{Sun})^{1/2}, \Phi^\gamma_{StW} / \Phi^\gamma_{SW} = u_1^2 / u_{1St}^2 \quad (32)$$

Here we assume that conditions around the star (CR and magnetic field pressure) are the same as near the Sun. Eq. (32) shows that dimension of hole in GR emissivity increases with increasing of stellar wind and speed of mass loosing in degree 1/2, and deeps of GR hole increases with stellar wind velocity in square. For example, for star with $\dot{M}_{St} \approx 10^{-8} M_{Sun}/year$ and $u_{1St} \approx 10^8 \, cm/sec$, we obtain $n_{1St} \approx 2 \times 10^6 \, cm^{-3}$, $r_{oSt} \approx 1.5 \times 10^5$ AU and $\Phi^\gamma_{StW}/\Phi^\gamma_{SW} \approx 1/6$. Let us note that in this case observation along line near the star (see Table 1) will give GR flux corresponded to background GR from about $3 \times 10^{22} \, cm$ interstellar medium what is much bigger than background GR in all directions from the galactic disc; this variable GR flux can be measured by present GR telescopes (if the modulation is not very big).

8. ACKNOWLEDGMENTS

I would like to thank Prof. Maurice Shapiro and Prof. John Wefel for the possibility to attend the school of cosmic ray astrophysics, for very interesting discussions and fruitful comments, and the staff of Ettore Majorana Center for very kind hospitality. This research was supported by Israel Cosmic Ray Center and Emilio Segre' Observatory, affiliated to Tel Aviv University, Technion, and Israel Space Agency.

9. REFERENCES

1. Dorman, L.I. (1996) Cosmic ray nonlinear processes in gamma-ray sources, *Astronomy and Astrophysics*, Suppl. Ser., **120**, No. 4, 427-435.
2. Dorman, L.I. (1997) Angle distribution and time variation of gamma ray flux from solar and stellar winds, 2. Generation by galactic cosmic rays, in C.D. Dermer, M.S. Strickman, and J.D. Kurfess (eds.), *Proc. 4th Compton Symposium*, AIP Conference Proceedings 410, Williamsburg, VA, Part 2, 1183-1187.
3. Berezinskii, V.S., Bulanov, S.V., Ginzburg, V.L., Dogiel, V.A., and Ptuskin, V.S. (1990) *Cosmic Ray Astrophysics*, Fyzmatgiz, Moscow.
4. Dorman, L.I., Ptuskin, V.S. and Zirakashvili, V.N. (1990) Outer Heliosphere: pulsations, cosmic rays and stream kinetic instability, in S. Grzedzielski and D.E. Page (eds.) *Physics of the Outer Heliosphere*, Pergamon Press, pp. 205-209.
5. Zirakashvili, V.N., Dorman, L.I., Ptuskin, V.S. and Babayan, V.Kh. (1991) Cosmic ray nonlinear modulation in the outer Heliosphere. *Proc.22-th Intern. Cosmic Ray Conf.*, Dublin, Vol.3, pp 585-588.
6. Dorman, L.I. (1995) Cosmic ray nonlinear effects in space plasma, 2. Dynamic Heliosphere, in M.M. Shapiro, R. Silberberg and J.P.Wefel (eds.). *Currents in High Energy Astrophysics*, Kluwer Academic Publishers., Dordrecht/Boston /London, NATO ASI Serie, Vol. 458, pp. 193-208.
7. Le Roux, J.A. & Fichtner, H., (1997) The influence of pickup, anomalous, and galactic cosmic ray protons on the structure of the heliospheric shock: a self consistent approach, *Ap. J*, **477**, L115-L118.
8. Stecker, F.W. (1971) *Cosmic Gamma Rays*, Mono Book Co, Baltimore.
9. Dermer, C.D. (1986) Secondary production of neutrl pi-mesons and the diffuse galactic gamma radiation, *A&A*, **157**, No. 2, 223-229.

10. Dermer, C.D. (1986) Binary collision rates of relativistic thermal plasmas, II-Spectra, *Ap. J,* **307**, 47-59.
11. Dorman, L.I. (1957) *Cosmic Ray Variations,* Gostekhtheorizdat, Moscow (English translation published in 1958 by Research Division of Ohio Air-Force Base, USA).
12. Dorman, L.I. (1963) *Cosmic Ray Variations and Space Research,* Nauka, Moscow.
13. Dorman L.I. (1963) *Astrophysical and Geophysical Aspects of Cosmic Rays,* in J.G. Wilson and S.A.Wouthuysen (eds.) *Progress of Cosmic Ray and Elementary Particle Physics,* North-Holland Publ. Co., Amsterdam, Vol. 7.
14. Dorman, L.I. (1974) *Cosmic Rays: Variations and Space Exploration.* North-Holland Publ. Co., Amsterdam.
15. Dorman, L.I. (1975) *Variations of Galactic Cosmic Rays,* Moscow University Press, Moscow
16. Parker, E.N. (1963) *Interplanetary Dynamically Processes,* Intersci. Publ., New York-London.
17. Dorman, L.I. (1959) To the theory of cosmic ray modulation by solar wind, *Proc. of 6-th Intern. Cosmic Ray Conf.,* Moscow, Vol. 4, pp. 328-334.
18. Simpson J.A., 1983, Ann. Rev. Nucl. Particle Physics, **33**, 323.
19. Dorman, L.I. (1975) *Experimental and Theoretical Foundations of Cosmic Ray Astrophysics,* Fizmatgiz, Moscow.
20. Belov, A.V., Gushchina, R.T., Dorman, L.I., and Sirotina, I.V. (1988) Rigidity dependence of cosmic ray modulation parameter in the different epoch of solar activity cycle, *Izvestia Ac. of Sci. of USSR,* Ser. Phys., **52**, N 12, pp. 2334-2337.
21. Belov, A.V., Gushchina, R.T., Dorman, L.I., and Sirotina, I.V. (1990) Rigidity spectrum of cosmic ray modulation, *Proc. 21-th Intern. Cosmic Ray Conf.,* Adelaide, Vol. 6, pp. 52-55.
22. Dorman, I.V., and Dorman, L.I., (1967) Solar wind properties obtained from the study of the 11-year cosmic ray cycle, *J. Geophys. Res.* 72, 1513-1520.
23. Dorman, I.V. & Dorman, L.I. (1967) Propagation of energetic particles through interplanetary space according to the data of 11-year cosmic ray variations. *J. Atmosph. and Terr. Phys.,* **29**, No.4, 429-449.
24. Dorman, I.V. & Dorman, L.I. (1968) Hysteresis phenomena in cosmic rays, properties of solar wind and energetic spectrum of different nuclei in the Galaxy, *Proc. 5-th All-Union Winter School on Cosmophysics,* Apatity, pp.183-196.
25. Alania, M.V., and Dorman, L.I. (1981) *Cosmic Ray Distribution in the Interplanetary Space.* Metsniereba, Tbilisi.
26. Alania, M.V., Dorman, L.I., Aslamazashvili, R.G., Gushchina, R.T., and Dzhapiashvili, T.V. (1987) *Galactic Cosmic Ray Modulation by Solar Wind.* Metsniereba, Tbilisi.
27. Dorman, L.I., Villoresi, G., Dorman, I.V., Iucci, N., and Parisi, M. (1997) High rigidity CR-SA hysteresis phenomenon and dimension of modulation region in the Heliosphere in dependence of particle rigidity, *Proc. of 25- th Intern. Cosmic Ray Conference,* Durban (South Africa), Vol. 2, pp 69-72.

28. Dorman, L.I., Villoresi, G., Dorman, I.V., Iucci, N., and Parisi, M. (1997) Low rigidity CR-SA hysteresis phenomenon and average dimension of the modulation region and Heliosphere, *Proc. of 25- th Intern. Cosmic Ray Conference,* Durban (South Africa), Vol. 2, pp 73-76.

29. Dorman, L.I. (1975) *Variations of Galactic Cosmic Rays*, Moscow Univ. Press, Moscow

30. Zusmanovich, A.G., (1986) *Galactic Cosmic Rays in the Interplanetary Space*, Nauka, Alma-Ata

31. Dorman, L.I. (2001) Cosmic ray long-term variation: even-odd cycle effect, role of drifts, and the onset of cycle 23, *Advances in Space Research* , D1.1-0037.

32. Dorman, L.I., Dorman, I.V., Iucci, N., Parisi, M., and Villoresi, G. (2001) Solar activity-cosmic rays hysteresis phenomenon during cycle 22 and determination of drift effects role and modulation region dimension, *Advances in Space Research* , Paper D1.1-0038.

33. Burger, R.A., and Potgieter, M.S. (1999) The effect of large heliospheric current sheet tilt angles in numerical modulation models: a theoretical assessment, *Proc. of 26-th Intern. Cosmic Ray Conference* , Salt Lake City, Vol. 7, pp. 13-16.

COMETS AS X-RAY OBJECTS

S. IBADOV
Institute of Astrophysics
Dushanbe 734042, Tajikistan

The discovery of soft X–ray radiation from comet Hyakutake by space telescope ROSAT in March 1996 opened a new field of research. The measured X–ray luminosities of comets significantly exceed the value predicted. Possibility of production of the observable X–rays in comets due to generation of high–temperature plasma blobs from high–velocity collisions between cometary and interplanetary dust particles is considered.

1. Introduction

Rapid progress of X–ray astronomy in the last fifty years including the discovery of solar X–rays in 1948–1949 (integral flux of the order of 0.1 erg cm^{-2} s^{-1} at a heliocentric distance R = 1 AU), the discovery of the first extrasolar X–ray object SCO X–1 in 1962 at the flux level 10^{-7} erg cm^{-2} s^{-1} and extremely high detecting possibilities of modern space telescopes like ROSAT, the sensitivity threshold 10^{-14} erg cm^{-2} s^{-1} for 0.1–2 keV soft X–rays [1], strongly stimulated also searches for X–ray generating processes in comets [2, 3, 4].

Soft cometary X–ray emission was discovered first in comet Hyakutake (C/1996 B2) with ROSAT in 27 March 1996 [5] and the results of a theoretical approach [4] served as a motive for that observations [6]. It is now well established that comets emit X–rays regularly and, moreover, the measured X–ray luminosities of comets significantly exceed the value predicted [7–11].

In order to explain this phenomenon different theoretical models were proposed [12–20]. However, the problem of basic mechanisms for generation of cometary X–rays remains open, so that emission differences due to various factors (cometary dynamical class, anisotropy in distributions of cometary and interplanetary dust particles, size distributions of the dust particles, bandpass of photon detectors, etc.) have yet to be understood and corresponding theoretical models for production of energetic photons in comets should be developed and improved (cf. [11]).

M. M. Shapiro et al. (eds.), Astrophysical Sources of High Energy Particles and Radiation, 245–250.
© 2001 *Kluwer Academic Publishers. Printed in the Netherlands.*

Meanwhile, the expected X–ray luminosity of comets due to generation of hot plasma blobs from high–velocity collisions between cometary and interplanetary dust particles calculated [4] corresponds only to radiation from the collisionally thick zone of the cometary coma. Moreover, the calculations have been made without taking into account the presence in the comae of comets a dense cloud of very small dust particles (VSDPs) detected in the coma of comet Halley 1986 III on the basis of the in situ measurements [21].

In the present paper the possibility of production of observable X–rays in dusty comets due to generation of high–temperature plasma in collisions between cometary and interplanetary dust particles is considered.

2. Production of High Energy Photons in Dusty Comets

Interaction between cometary gas–dust comas and interplanetary dust particles includes two principally different mechanisms: meteor–like mechanism connected with irradiation of interplanetary dust particles by cometary molecules and explosion–type mechanism due to collisions between cometary and interplanetary dust particles. Dusty comets like comet Halley 1986 III are characterized by the dust to gas production rate ratio more than 0.1 [22, 23]. In such comets the main interaction mechanism between the cometary gas–dust coma and interplanetary dust is an explosion–type mechanism resulting in production of high–temperature expanding plasma blobs and high energy photons [3].

2.1. ENERGY OF PHOTONS

The most probable energy of photons emitted from the plasma blobs, E_m, is determined by the initial temperature of the blobs, T_o, namely

$$E_m = 2.8 k T_o = \frac{A m_p f_p V^2}{4\left(1 + z + 2 x_1 f_c / 3\right)},$$
(1)

where k is the Boltzmann constant, A is the mean mass number of atoms in colliding dust particles, m_p is the proton mass, f_p is the efficiency of plasma blob production, V is the relative velocity of colliding particles, z iz the mean multiplicity of charge of the plasma ions, x_1 is the mean relative ionization energy, f_c is the coefficient for dissociation energy of dust particle molecules and excitation energy of the plasma ions [4].

The ecliptic concentrated main component of interplanetary dust cloud, as known (see, e.g., [24]), consists of dust particles on quasi–circular orbits and prograde motion with Keplerian orbital velocity $V_m = 30 R^{-1/2}$. Orbital velocity of comets near inner heliosphere may be approximated by a parabolic velocity $V_c = 42 R^{-1/2}$ km/s. Hence, the relative velocity between colliding cometary and interplanetary dust particles may be presented as

$$V = V_1 R^{-1/2} .$$
(2)

Here V_1 is the value of V at R=1 AU, i.e. V_1 = 20, 50 and 70 km/s for comets with near–ecliptic, near–polar and retrograde orbits, respectively.

Accepting A=20, f_p=0.8, z=1, x_1=1 and f_c=1.3 according to (1) and (2) we get E_m(R=1AU) = 5, 30 and 60 eV for the three characteristic types of comets.

Thus, in the inner heliosphere, R ≤ 1 AU, dusty comets with retrograde and near–polar orbits like comet Halley (inclination i=162.2⁰) and comet Hyakutake (i=124.9⁰) become sources of extreme ultraviolet (EUV) and soft X–rays due to collisions between cometary and interplanetary dust particles.

2.2. EUV/X–RAY LUMINOSITY OF COMETS DUE TO DUST–DUST COLLISIONS

The EUV/X–ray luminosity of comets due to generation of hot plasma blobs in collisions of cometary and interplanetary dust particles may be presented as the sum of the luminosities of the collisionally thick and collisionally thin zones of the cometary dust coma, i.e. Lx = Ls + Lv, where

$$L_s = k_x f_r \frac{\rho V^3}{4} \left(\pi r_x^2 \right),$$ (3)

$$L_v = k_x f_r \rho V^3 r_a r_x \ln \frac{r_a}{r_x} = L_s \frac{4 r_a}{\pi r_x} \ln \frac{r_a}{r_x},$$ (4)

$$f_r = \frac{15 y_1^3}{\pi^4 e^{y_1}} \left(1 - \frac{3}{y_1} + \frac{6}{y_1^2} - \frac{6}{y_1^3} \right),$$ (5)

k_x is the efficiency of conversion of the kinetic energy of colliding dust particles into the radiation energy [4], f_r is the photon detector regime/bandpass factor, $y_1=E_o/kT_o$, E_o is the lower limit of the photon detector bandpass, ρ is the spatial mass density of interplanetary dust [24], r_a is the aperture radius of the photon detector, r_x is the radius of the collisionally thick zone of the cometary dust coma, i.e. the cometocentric distance wihin which the flux of interplanetary dust grains penetrating into the cometary coma fully transforms into plasma blobs due to collisions with cometary dust particles; $f_r(E_o = 60$ eV, R=1 AU) = 5.4 x 10^{-11}, 6.9 x 10^{-2} and 0.1 for V = 20, 50 and 70 km/s , respectively.

The value of r_x depends on the value of the differential size index, p, in the power law for the size spectrum of cometary dust particles:

$$r_{x1} = \frac{3 M_d \ln\left(a_2 / a_1 \right)}{4 d V_d a_2} \qquad \text{for } p = 3,$$ (6)

$$r_{x2} = \frac{3M_d}{4dV_d a_1 \ln\left(a_2 / a_1\right)} \qquad \text{for } p = 4, \qquad (7)$$

$$r_{x3} = \frac{3(p-4)M_d}{4(p-3)dV_d a_1} \qquad \text{for } p > 4, \qquad (8)$$

where M_d is the dust production rate of the cometary nucleus, d and V_d are the density and outflowing velocity of dust particles, a_1 and a_2 are the minimum and maximum radii of dust particles emitted from the cometary nucleus; high values of the differential size index, p=5–7, were detected in situ by Giotto for particles from very active parts of the nucleus of comet Halley ([23] p.341).

Of the wide range of heliocentric distances of comets, $R \geq 0.01$ AU, only the outer region, R=1–3 AU, is covered at present by X–ray observations (see, e.g., [11]). Particularly, for comet Hyakutake at R = 1 AU with r_a= 1.2 x 10^5 km the measured luminosity is $L_{x,m}$ ($E_o \geq 60$ eV) = 4 x 10^{15} erg s^{-1} [5].

For Hyakutake–like comets according to (1)–(8) with R=1 AU, V = 5 x 10^6 cm s^{-1}, k_x = 0.03, f_r(E_o = 60 eV)= 6.9 x 10^{-2}, ρ = 5 x 10^{-22} g cm^{-3} , r_a = 1.2 x 10^{10} cm, M_d = 10^7 g s^{-1}, d = 0.5 g cm^{-3} , V_d = 5x10^4 cm s^{-1}, a_1 = 10^{-6} cm, a_2 = 0.1 cm we have L_x ($E_o \geq 60$ eV) = 6.8 x 10^{11}, 2.0 x 10^{14} and 9.7 x 10^{14} erg s^{-1} for p = 3, 4 and 5, respectively.

At the same time the data [16] for VSDPs M_d = 10^8 g s^{-1}, m_1 = 4 x 10^{-19} g, m_2 = 10^{-17} g, a_1 = 5.7 x 10^{-7} cm, a_2 = 1.7 x 10^{-6} cm, d=1 g cm^{-3} and p=4 result in L_x ($E_o \geq 60$ eV) = 6 x 10^{15} erg s^{-1}.

Hence, encounters of dusty comets with high–velocity interplanetary meteoroid swarms providing V > 50–70 km s^{-1} and ρ = 10^{-19} –10^{-21} g cm^{-3} [25] can produce observable enhancements of the soft component of cometary X–ray emission. These variations may be considered as an addition to cometary X–ray radiation due to charge exchange of solar wind multicharge ions on cometary molecules [15, 19].

3. Conclusions

High–velocity comets like comet Halley and comet Hyakutake having retrograde orbital motion and dense cloud of very small dust particles (VSDPs) in their comas become effective sources of soft X–ray radiation due to high–velocity collisions between cometary and interplanetary dust particles at small heliocentric distances, $R \leq 1$ AU. Encounters of such comets with dense fluxes of interplanetary meteoroids will produce enhancements of the soft component of the cometary X–ray emission.

X–ray observations of high–velocity dusty comets in the inner heliosphere are also of interest for studying high–temperature plasma generation by high–velocity, V > 70–100 km/s, dust–dust collisions.

Acknowledgements

The author is grateful to Prof. John P. Wefel for invitation to the NATO Advanced Study Institute "Astrophysical Sources of High Energy Particles and Radiation" (Erice, Sicily, Italy, 11–21 November 2000) and Travel Grant, to Staff of the Ettore Majorana Center for hospitality, to Prof. J. Trumper for stimulating interest, to Dr. K. Dennerl and F.S. Ibodov for cooperation.

References

1. Trumper, J. (1980) The Rontgen Satellite (ROSAT), Preprint MPI f. Phys. Astrophys., Munchen.
2. Ibadov, S. (1985) On two mechanisms of X-ray generation in comets, in R.H. Giese and P. Lamy (eds.), Properties and Interactions of Interplanetary Dust, Reidel Publishing Company,Dordrecht, pp. 365–368.
3. Ibadov, S. (1987) Interaction mechanisms of comets with the Zodiacal dust cloud, ESA SP–278, 655–656.
4. Ibadov, S. (1990) On the efficiency of X-ray generation in impacts of cometary and zodiacal dust particles, Icarus 86, 283–288.
5. Lisse, C.M., Dennerl, K., Englhauser J., Harden, M., Marshall, F.E., Mumma, M.J., Petre, R., Pye, J.P., Ricketts, M.J., Schmitt, J., Trumper, J., and West, R.G. (1996) Discovery of X-ray and EUV emission from comet Hyakutake C/1996 B2, Science 274, 205–209.
6. Dennerl. K., Lisse, C.M., and Trumper, J. (1998) Private communications to S. Ibadov.
7. Dennerl, K., Englhauser, J., and Trumper, J. (1997) X-ray emission from comets detected in the Rontgen X-ray satellite all-sky survey, Science 277, 1625–1629.
8. Mumma, M.J., Krasnopolsky, V.A., and Abbott, M.J. (1997) Soft X-rays from four comets observed by EUVE, Astrophys. J. 491, L125–L128.
9. Krasnopolsky, V.A., Mumma, M.J., Abbott, M., et al. (1997) Discovery of soft X-rays from Comet Hale-Bopp using EUVE, Science 277, 1488–1491.
10. Owens, A., Parmar, A.N., Oostrbroek, T., et al. (1998) Evidence for dust- related X-ray emission from Comet C/1995 O1 (Hale-Bopp), Astrophys. J. 493, 47–51.
11. Lisse, C.M., Christian, D., Dennerl, K., et al. (1999) X-ray and EUV emission from comet P/Encke 1997, Icarus 141, 316–330.
12. Brandt, J.C., Lisse, C.M., and Yi, Y. (1996) Small current sheets in the solar wind as the cause of X-ray emission in comets, Bull. Amer. Astron. Soc. 29, 728.
13. Wickramasinghe, N.C., and Hoyle, F. (1996) Very small dust particles (VSDPs) in comet C/1996 B2 (Hyakutake), Astrophys. Space Sci. 239, 121–123.
14. Bingham, R., Dawson, J.M., Shapiro, V.D., Mendis, D.A., and Kellet, B.J. (1997) Generation of X-rays from comet Hyakutake C/1996 B2, Science 275, 49–51.
15. Cravens, T.E. (1997) Comet Hyakutake X-ray source: Charge transfer of solar wind heavy ions, Geophys. Res. Lett. 24, 105–109.
16. Haberli, R.M., Gombosi, T.I., deZeeuw, D.L., Combi, M.R., and Powell, K.G. (1997) Modeling of cometary X-rays caused by solar wind minor ions, Science 276, 939–942.
17. Ip, W.-H., and Chow, V.W. (1997) On hypervelocity impact phenomena of microdust and nano X-ray flares in cometary comae, Icarus 130, 217–221.
18. Krasnopolsky, V. (1997) On the nature of soft X-ray radiation in comets, Icarus 128, 368–385.
19. Wegmann, R., Schmidt H.U., Lisse, C.M., Dennerl, K., and Englhauser, J. (1998) X-rays from comets generated by energetic solar wind particles, Planet. Space Sci. 46, 603–612.
20. Uchida, M., Morikawa, M., Kubotani, H., and Mouri, H. (1998) X-ray spectra of comets, Astrophys. J. 498, 863–870.
21. Utterback, N.G., and Kissel, J. (1990) Attogram dust cloud a million kilometers from comet Halley, Astron. J. 100, 1315–1322.

22. Sagdeev, R.Z., Blamont, J., Galeev, A.A., et al. (1986) Vega spacecraft encounters with comet Halley, Nature 321, 259–262.

23. McDonnel, J.A.M., Alexander, W.M., Burton, W.M., et al. (1986) Dust density and mass distribution near comet Halley from Giotto observations, Nature 321, 338–341.

24. Mann, I., and Grun, E. (1996) Dust studies on a solar probe, Adv. Space Sci. 17, (3)99–(3)102.

25. Lebedinets, V.N. (1984) Visibility of meteoroid swarms on the background of zodiacal light, Solar System Res. 18, 23–29.

COSMIC RAY ORIGIN: GENERAL OVERVIEW

V.S. PTUSKIN
Institute for Terrestrial Magnetism, Ionosphere and Radio Wave Propagation of the Russian Academy of Sciences (IZMIRAN), Troitsk, Moscow region 1421190, Russia

A summary of current theoretical results on the acceleration and propagation of cosmic rays in the Galaxy is presented. These topics will be discussed in greater detail in other papers in this volume.

1. Introduction

Cosmic rays present a remarkable example of naturally generated relativistic particles detected over an enormous energy range from about 10^7 eV up to more than 10^{20} eV. They are associated with the most energetic events and active objects in the Universe: supernova bursts, pulsars, relativistic jets, active galactic nuclei etc. To a first approximation, the energy spectrum of cosmic rays can be described by a power law on more than 10 decades on particle energy, so that the dependence of cosmic ray intensity on particle energy is close to $I \propto E^{-2.7}$. There is a turning down of the observed spectrum at energies less than about 10 GeV that is mainly due to the solar wind action. Closer examination reveals some additional structure in the galactic cosmic ray spectrum that includes the flattening of the interstellar spectrum at energies below 2×10^9 eV/nucleon, the knee at 3×10^{15} eV, and the ankle at 10^{19} eV. This structure is schematically presented in Figure 1. Below we discuss possible interpretation of cosmic ray observations. The extensive reviews of theoretical and experimental results on cosmic ray astrophysics were given in [1,2].

2. Supernova Remnants as Sources of Galactic Cosmic Rays

If we use the energy requirements and the nonthermal radiation as a guideline, then the most powerful accelerators in the Galaxy should be supernovae and supernova remnants, pulsars and neutron stars in close binary systems. The balance condition for cosmic ray energy density in the Galaxy gives estimate of the total power of the sources $Q = wV/T = wvM_g/X = 5 \times 10^{40}$ erg/s, where $w = 3 \times 10^{-12}$ erg/cm^3 is the energy density of cosmic rays; $v \approx c$ is the particle velocity; V is the

M. M. Shapiro et al. (eds.), Astrophysical Sources of High Energy Particles and Radiation, 251–262.
© *2001 Kluwer Academic Publishers. Printed in the Netherlands.*

volume of the Galaxy occupied by cosmic rays; T is the leakage time of cosmic rays from the Galaxy; $M = 5 \times 10^{42}$ g is the total mass of the interstellar gas in the Galaxy; $X = v\rho T \approx 10$ g/cm^2 is the mean matter thickness traversed by cosmic rays in course of their leakage out of the Galaxy, see also below (ρ is the mean gas density in the volume V). For the acceleration by supernovae, this estimate implies the release of energy in the form of cosmic rays approximately 10^{50} erg per supernova if the supernovae rate in the Galaxy is $1/(30$ years$)$. This value comes to about 10 % of the kinetic energy of the ejecta which does not contradict the prediction of the theory of diffusive shock acceleration, see e.g. [3]. The acceleration of cosmic rays is produced by the shock, which results from the supernova explosion and propagates in the interstellar medium or in the wind of the progenitor star.

The rotational energy of a young pulsar with period P which is left behind the supernova is estimated as $2 \times 10^{50}(P/10$ ms$)^{-2}$ erg. This is an additional energy reservoir for particle acceleration.

The data on the nonthermal radioemission of supernova remnants reinforce the above energy estimates. The nonthermal radioemission is due to the synchrotron radiation of relativistic electrons moving in magnetic field. Assuming the equipartition between cosmic ray and magnetic field energy densities and with supposition of electron to proton ratio equals to 10^{-2}, one can find the total energy of relativistic electrons 2×10^{48} erg in the remnant Cas A and $5 \times 10^{48} - 10^{49}$ erg (150 MeV - 30 GeV; 2×10^{-5} G) in the remnants IC 443 and Cygnus Loop, see [4]. The shell-type supernova remnants exhibit a broad range of spectral indices, centered roughly on $\alpha \sim 0.5$. This can be related to the electron power law index through the equation $\gamma = 2\alpha + 1 \sim 2.0$ but the range is from about $\gamma \sim 1.4$ to $\gamma \sim 2.6$. The interpretation of nonthermal radio emission from the external galaxy M33 [5] showed that supernova remnants are the sites of the acceleration of relativistic electrons with the same efficiency which is needed to provide the observed intensity of galactic cosmic-ray electrons.

Gamma-ray emission associated with a few bright supernova remnants has been found using the EGRET catalogue of gamma-ray sources at $E > 30$ MeV [6,7]. The gamma-ray fluxes from the two most prominent sources γ Cigni and IC443 indicate an energy of about 3×10^{49} erg for relativistic protons and nuclei confined in each envelope, assuming that gamma rays are generated through pp interactions and subsequent $\pi^0 \rightarrow 2\gamma$ decays.

The detection of non-thermal X-rays radiation from the supernova remnant SN1006 [8] and some other remnants (G347.3-0.5, IC443, and Cas A) is considered as evidence of the synchrotron emission of electrons with energies up to 10^{14} eV. The inverse Compton scattering of background photons by these electrons is also the most probable mechanism of the emission of TeV gamma rays detected from SN1006 [9]. There is yet no confirmation of the presence of very high energy protons and nuclei in supernova remnants that should manifest itself through gamma rays generated via π^0 channel. This could be a result of a relatively steep proton source spectrum ($\gamma \sim 2.3$) or low maximum energy of accelerated particles in studied supernova remnants. The models of radio, X-ray, and gamma ray emission from supernovae were developed in [10,11,12].

As was mentioned above, the diffusive shock acceleration at a supernova blast wave is the most probable mechanism of particle acceleration in supernova remnants. The acceleration of a fast particle diffusing near the shock front is a version of the first-order Fermi acceleration. It is due to repeated crossing of the shock front in a random particle walk and to an energy gain in the head on collisions with scattering centers embedded in the background plasma. The characteristic time of particle acceleration is $3(D_1/u_1 + D_2/u_2)/(u_1 - u_2)$, where u is the gas velocity and the subscripts $1, 2$ refer to the upstream and downstream regions of the gas flow separated by the standing shock front. The stationary spectrum of accelerated test particles in the downstream region has a power-law form $N(p)dp \propto p^{-(r+2)/(r-1)}dp$, where $r = u_1/u_2$ is the compression ratio in the shock. For strong shocks $r = 4$ and $N(p) \propto p^{-2}$. The efficiency of acceleration can be so high that the back reaction of the pressure of accelerated particles changes the shock structure and cosmic rays can not be considered as test particles. The modification of the shock results in the deviation of the particle energy spectrum from the shape typical for the shock with a step-function profile. However, the numerical simulation [13] showed that even in this case the time-average spectrum of energetic particles injected from the supernova remnant into the interstellar medium can be approximately described as a power law with $\gamma \sim 2.1$.

The maximum energy of accelerated particles is determined either by losses, or by the finite size and age of the shock wave. For the spherical shock front with a radius R, the acceleration occurs under the condition $g = Ru_1/D_1 > 1$ (a numerical factor is omitted here, and $D_1 > D_2$ is assumed). The diffusion coefficient usually increases with energy and the acceleration becomes not efficient at high enough energies when $g \lesssim 1$. The stream instability of accelerated particles creates the enhanced turbulence in the vicinity of the shock. This can give the diffusion coefficient close to the Bohm value $vr_g/3$ ($r_g = pc/ZeB$ is the particle Larmor radius) and provide the maximum particle energy up to 10^{14} eV/n for the interstellar magnetic field 5×10^{-6} G and the age of the remnant about 10^3 yrs.

Actually, for the core collapse supernovae (SNIb and SNII types), the blast wave propagates in the stellar wind of a presupernova star rather than in the interstellar medium. The estimated maximum energy of accelerated particles is about 10^{16} eV/n in the winds of the red giants and the Wolf-Rayet stars [14]. The maximum energy is gained at about 1 yr after explosion and does not change considerably for another few years. The maximum energy of accelerated particles may reach 3×10^{18} eV if the magnetic field on the surface of the Wolf-Rayet stars is as high as 3×10^3 G, i.e. ~ 100 times greater than usually accepted [15].

3. Cosmic Ray Propagation in the Galaxy

The interaction of relativistic charged particles with galactic magnetic fields provides high isotropy and relatively large confinement time of cosmic rays in the Galaxy. It is accepted that the diffusion approximation gives an adequate description of cosmic ray propagation in the Galaxy at energies up to about 10^{17} eV. The diffusion model provides a basis for the interpretation of cosmic ray data as well as related radio-astronomical, X-ray and gamma-ray observations [1,16].

The modeling of cosmic-ray diffusion in the Galaxy includes the solution of a transport equation with a given source distribution and boundary conditions for all cosmic-ray species. The transport equation describes diffusion, convection by the hypothetical galactic wind, and changes of energy (energy losses and possible distributed acceleration in the interstellar medium). In addition, nuclear collisions with interstellar gas atoms resulting in the production of secondary energetic particles should be taken into account when considering proton-nucleus component. Hundreds of stable and radioactive isotopes are included in the calculations of nuclear fragmentation and transformation of energetic nuclei in the course of their interaction with the interstellar gas.

The homogeneous leaky box model is a well-accepted approximation to the diffusion model. The exit of particles from the Galaxy is described in this simple model by some effective escape time T_{lb} and corresponding escape length X_{lb} which do not depend on position. The relation between parameters of the diffusion model with static halo and the leaky box model is given by the following approximate equation valid for an observer located in the galactic disk [17]:

$$X_{lb} = \mu \beta c H / (2D) \qquad (1)$$

where $\mu = 2.4 \times 10^{-3}$ g/cm^2 is the surface gas density of the galactic disk, $\beta = v/c$, H is the size of the galactic cosmic ray halo. It is assumed that the cosmic ray halo is flat, so that $R \gg H$, where $R \sim 20$ kpc is the radius of cylindrical halo. The value of H is defined as the distance from the galactic midplane to the halo boundary where cosmic rays freely escape from the Galaxy and their number density is negligibly small. The cosmic ray sources are distributed in a thin galactic disk. The equivalence of the diffusion and the leaky box models with the relation (1) between their parameters holds for not very heavy stable nuclei. The diffusion and the leaky box models are not equivalent for high-energy electrons with strong energy losses and for rapidly decaying radioactive isotopes which have very inhomogeneous spatial distributions in the Galaxy.

The data on abundance of stable secondary nuclei B and Sc+Ti+V in cosmic rays allows finding the value of the escape length. According to [18]

$$X_{lb} = 11.8\beta \frac{g}{cm^2}, R < 4.9 GV; \quad X_{lb} = 11.8\beta \, (R/4.9GV)^{-0.54} \frac{g}{cm^2}, R \geq 4.9GV.$$

$$(2)$$

Here $R = pc/Ze$ is the particle magnetic rigidity. This parametrization is valid for particles in the interstellar medium with energies from about 0.4 GeV/n to 300 GeV/n where data on secondary nuclei are available. The differential source spectrum found from the fit to the observed spectra of primary nuclei has a power law form on rigidity $Q \propto R^{-\gamma_s}$, $\gamma_s = 2.35$. The found escape length implies constant diffusion coefficient at rigidity $R < 4.9$ GV and gives rigidity-dependent diffusion at higher rigidities: $D = 2.0 \times 10^{28} (H/5 \text{ kpc})(R/1 \text{ GV})^{0.54}$ cm^2/s, $R \geq 4.9$ GV.

The information on diffusion coefficient can be also obtained from the data on radioactive secondary isotopes with β-decay. The isotopes ^{10}Be (the decay lifetime of an ion at rest is $\tau = 2.3 \times 10^6$ yr), ^{26}Al (1.3×10^6 yr), ^{36}Cl (4.3×10^5 yr), ^{54}Mn(9.1×10^5 yr) are usually employed. The content of these isotopes is determined by the conditions of cosmic ray propagation in the galactic vicinity $\sqrt{D\tau} \sim 300$ pc about an observer position. The surviving fraction of radioactive isotopes derived from the low-energy observations gives the value of the diffusion coefficient $\sim 3.4 \times 10^{28}$ cm^2/s at energy 0.4 GeV/n in the interstellar space and the size of cosmic ray halo $H \sim 4$ kpc, see [19]. This value of H is in agreement with the radioastronomical observations [20] which indicate the presence of a thick nonthermal galactic radio disk. The synchrotron radio emission of this disk is generated by the electron component of cosmic rays moving in galactic magnetic field.

The value of the diffusion coefficient found in the frameworks of the empirical diffusion model can be compared with the prediction of the theory of charged particle motion in random and regular galactic magnetic fields. The typical strength of magnetic field in the galactic disk is approximately $B \sim 5 \ \mu$G [21]. The random component of the field exceeds the regular (average) component. The last is predominantly azimuthal and has the value $B_0 \sim 2 - 3$ μG. The observed power spectrum of random field has a form $W(k)dk \propto k^{-2+a}dk$, $a \sim 0.2 - 0.6$ in a wide range of wave numbers $1/(10^{20}$ cm$) < k < 1/(10^8$ cm$)$. Note that $a = 1/3$ corresponds to Kolmogorov spectrum.

The kinetic theory of cosmic ray transport in the galactic magnetic fields is constructed similarly to the well-studied case of cosmic ray transport in the interplanetary magnetic fields, see e.g. [1,22]. A charged particle with Larmor radius r_g is mainly scattered by magnetic irregularities of the size $1/k \sim r_g$. This resonant scattering leads to the spatial diffusion of cosmic rays. For typical parameters of the interstellar magnetic field, the diffusion coefficient can be estimated as

$$D = \frac{1}{3}vrg\left(\frac{B_{tot}^2}{\delta B_{res}^2}\right) = 3 \times 10^{28}\,(R/1GV)^a\,cm^2/s,\, 0.2 \lesssim a \lesssim 0.6. \quad (3)$$

Here δB_{res} is the value of random field fluctuations at the resonant scale, B_{tot} is the total magnetic field (the inequality $\delta B_{res} \lesssim B_{tot}$ is assumed). The diffusion is anisotropic locally and goes predominantly along the magnetic field but the large scale wandering of magnetic field lines makes diffusion close to isotropic on scales $\gtrsim 100$ pc.

The estimate (3) is compatible with the high-energy asymptotics found from the empirical diffusion model. Nonetheless, the resonant scattering on random field with a power law spectrum does not provide an independent on energy diffusion at low energies that follows from the scaling $X_{lb} \propto \beta$ at $R < 4.9$ GV. It was suggested [23] that the large-scale convective motion of the interstellar medium may dominate at low energies. The convective transport of cosmic rays may have the form of the galactic wind with velocity $20 - 30$ km/s or the form of the turbulent diffusion with diffusion coefficient of the order of 3×10^{28} cm^2/s.

The alternative explanation is offered by the model with reacceleration of cosmic rays in the interstellar medium [24,25]. The distributed reacceleration of cosmic rays after their exit from the compact sources (supernova remnants) changes the shape of particle spectra. In particular, it leads to the increase of secondary/primary ratios with energy at $R < 4.9$ GV where reacceleration is relatively strong, and to the steep decrease of these ratios at $R > 4.9$ GV where the efficiency of reacceleration is suppressed by fast escape of particles from the Galaxy. This effect can mimic the turning down of the leaky box escape length (2) at $R < 4.9$ even if the diffusion coefficient is monotonic function of energy. In a minimal model, the stochastic reacceleration occurs as a result of scattering on the same random magnetohydrodynamic waves which are responsible for the spatial diffusion of cosmic rays. The rate of reacceleration is determined by the particle diffusion coefficient on momentum $D_{pp} \sim p^2 V_a^2 / D$, where $V_a \sim 30$ km/s is the Alfven velocity. As this reacceleration takes place, the secondary/primary ratios are reproduced with the diffusion coefficient equals to

$$D = 5.9 \times 10^{28} \beta \, (H/5kpc) \, (R/1GV)^{0.3} \, cm^2/s. \tag{4}$$

This equation is valid for all rigidities and corresponds to the scattering on random field with a Kolmogorov spectrum.

More weak dependence of diffusion on rigidity in the model with reacceleration, $D \propto R^{0.3}$, compared with the model without reacceleration, $D \propto R^{0.54}$, should manifest itself in a more weak energy dependence of secondary/primary ratios at $E \gg 20$ GeV/n where reacceleration is not essential. The currently available data do not allow yet distinguishing between the models.

As was mentioned above, the regular convection of cosmic-ray particles by the galactic wind may work simultaneously with diffusion in a process of cosmic ray transport in the Galaxy. The wind velocity does not greatly exceed 30 km/s at the height 1 kpc above the galactic plane but could reach a few hundreds km/s beyond few dozens kpc. Different wind models were considered in the context of cosmic-ray propagation [1,26,27]. In the most advanced version of this model [28,29] the cosmic-ray pressure drives the galactic wind. Cosmic-ray sources are distributed in the galactic disk and the streaming of relativistic particles along the spiral magnetic field out of the Galaxy generates Alfvenic turbulence sufficient to provide cosmic-ray diffusion. This mechanism of wave generation works at distances larger than about 1 kpc above the galactic plane since the ion-neutral collisions prevent the development of stream instability at smaller distances. The geometry of the region of cosmic-ray propagation and the value of cosmic-ray diffusion coefficient in the upper halo are not prescribed in this model but are self-consistently determined by the cosmic-ray dynamics. In particular, the scaling $D \propto \beta R^{\gamma_s - 1}$ is predicted in this model if the nonlinear Landau damping on thermal ions balances the cosmic ray stream instability.

The calculated wind velocity u is approximately linear function of distance z from the galactic disk, $u = wz$, $w =$const, at $z \lesssim 20$ kpc. Diffusion in this case is more important than convection for cosmic ray transport at distances $z < z_c$, $z_c = (D/w)^{1/2}$. Convection dominates at $z > z_c$. The critical distance z_c depends on particle energy through the dependence of the diffusion coefficient D on energy. (Note that the distance z_c is defined by the condition that the corresponding Peclet number is equal to unity: $u(z_c)z_c/D = 1$.) Very roughly, the wind model for an observer at the galactic disk can be approximated by the pure diffusion model with the effective size of the halo equal to z_c. In particular, H should be substituted by z_c in eq. (1).

The existence of wind in our Galaxy stands unproved. New cosmic ray data together with other astronomical observations can clarify the problem. For example, the energy dependence of surviving fraction of decaying secondary isotopes at $1 - 20$ GeV/n is sensitive to $D(E)$ and $z_c(E)$ dependence and might be used to verify the presence of the Galactic wind [30].

4. Ultra-High Energies

The cosmic rays spectrum becomes steeper by $\delta\gamma \sim 0.5$ at energy 3×10^{15} eV [31]. The "knee" is the main reliably established feature in high energy cosmic-ray spectrum. In principal, it may reflect either the break of the source spectrum or the more rapid leakage of ultra-high energy cosmic rays out of the Galaxy. The needed shape of the source spectrum could be produced in a two-stage model where individual supernova remnants accelerate particles up to the knee and subsequent energy gain is due to collective reacceleration on many shocks produced by other supernovae [32,33]. In principal, the spectrum with knee could be a superposition of sources of different nature with different spectra that implies the surprisingly fine tuning of fluxes from independent sources needed to build the pronounced knee in the total spectrum. Increased probability of escape from the Galaxy for very high energy particles may explain the knee in the observed cosmic ray spectrum even if the source spectrum is featureless. In particular, the knee might occur as a result of interplay between the diffusion of cosmic rays along magnetic field lines and the drift (Hall diffusion) perpendicular to the regular, predominantly azimuthal, galactic magnetic field [34]. The drift has more strong dependence on energy than diffusion and can dominate at high energies. This might make the cosmic ray spectrum steeper above the knee.

Most probably, the ultimate explanation of the knee will be obtained when the principal mechanism of cosmic ray acceleration up to ultra-high energies will be definitely established. An essential common feature of the processes of particle reacceleration and diffusion is their dependence on magnetic rigidity. So, one expects the increase of the abundance of heavy nuclei in cosmic rays as a function of total energy per particle when energy goes through the knee. This prediction is not reliably checked experimentally.

The origin of the highest energy cosmic rays is not established yet, see reviews [1,2,35]. The cosmic ray spectrum features a flattening at energies larger than $3 \times 10^{18} - 10^{19}$ eV. The maximum energy of detected events is 3×10^{20} eV. The elemental composition is heavy (may be dominated by iron nuclei) at $10^{17} - 10^{18}$ eV but is probably more light at $E > 10^{18}$ eV. The amplitude of cosmic ray anisotropy at 10^{18} eV is ~ 4 % with a broad cosmic-ray flow from the directions of Galactic Center and the Cygnus region. No significant large-scale anisotropy was found at higher energies, and the upper limit is about 30 % at 10^{19} eV. One may try to explain these data in terms of the two-component empirical model [36] where heavy galactic component turns down at $E > 3 \times 10^{18}$ eV. The extragalactic proton component has flat spectrum and dominates at $E > 3 \times 10^{18} - 10^{19}$ eV. An alternative interpretation assumes pure galactic origin for all particles. The changes in energy spectrum and elemental composition are explained by decreasing with magnetic rigidity probability to confine particles in large galactic magnetic corona, see [37].

The mere fact that observed cosmic rays (nuclei) have energies up to 3×10^{20} eV limits their age in the universal microwave background radiation to $T \lesssim 10^8$ yr because of photo-pion production and nuclear photo-disintegration. This implies that the distance to the sources of highest energy cosmic rays does not exceed 30 Mpc. These particles should come from the local Metagalactic vicinity, say from the Local Supercluster, or even have a Galactic origin. In the case of cosmic ray production homogeneously distributed on the Hubble scale $c/H_0 = 3 \times 10^3 h^{-1}$ Mpc, the particle spectrum is expected to have a characteristic cutoff at about 3×10^{19} eV, the so called Greisen-Zatsepin-Kuzmin cutoff. The data do not show the cutoff [38] but this may be not in contradiction with a universal extragalactic model of cosmic ray origin since the galaxy distribution in the Universe is not uniform and sometimes is described as a fractal [39]. The number of galaxies within a distance R from an observer scales as $N(< R) \propto R^D$ for a fractal correlation dimension D. The value $D = 3$ corresponds to the uniform (not fractal) distribution whereas $D \sim 2$ is observed up to the maximum scale $L \sim 100$ Mpc. The deficit of distant sources leads to the depression of cosmic ray intensity at relatively low energies and straightens the overall cosmic ray spectrum. Figure 2 illustrates the effect of clumpiness of source distribution [40]. 6

The estimates based on the total energy release and the spatial density of astronomical objects of different types showed that the galaxies with active nuclei and the Virgo cluster of galaxies, are the most probable extragalactic sources capable to maintain the observed intensity of highest-energy cosmic rays [1]. The interacting galaxies were also proposed [42]. A possible common origin of ultra high energy particles and Gamma Ray Bursts (GRB) was suggested in [43,44].

The difficulties with the acceleration of cosmic rays to the highest observed energies stimulated interest in the so-called top-down mechanisms. Production of extremely energetic particles with energies up to $\sim 10^{23} - 10^{25}$ eV is possible in the course of annihilation or collapse of cosmological topological defects such as monopoles, cosmic strings etc. [45], and decays of the hypothetical superheavy long-lived relic particles accumulated in the halo around our Galaxy [46]. The characteristic feature of top-down scenarios is an excess of gamma ray flux over the nucleon flux. In this connection, it should be noted that the analysis of the highest energy event $\sim 3 \times 10^{20}$ eV detected by the Fly's Eye installation disfavors its not proton production.

References

Berezinskii, V.S. et al. (1990) *Astrophysics of Cosmic Rays* , North Holland.

Nagano, M., Watson, A.A. (2000) Observations and implications of the ultrahigh-energy cosmic rays , *Rev. Modern Phys.* **72**, 690-732.

Jones, F.C., Ellison, D.C. (1991) The plasma physics of shock acceleration, *Space Sci. Rev.*, **58**, 259-346.

Lozinskaya, T.A. (1992) *Supernovae and Stellar Wind: The Interaction with the Interstellar Medium*, AIP.

Duric, N.F. et al.(1993) The VLA-WSRT survay of M33 - Statistical properties of a sample of optically selected supernova remnants, *A&A Suppl.* **99**, 217-255.

Sturner, S.J., Dermer, C.D. (1995) Association of unidentified, low latitude EGRET sources with supernova remnants, *A&A* **293**, L17-L20.

Esposito, J. et al. (1996) EGRET observations of radio-bright supernova remnants, *ApJ* **461**, 820-827.

Koyama, K. et al. (1995) Evidence for shock acceleration of high-energy electrons in the supernova remnant SN1006, *Nature* **378**, 255-258.

Tanimori, T. et al. (1998) Discovery of TeV gamma rays from SN 1006: further evidence for the supernova remnant origin of cosmic rays, *ApJ* **497** , L25-L28.

Reynolds, S.P. (1998) Models of synchrotron X-rays from shell supernova remnants, *ApJ* **493**, 375-396.

Gaisser, T.K. et al. (1998) Gamma-ray production in SN remnants, **ApJ 492**, 219-230.

Baring, M.G. et al. (1999) Radio to gamma-ray emission from shell-type supernova remnants: predictions from nonlinear shock acceleration models, *ApJ*, **513**, 311-338.

Berezhko, E.G. et al. (1996) Cosmic ray acceleration in SN remnants, *JETP* **82**, 1-21.

Völk, H.J., Biermann, P.L. (1988) Maximum energy of cosmic-ray particles accelerated by supernova remnant shocks in stellar wind cavities, *ApJ* **333**, L65-L68.

Biermann, P.L., Cassinelli, J.P. (1993) Cosmic rays. II. Evidence for a magnetic rotator Wolf-Rayet star origin, A&A **277**, 691-706.

Strong, A.W., Moskalenko, I.V. (1998) Propagation of cosmic ray nucleons in the Galaxy, *ApJ* **509**, 212-228.

Ginzburg, V.L., Ptuskin, V.S. (1976) On the origin of cosmic rays: some problems of high-energy astrophysics, *Rev. Mod. Phys.* **48**, 161-189.

Jones, F.C. et al. (2001) The modified weighted slab technique: models and results, *ApJ*, in press (astro-ph/0007293).

Ptuskin, V.S., Soutoul, A. (1998) Cosmic ray clocks, *Space Sci. Rev.* **85**, 223-236.

Beuermann, K. et al. (1985) Radio structure of the Galaxy, *A&A* **153**, 17-34.

Ruzmaikin, A.A. et al. (1998) *Magnetic Fields of Galaxies*, Kluwer Acad. Publ., Dordrecht.

Jokipii, J.R. (1971) Propagation of cosmic rays in the solar wind, *Rev. Geophys. Space Phys.* **9**, 27-87.

Jones, F.C. (1979) The dynamical halo and the variation of cosmic-ray path length with energy, *ApJ* **229**, 747-752.

Simon, M. et al. (1986) Propagation of injected cosmic rays under distributed reacceleration, *ApJ* **300**, 32-40.

Seo, E.S., Ptuskin, V.S. (1994) Stochastic reacceleration of cosmic rays in the interstellar medium, *ApJ* **431**, 705-714.

Webber, W.R. et al. (1992) Propagation of cosmic-ray nuclei in a diffusing galaxy with convective halo and thin matter disk, *ApJ* **90**, 96-104.

Bloemen, J.B.G.M. et al. (1993) Galactic diffusion and wind models of cosmic ray transport, *A&A* **267**, 372-387.

Zirakashvili, V.N. et al. (1996) Magnetohydrodynamic wind driven by cosmic rays in a rotating galaxy, *A&A* **311**, 113-137.

Ptuskin, V.S. et al. (1997) Transport of relativistic nucleons in a galactic wind driven by cosmic rays, *A&A* **321**, 434-443.

Ptuskin, V.S. (2000) The cosmic ray transport in the Galaxy, In: *Acceleration and Transport of Energetic Particles*, AIP, ed. R.A.Mewaldt et al., p. 391- 396.

Kulikov, G.V., Khristiansen, G.B. (1958) On the size spectrum of extensive air showers, *JETP* **35**, 441-447.

Axford, W.I. (1994) The origins of high-energy cosmic rays, *ApJ Suppl.* **90**, 937-944.

Bykov, A.M., Toptygin, I.N. (1993) Kinetics of particles in the strongly turbulent plasmas, *Physics Uspekhi* **36**, 1020-1052.

Ptuskin, V.S. et al. (1993) Diffusion and drift of very high energy cosmic rays in galactic magnetic fields, *A&A* **268**, 726-735.

Bhattacharjee, P., Sigl, G. (2000) Origin and propagation of extremely high-energy cosmic rays, *Phys. Report* **327**, 109-247.

Bird, D.J. et al. (1995) Detection of a cosmic ray with measured energy well beyond the expected spectral cutoff due to cosmic microwave radiation, *ApJ*, **441**, 144-150.

Zirakashvili, V.N. et al. (1998) Propagation of ultra high energy cosmic rays in Galactic magnetic fields, *Astron. Lett.* **24**, 139-144.

Takeda, M. et al. (1998) Extension of the cosmic-ray energy spectrum beyond the predicted Greisen-Zatsepin-Kuz'min cutoff, *Phys. Rev. Lett.* **81**, 1163-1166.

Mandelbrot, B.B. (1998) in *Current Topics in Astrophundamental Physics: Primordial Cosmology*, eds. N. Sanchez and A. Zichichi, Kluwer, p. 583.

Ptuskin, V.S. et al. (1999) Cosmic ray spectrum above the Greizen- Zatsepin-Kuzmin cutoff, *26th Intern. Cosmic Ray Conf.*, Salt Lake City 4, 271- 274.

Wu, K.K.S. et al. (1999) The smoothnes of the Universe, *Nature* **397**, 225-230.

Cesarsky, C. J., Ptuskin, V.S. (1993) Acceleration of highest-energy cosmic rays in galaxy collisions, *23rd Intern. Cosmic Ray. Conf.*, Calgary 2, 341-344.

Waxman, E. (1995) Cosmological gamma-ray bursts and the highest energy cosmic rays, *Phys. Rev. Lett.* **75**, 386-389.

Vietri, M. (1995) The acceleration of ultra-high-energy cosmic rays in Gamma-Ray Bursts, *ApJ* **453**, 883-889.

Hill, C.T. et al. (1987) Ultra-high-energy cosmic rays from superconducting cosmic strings, *Phys. Rev.* **D36**, 1007-1016.

Berezinsky, V.S., Vilenkin, A. (1997) Cosmic necklaces and ultrahigh energy cosmic rays, *Phys. Rev. Lett.* **79**, 5202-5205.

GALACTIC COSMIC RAY COMPOSITION:
FROM THE ANOMALOUS COMPONENT TO THE KNEE

M.I. PANASYUK
Skobeltsyn Institute of Nuclear Physics of Moscow State University
119899, Vorob'ovy Gory, Moscow, Russia

Abstract. The experimental data on the chemical composition of galactic cosmic rays (GCR) are analysed in the energy range between ~10 MeV/nucl and ~ PeV. The lower part of the energy range corresponds to the so-called 'anomalous cosmic rays' (ACR); the difference between ACR and more energetic GCR is their charge state which is close to +1, as well as their chemical composition. While the issue of ACR origin is practically resolved, the mechanism of nuclear component acceleration in the energy above hundreds GeV/nucl is still subject to discussion due to the ambiguity of the available experimental data.

1. Introduction

The second half of the last century was marked by two discoveries concerning cosmic rays with energies below PeV's, which contributed immensely to the understanding of the fundamental problem of cosmic ray physics, and namely, their origin and acceleration mechanism.

In 1958 Russian scientists discovered the 'knee' in the all-particle energy spectrum at energies ~3 PeV [1]. It was revealed, that this point in the energy spectrum is close to the threshold energy for particle acceleration in supernova blasts (see e.g. [2]). It is in this region that a sharp change in the chemical composition of GCR caused by transition from one acceleration mechanism (acceleration in supernova remnants) to another source (e.g. an extra-galactic) can be observed. However, there are certain arguments in favour of a more gradual change of the GCR chemical composition in this region: e.g. due to the diffusion character of the cosmic ray propagation mechanism.(see survey [3]). Experimental data on the chemical composition of GCR at energies below the knee could play a key role in the identification of the actual mechanism of particle acceleration and source, however, their ambiguity on the one hand, and disagreement with experimental data beyond the 'knee' (at energies above PeV) do not permit to draw unambiguous conclusions. The problem is aggravated by obvious experimental difficulties: at energies below the 'knee' direct cosmic ray measurement techniques on satellites and balloons are possible whereas beyond the 'knee' ground techniques, involving a vast amount of simulation (of particle penetration through the atmosphere) have to be used.

M. M. Shapiro et al. (eds.), Astrophysical Sources of High Energy Particles and Radiation, 263–273.
© 2001 *Kluwer Academic Publishers. Printed in the Netherlands.*

The second discovery concerns the least energetic part of the GCR spectrum, the region, which is subject to solar modulation . It was discovered that at energies ~10 MeV/nucl during years of minimum solar activity a local maximum appears in the spectra of certain nuclei (He, N, O, and others) [4]. Further experimental research (see e.g. the surveys [5] and [6]) confirmed the theoretical prerequisite [7] on their origin from neutrals of the local interstellar medium (LISM) and their fundamental difference from more energetic GCR: it was discovered, that their charge state is close to +1, unlike energetic GCR which are fully ionised.

2. The Anomalous Cosmic Ray Component

The term 'anomalous cosmic rays' was first introduced in 1973 after the discovery of a local maximum at $E \sim 10$ MeV/nucl in the energy spectrum of such cosmic ray elements as ^4He and ^{16}O [4],[8]. This maximum appeared during solar activity minimum (in 1976-1977) at about 10 MeV/nucl, i.e. at energies between those of particles of solar origin - solar energetic particles (<10 MeV/nucl) and traditional galactic cosmic rays (>100 MeV/nucl).

After the discovery . of the GCR anomalous component Fisk, Kozlovsky and Ramaty [7] suggested a theory describing their origin. According to their hypothesis ACR are neutral atoms of the LISM which penetrate inside the heliosphere where they are ionised by solar ultraviolet radiation or due to charge-exchange with ions of the solar wind; then they are picked up by the solar wind and carried away towards the heliopause, where they are accelerated to energies of ~10 MeV/nucl, and finally return back to the Sun. Later it was shown, that this process can be multiple [9]. The charge state of ACR ions can be 1+ or >2+.

The suggested mechanism of ACR origin and acceleration assumes a relative increase of intensity for the elements with a high ionisation potential, whereas for elements with a low ionisation potential (e.g. for Mg, Si, Fe) there should be no 'anomalous' increase of the flux. Fig.1 [10]
shows the energy spectra of He, C, N, and O observed by Voyager-2 at the distance of 23 AU during the year of minimum solar activity. The maximum of ACR fluxes is revealed at energies < 50 MeV/nucl. At larger energies GCR particles start to dominate.

It was experimental proof of ACR ions having charge states close to +1, that could serve as the final argument in favour of the above described model of ACR propagation and acceleration in the heliosphere.

The most convincing proof that ACR have charge states close to 1+ were the results of a series of experiments which used the effect of ion separation by the Earth's magnetic field. (see e.g. [10]). This technique was based on the comparison of ^{16}O fluxes, observed by the IMP-8 satellite in the interplanetary medium and simultaneous measurements at low altitudes (below 350 km) on satellites of the 'Cosmos' series. Detailed analysis of these results is given in [10]. Unambiguous determining of the charge state of ^{16}O was the key result of these experiments and proof of the validity of the Fisk, Kozlovsky and Ramaty hypothesis. Fig.2 [9] shows the results of comparison of the ^{16}O ion data (measured on 'Cosmos' and 'IMP-8' satellites) and model calculations, based on the penetration ^{16}O ions with charge values $Q=+1$ and $Q=+8$ into the inner magnetosphere. The mean charge state for ^{16}O was found to be $Q = 0.9 \pm^{0.3}_{0.2}$.

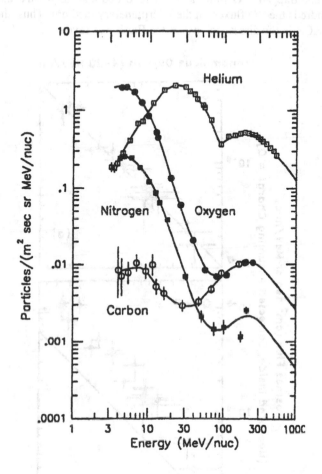

Figure 1. The ACR energy spectra: He, C, N, O , according to the Voyager-2 observations at ~23 AE in 1987 .

Studies of ACR are very important for cosmic ray physics. It is known that while propagating through the interstellar medium cosmic rays (ions with energies above ~10 MeV/nucl) soon become fully stripped. In order to lose their orbital electrons the ions need to travel through just several tens of mg/cm^2 of matter. Therefore, the presence of orbital electrons in the low energy cosmic ray component atoms can serve as evidence that their source is located in the direct vicinity of the solar system.

Another important result of ACR studies carried out during the 90-ies was proof that ions can be trapped by the Earth's geomagnetic field. This idea was suggested by Blake and Friesen in 1977 [11]. According to this theory a singly-ionised ACR ion, penetrating inside the geomagnetic field, is stripped by the residual atmosphere at altitudes of ~300 km, becomes multiply-charged and is trapped by the geomagnetic field.

This theory was confirmed in the experiments onboard satellites of the 'Cosmos' series [12]. The trapped ^{16}O were actually recorded and their flux exceeded by a factor of hundreds the ^{16}O fluxes in the interplanetary medium. Thus, the possibility of studying ACR in the direct vicinity of the Earth was revealed.

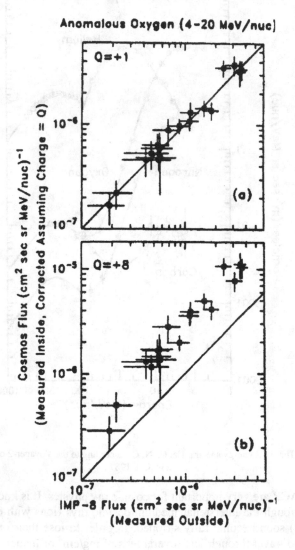

Anomalous Oxygen (4-20 MeV/nuc)

Figure 2. Comparison of anomalous oxygen intensities measured inside the magnetosphere ('Cosmos' data) and outside the magnetosphere (IMP-8 data) for supposed ^{16}O charged states of +1 and +8.

Starting from 1992 large-scale tudies of penetrating and trapped ACR began on 'SAMPEX'. It was confirmed, that besides ^{16}O, and ^{14}N ACR also contain Ne and

Ar. The different efficiency of ACR trapping and losses in the geomagnetic field effect their energy spectra: they become softer in comparison to the interplanetary ones [13]. As a consequence, their relative composition also changes. Table 1 shows the relative abundances for the individual components of ACR for trapped and interplanetary ions according to 'SAMPEX' data [13]. The discrepancies are obvious.

TABLE 1. The relative abundances of ACR according to 'SAMPEX' data [13].

Ratio	Trapped 16-45 MeV/nucl	Interplanetary > 17 MeV/nucl
C/O	~0.0004	0.014±0.009
N/O	0.09±0.01	0.19±0.03
Ne/O	0.04±0.01	0.06±0.02

Measurements of ACR during the last two solar activity cycles show that this GCR component is extremely sensitive to solar modulation. The flux of anomalous ^{16}O with $E \sim 10$ MeV/nucl in the interplanetary medium varies within a factor of ~100. The behaviour of trapped anomalous ^{16}O is similar.

Hence, satellite studies of anomalous cosmic rays carried out during the 70-ies-90-ies led to the understanding that the most low-energy component of GCR, observed in the heliosphere, consists of particles which originate in the LISM - the part of the Galaxy which is located close to us, and the limited time of their propagation and acceleration inside the heliosphere is responsible for their low charge state.

Unlike ACR, GCR with energies exceeding hundreds of MeV/nucl are not subject to modulation by solar magnetic fields and their energy spectra are thought to be stable.

3. Galactic Cosmic Rays Before the 'Knee'

The energy range prior to the 'knee', i.e. before several PeV is mainly accessible by 'direct' experiments on satellites and balloons. This research began with space radiation studies onboard soviet satellites of the 'Proton' series [14] and are being continued to date. Detailed analysis of this energy range is given in many surveys (see e.g. [15], [16], [17]). Here we will restrict ourselves to the analysis of experimental data, concerning protons and particles with $z > 2$.

Table 2 shows the currently available experimental data on protons, obtained in satellite and balloon experiments using X-ray emulsion chambers (XEC) and calorimeter techniques.

Table 2. The spectral indices for protons according to data of different experiments
(the references are given in [17]).

#	Experiment	Energy range, eV	$|\gamma_p|$
1.	'Proton'	$10^{12} \div 10^{13}$	$\gamma_p - 1 = 2.2 \div 2.3$
2.	SOKOL	$5 \cdot 10^{12} \div 10^{14}$	2.85 ± 0.14
		$4 \cdot 10^{12} \div 2 \cdot 10^{14}$	$\gamma_p - 1 = 2.11 \pm 0.15$
3.	MUBEE	$10^{13} \div 2 \cdot 10^{14}$	3.14 ± 0.08
4.	JACEE	$8 \cdot 10^{12} \div 8 \cdot 10^{14}$	2.8 ± 0.04
5.	RUNJOB	$3 \cdot 10^{12} \div 3 \cdot 10^{13}$	~2.8
6.	Kawamura et al.	$> 5 \cdot 10^{12}$	$\gamma_p - 1 = 1.82 \pm 0.13$

It is also possible to estimate the proton spectra in the energy range before the 'knee' by 'indirect' techniques. The data on the muon component discussed in the survey [18] indicates, that γ_p lies within $2.64 \div 2.8$. The results of the high mountain studies of the proton spectrum cited in [17] indicate, that the proton spectrum index γ_p is ≥ 2.7: $\gamma_p = 2.72 \pm 0.04$ (CANGAROO); $\gamma_p = 2.72 \pm 0.04$ (Chakaltaya).

Measurements of the proton spectrum in the pioneering experiment on the 'Proton' satellite [14] show indications of a 'bend' in the proton spectrum: γ_p changes from 2.7 to 3.2 at energies around 1-2 TeV. The author (Grigorov) in a number of papers (see e.g. [19]) gives a number of additional arguments in favour of this 'bend', however, at present there is no other experiment in this energy range, in which the authors would confirm this point of view.

The number of experimental results for the heavy component of cosmic rays ($z > 2$) in the energy range before the 'knee' is even more restricted than for the proton component. These results are shown in Table 3.

Table 3. The spectral indices for particles with $z > 2$ according to different experiments. (The references can be found in [17]).

#	Experiment	Energy range, eV	$|\gamma_{z>2}|$
1.	CRN	$7 \cdot 10^{10} \div 10^{12}$	~2.4
2.	SOKOL	$1.5 \cdot 10^{12} \div 1.5 \cdot 10^{13}$	2.58 ± 0.07
3.	MUBEE	$5 \cdot 10^{11} \div 2 \cdot 10^{12}$	2.4 ± 0.16 (CNO)
			2.3 ± 0.57 (Fe)
4.	JACEE	$1.9 \cdot 10^{12} \div 3 \cdot 10^{14}$	2.38 ± 0.11
5.	RUNJOB	$3 \cdot 10^{12} \div 2 \cdot 10^{13}$	~2.5

The data shown in Table 3, in spite of their limited statistics, undoubtedly indicate a harder spectrum for these .particles in comparison to protons. Fig.3. shows the energy spectra for protons and nuclei of the Fe group which are the result of compilation from the data of Tables 2 and 3.

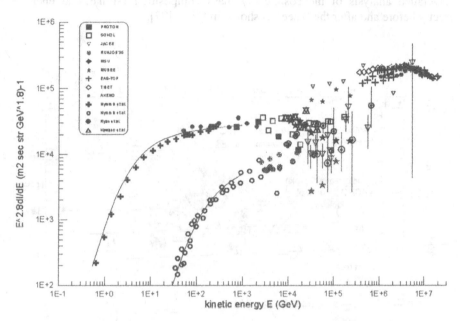

Figure 3. The energy spectra of protons, Fe group ions, and all-particle spectra from different experiments.

In spite of limited statistics for the nuclear component measurements in the energy range above 1 TeV there is no doubt that there is a monotonous hardening of the heavy cosmic ray component spectrum relatively to the proton one.

This is the most important result in the studies of cosmic rays at energies above the 'knee' over the last several decades.

As it can be seen from Fig.3. the chemical composition of cosmic rays gets heavier starting from energies which are quite far from the 'knee' (not smaller than ~1 TeV). This is a key experimental result which can be used to verify any theories of cosmic ray origin and acceleration in this range of energies. Detailed comparison with theoretical models lies outside the scope of this paper, however, we can state, that the latest versions of theoretical models [20,21] convincingly demonstrate that is possible to explain the experimental data on the cosmic ray chemical composition (up to at least ~ 100 TeV) as the result of their acceleration on supernova remnants.

The most important question is: what is the chemical composition of cosmic rays at larger energies, i.e. around the 'knee' and what is the acceleration limit which can be achieved in supernova blasts.

4. Galactic Cosmic Rays Around the 'Knee'

Studies of the chemical composition of cosmic rays with energies ≥ 100 TeV on satellites and balloons using 'direct' techniques remain impossible. In this range of energies estimates of the cosmic ray composition, based on the analysis of EAS development is muon, electromagnetic and hardron component as well as cherenkov, fluorescent emission and the XEC technique, are used.

Detailed analysis of the cosmic ray mass composition estimates at energies directly before and after the 'knee' is shown in Fig.4. [22].

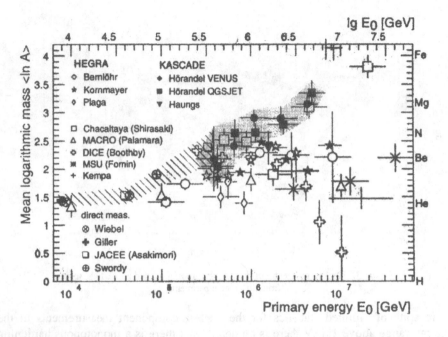

Figure 4. The energy dependence of mean galactic cosmic ray mass <A> according to different experimental data.

It can be noted, that the main conclusion (cosmic ray mass composition becoming heavier with increasing energy) is confirmed only by EAS measurements of the KASCADE set-up for the hardron component with utilisation of the VENUS and QGSJET high-energy hardron interaction models. A similar result can be obtained from the Chakaltaya XEC set-up data, but using the predetermined super-heavy cosmic ray composition.

There are other results which contradict those mentioned above (see Fig.4.). Furthermore, studies of the mass composition on the same set-up but employing different interaction models can lead to different results (see, e.g. [23,22]).

All these considerations show, that calculations of the chemical composition using measurements of the cascade process developing inside the atmosphere or detector (for XEC) are essentially model-dependent.

The absence of adequate models describing interactions of high-energy particles with matter is probably the main reason for the discrepancies in the experimental results around the 'knee'. However, the 'knee' itself as the only kink in the all-particle energy spectrum at energies ~ 3 TeV is subject to doubt in some papers. (see [24]).

Additional results of EAS cherenkov radiation studies (made by the TUNKA ground set-up) and measurements of the cherenkov light reflected by the snow-covered surface (the SPHERE balloon experiment) can be mentioned as evidence of the fine structure of the 'knee'. Both these experiments gave similar results [25]: around several tens of PeV there is a statistically significant peak in the intensity (Fig.5.)

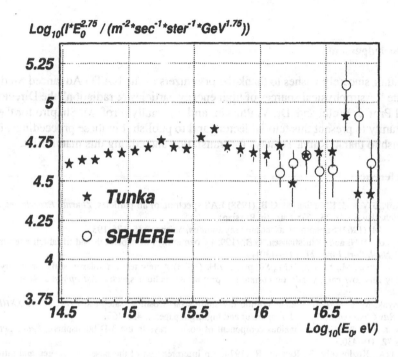

Figure 5. The all-particle energy spectra according to the data of 'TUNKA" and 'SPHERE' experiments.

According to [24] this is evidence of the appearance of a new source of cosmic rays - a local source, generating a z-dependent spectrum against the background of a monotonous all-particle spectrum from supernova.

5. Conclusions

Studies of the chemical composition of cosmic rays plays a fundamental role in our understanding of cosmic ray origin and acceleration. In the present report attention was mostly paid to the most low energy component of galactic cosmic rays - the anomalous component and the 'knee' region. In the first case knowledge of the cosmic ray chemical composition obtained on satellites using 'direct' techniques helped to establish adequate understanding of their origin and acceleration. In the second case difficulties in using 'direct' techniques in the PeV energy range, primarily associated with the enormous dimensions of the required satellite instruments, and, on the other hand, insufficient knowledge of particle interaction processes at such high energies, makes it currently impossible to establish the origin of these cosmic rays.

Acknowledgements

The author sincerely wishes to thank the organizers of the NATO Advanced Study Institute " Astrophysical sources of high energy particles & radiation" the Directors of ASI Prof. J. Wefel and Dr. V. Ptuskin and especially Prof. M. Shapiro for the opportunity to present this tutorial lecture and to publish it in these proceedings. I also wish to thank Katya Tolstaya for translation and editing this manuscript.

6. References

1. Kulikov, G.V. and Khristiansen, G.B. (1958) EAS spectrum of all-particles, *Zhurnal Experimental' noi i Teoreticheskoi Fiziki* **35**, 635-640. (in Russian).

2. Shapiro, M. (1962) Supernovae as cosmic ray sources, *Science* **135**, 175-193.

3. Kalmykov, N.N. and Khristiansen, G.B. (1995) Cosmic rays of superhigh and ultrahigh energies, *J. Phys. G: Nucl. Part. Phys.* **21**, 1279-1301.

4. Garcia-Munoz, M., Mason, G.M., Simpson, J.H. (1973) A new test for solar modulation theory: the 1972 May-July low energy galactic cosmic ray proton and helium spectra, *Astrophys. J. (Lett.).* **182**, L81-L84.

5. Panasyuk, M. (1993) Anomalous cosmic ray studies on 'Cosmos' satellites, *Proc. of the XXIII Int. Cosmic Ray Conference.* (Invited, rapporteur and highlight papers), 455-463.

6. Klecker, B. (1995). The anomalous component of cosmic rays in the 3-D heliosphere, *Space Science Reviews* **72**, 419-430.

7. Fisk, L.A., Kozlovskiy, B., Ramaty, R. (1974) An interpretation of the observed oxygen and nitrogen enhancements in low-energy cosmic rays. *Astrophys. J. (Lett.)* **190**, L35-L38.

8. Hovestadt, D.O., Valmer, O., Gloeckler, G., Fan, C. (1973) Differential energy spectra of low-energy (8.5 MeV per nucleon) heavy cosmic rays during solar quiet times, *Phys. Rev. Lett.* **31**, 650-667.

9. Mewaldt, R.A., Selesnick, R.S., Cummings, J.R., Stone, E.C., Von Rosenvinge, T.T. (1996) Evidence for multiply charged anomalous cosmic rays, *Astrophys. J.* **466**, L43-L46.

10. Adams, J.H., Garcia-Munoz, M., Grigorov, N.L., Klecker, B., Kondratyeva, M.A., Mason, G.M., McGuire, E., Mewaldt, R., Panasyuk, M.I., Tretyakova, Ch.A., Tylka, A.J., Zhuravlev, D.A. (1991) The charge state of the anomalous component of cosmic rays, *Astrophys J.(Lett)* **375**, L45-L48.

11. Blake, J.B. and Friesen, L.M. (1977). A technique to determine the charge state of the anomalous low energy cosmic rays., *Proc. of the 15th ICRC* **2**, 341-346.

12. Grigorov, N.L., Kondratyeva, M.A., Panasyuk, M.I., Tretyakova, Ch.A., Adams, J.H., Blake, J.B., Shultz, M., Mewaldt, R.A., Tylka A.J. (1991) Evidence for trapped anomalous cosmic ray oxygen ions in the inner magnetosphere, *Geophys. Res.. Letters* **18**, 1959-1962.

13. Selesnik, R.S., Cummings, A.C., Cummings, J.R., Mewaldt, R.A., Stone, E.C., Von Rosenvinge, T.T. (1995). Geomagnetically trapped anomalous cosmic rays, *J.Geophys. Res.* **100**, 9503-9509.

14. Grigorov, N.L., Nesterov, V.E., Rappoport, I.D. (1970) Measurements of particle spectra on the 'Proton-1,2,3' satellites, *Yadernaya fizika* **11**, 1058-1067. (in Russian).

15. Shibata, T. (1996) Cosmic Ray spectrum and composition: direct observation, *Nuovo Cimento* **19**, 713-721.

16. Watson, A.A. (1998) Charged cosmic rays above 1 TeV, *Invited, Rapporteur and Highlight Papers of the 25th ICRC* **8**, 257-280.

17. Panasyuk, M.I. (1998) Galactic Cosmic Ray Composition, *Proc. 16th European Cosmic Ray Symposium*, 235-244.

18. Ryazhskaya, O.G. (1996) Muons and neutrinos in the cosmic radiation, *Nuovo Cimento* **19**, 655-670.

19. Grigorov, N.L., and Tolstaya, E.D. (1998) How many 'knees' does the galactic cosmic ray spectrum have?, *Preprint INP MSU* -**98-8-509**, Skobeltsyn Institute of Nuclear Physics.

20. Berezhko, E.G. (1996) Maximum energy of cosmic rays accelerated by supernova shocks, *Astroparticle Physics* **5**, 367-378.

21. Ellison, D.C., Drury, L., Meyer, J.P. (1997) Galactic cosmic rays from supernova remnants, II. Shock acceleration of gas and dust, *Astrophys.J.* **487**, 197-217.

22. Hörandel, J.R. (1998) Cosmic-ray mass composition in the PeV region estimated from the hardronic component of EAS, *Proc. of the 16th European Cosmic Ray Symposium* , 579-582.

23. Weber, J.H. (1998) Estimation of the chemical composition in the 'knee' region from the muon/electron ratio in EAS., *Proc. of the 16th European Cosmic Ray Symposium*, 567-570.

24. Erlykin, A.D., Wolfendale, A.W. (1998) High-energy cosmic ray spectroscopy, *Astroparticle Physics* **8**, 265-281.

25. Kuzmichev, L.A., Antonov, R. A. (2000) Personal communication.

LOW ENERGY COSMIC RAY NUCLEI AND THEIR PROPAGATION IN THE GALAXY

MARIA GILLER

Division of Experimental Physics, University of Lodz
Pomorska 149/153, 90-236 Lodz, Poland
mgiller@kfd2.fic.uni.lodz.pl

Abstract   In these lectures we show what information can be obtained from studying fluxes of various isotopes in cosmic rays at low energies. The presence of secondary nuclei (produced in the interstellar matter from fragmentation of the primaries) allows, in principle, to derive the path length (in $g \cdot cm^{-2}$) distribution traversed by nuclei, accelerated in the sources, on their way to the Solar System. Secondary to primary ratios measured as a function of energy, testify in favour of a quicker leakage of CR from the Galaxy at higher energies. They also suggest an increasing role of particle convection (out of the Galactic disk) over diffusion with decreasing energy. We discuss a possibility of partial particle acceleration during propagation in the Galaxy and describe methods of verifying this hypothesis. It is also explained how to derive the cosmic ray age from the fluxes of some radioactive secondary isotopes.

1. Introduction

Cosmic Rays (CR) are fluxes of various nuclei arriving to the Earth atmosphere from the outside of the Solar System practically isotropically. Up to energies $\leq 10^{14}$ eV the fluxes of individual nuclei are large enough to be measured directly on high altitude balloons, satellites and recently also on spacecrafts.

When their chemical composition of low energies (1–10 GeV per nucleon) is compared with that of the Solar System and with the composition of meteorites (see Fig.1), a general similarity can be seen, together with some striking differences: when normalizing the relative abundances to magnesium ($Z = 12$), a medium charge nucleus, there is a rough agreement between most of even Z nuclei in the CR composition on one side and the two local ones on another, suggesting that the accelerated matter has a composition similar to that of the Solar System. However, hydrogen and helium are significantly underabundant in the CR flux, when compared with other elements.

275

M. M. Shapiro et al. (eds.), Astrophysical Sources of High Energy Particles and Radiation, 275–304.

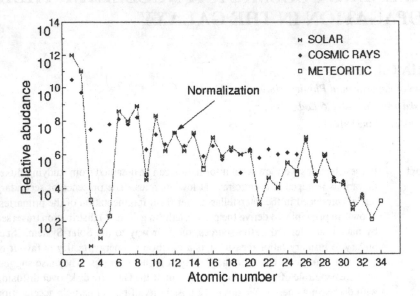

Figure 1. Composition of arriving cosmic rays compared with those in the solar photosphere (H abundance $\equiv 10^{12}$) and meteorites

The second main difference is that the CR elemental distribution is noticeably smoother than the other two, the two large dips at $Z = 3 - 5$ and $Z = 21 - 25$ being almost levelled out. The explanation of this fact is that CR nuclei from these two charge regions, practically absent in the source material, are produced by heavier CR nuclei in their collisions with the Galactic interstellar matter: Li, Be and B – mainly by carbon and oxygen and "sub-iron" elements by iron. Most of this lecture will by actually devoted to this observation and its implications on models of CR propagation in the Galaxy.

2. Experimental methods

In the GeV/n energy region, CR fluxes of individual nuclei range from several thousands of protons per $(m^2 \cdot s \cdot GeV)$ to several iron nuclei. For Z>26 (iron) the CR flux drops dramatically and for $Z > 40$ typically a few nuclei of a given element arrive to the Earth atmosphere for a million of these of iron. These are the so called ultra heavy elements and we shall not consider them in this lecture.

The experimental purpose is to measure energy, charge, mass and direction of each CR particle arriving at the detector. As we shall be concerned here mainly in the low energy region (below tens of GeV/n) measuring arrival directions serves in the first place for determination of particle trajectory within the detector for a more accurate derivation of its other parameters. Galactic anisotropy (if any) would be anyway blurred by the solar wind magnetic fields.

The easiest way to measure a high energy particle charge is to use its continuous interactions with the detector matter, producing measurable effects, usually proportional to Z^2. Scintillation light produced by a relativistic particle traversing a transparent scintillator is proportional to the particle energy loss for ionisation which in turn is $\sim Z^2 \cdot l$ (where l is the particle track length in the detector – necessary to be known, see above) and is very weakly energy dependent. Cherenkov light is $Z^2 \cdot l$ as well, if the particle energy is well above the threshold (factor of 3 gives 90% of the maximum signal). Thin semiconducting silicon detectors respond with signals $dE/dx \cdot l$, being $\sim Z^2 \cdot l$.

Nuclear emulsion detector reveal particle tracks with widths Z^2 (again an effect of many ionisation acts and production of δ-electrons). As there are fluctuations in the energy deposition for ionisation and, of course, experimental inaccuracies (in e.g. determination of l and/or the emitted light) the charge resolution ΔZ obtained in the present experiments is practically sufficient (as we shall see) up to $Z \simeq 30$. The bigger is the nucleus charge the better relative resolution is needed to distinguish it from its neighbors. Together with much lower fluxes of the ultra heavy nuclei, it is still impossible to resolve odd Z neighbors from its dominant even-Z elements.

The energy determination in the GeV/n region is usually achieved by stopping the particle in a thick detector (or several layers of it) and measuring the total energy deposited for ionisation. Once the particle charge is known one can calculate its energy by measuring its rigidity (from deflection in a magnetic spectrometer) and its velocity (from a time of flight – if it is a mildly relativistic particle). At higher energies the particle nuclear interaction can be used to measure total energy of the electromagnetic cascades developed in the detector (e.g. a sandwich of high Z material and nuclear or X-ray emulsion).

Not only the chemical composition and its energy dependence is important for understanding the CR origin. Also the isotopic contents of various elements, when compared with that of the local matter, can tell us a lot about CR acceleration and propagation conditions (we shall be more precise about it later). Thus, particle mass should be measured as well. In an experiment where particle velocity and momentum are measured (mentioned above), its mass can obviously be determined. This method, however, has some disadvantages: large weight of a necessary magnet and particle velocities close to c limit its broad application. A much more frequent method is based on the fact that for a given Z ionisation losses dE/dx are dependent on particle velocity only. Thus, measuring particle energy deposit ΔE in a thin detector (or better, in several of them) and the energy deposit in a thick detector (meaning – particle kinetic energy E) allows to determine its mass.

3.   Recent experiments

The first experiment we want to recall here is not very recent as the results were published in 1990. Accurate data on the CR chemical composition in a wide range of charges ($Z = 4 \div 30$) and energy ($0.6 \div 35$ GeV/n) were obtained by the French-Danish experiment C2 onboard the satellite HEAO-3 [1]. The detector consisted with 5 Cherenkov counters (with different refractive indices) – for Z and velocity determination and a hodoscope of 4 flash tube arrays (inserted between the counters) – for determination of particle track lenghts in the counters. The geometrical factor was 413 cm^2 · sr and the total number of "good" events (non interacting in the detector) was about $7 \cdot 10^6$.

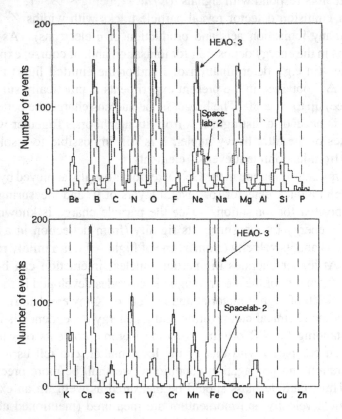

Figure 2.   CR charge distribution from HEAO-3 in a few GeV/n region compared with the results from Spacelab-2 above 70 GeV/n

How good was the charge resolution is presented on Fig.2, where the measured charge histogram is shown for energy range 2.35 – 6.4 GeV/n. ΔZ good enough even to resolve quite well Co (Z= 27) from the adjacent large peak of Fe ($Z = 26$). It was the best charge resolution obtained at that time. It allowed to determine the energy spectra of individual nuclei with a good precision. Also shown are the data for much higher energies ($\geq$ 70 GeV/n) obtained on the Space Shuttle Challenger in the Spacelab-2 mission [2]. Here, the transition radiation detectors were used for energy determination at the highest values, together with gas Cherenkov and scintilation counters for the lower energy range. It can be seen that the numbers of events are much lower and the charge resolution allows to separate clearly the odd-Z elements only for boron and nitrogen.

The results confirmed earlier conjectures that the energy spectra of elements present is the local matter (called primary CR) have almost the same shape. However, for those underabundant in the local matter (meaning – practically not present in the CR sources) but present in the CR flux (called the secondary nuclei, see also Introduction), the energy spectra are quite different. The most conspicuous example of this difference can be demonstrated by the boron to carbon flux (B/C) ratio as a function of energy (Fig.3a) where the HEAO-3 results are shown together with those of other experiments. The B/C ratio falls by about a factor of 3 if energy changes from $\sim$ 1 GeV per nucleon (GeV/n) to 30 GeV/n.

A decrease of any secondary to primary nuclear flux with energy is observed (see also Fig.3b. where the ratio of some sub-iron elements to iron is shown, as a function of energy). As we have already mentioned in the Introduction these observations constitute the main basis for constructing models of CR origin and propagation in the Galaxy. We shall devote to this issue the next chapter .

The chemical composition being well measured by HEAO-3 (also absolute fluxes have been determined with accuracies of a few percent), many experiments concentrated on the isotopic composition of individual CR elements. Here, we shall shortly describe only three of them and compare the obtained results. These are experiments onboard spacecrafts Voyager 1 and 2, Ulysses and ACE (Advanced Composition Explorer) on orbits far from the Earth. All are designed to measure CR particle charge and mass in a wide charge range and in the low energy region from tens to hundreds MeV/n. These particle parameters are determined by the dE/dx vs E method, described above. Fig.4 shows a cross-section view of the Ulysses High Energy Telescope (HET) detector elements [3] (see a more detailed description in the figure), being typical for the other experiments. A comparison of some characteristics of the there experiments is made in Table 1.

We can see that CRIS (Cosmic Ray Isotope Spectrometer) onboard the ACE spacecraft exceeds the other two by having a much larger geometry factor. We have also shown the number of ^{56}Fe events (published so far) and that of ^{10}Be. As we shall discuss it later, the amount of the latter isotope plays a major role in the determination of the cosmic ray age in the Galaxy. A comparison of mass distributions obtained by the three experiments for Be (an important light

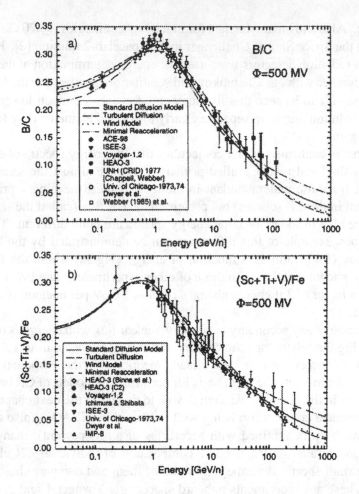

Figure 3. Compilation of two main secondary to primary ratio data: a) boron to carbon and b) (scandium + titanium + vanadium) to iron as a function of kinetic energy per nucleon. Lines are model predictions (see §4). Figure taken from [15]

Table 1. Comparison of some characteristics of three spacecraft experiments

Experiment	Energy range (MeV/n)	Geometry factor (cm$^2 \cdot$ sr)	Number of ^{56}Fe events	Number of ^{10}Be events
Voyager 1&2	30–200†	0.89–1.69	∼ 1400‡	14
HET on Ulysses	30–500	4–8	6035	53
CRIS on ACE	50–500	250	∼10000‡	∼200‡

† depending on charge
‡ numbers estimated from the published histograms

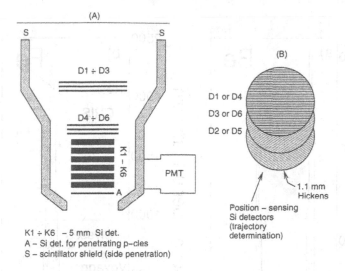

Figure 4. Cross section view of the Ulysses High Energy Telescope (HET) detector elements

element) [4, 5, 6] and Fe (an important heavy one) [7, 3, 8] is prescnted in Fig. 5a,b. The figures illustrate well the great progress that has been achieved in the determination of CR isotopic composition in the last few years. It is instructive to remember also that the 14 ^{10}Be events were being collected by the two Voyagers for 21 years, whereas it took less than two years to register about 200 of these nuclei by CRIS. However, up to a few years ago it were the Voyager's data that were most accurate in this energy region.

4. Propagation models

4.1. Source – propagation separation

Having the rich data about fluxes of individual CR isotopes it is possible to deduce quite a lot of information about their origin and the rest of this lecture will be devoted to these deductions.

First of all, let us come back to Fig. 3a,b, where two secondary to primary ratios are shown as a decreasing function of energy (above ~ 1 GeV/n). It turns out that all secondary isotopes have energy spectra steeper than primaries. This fact imposes the following picture: CR primaries are being accelerated in some compact objects (CR sources) and then propagate in the Galaxy through the interstellar gas without changing their energies (at least significantly), producing secondaries. The higher is the energy, the less secondaries are produced, thus the smaller is a typical path length (in $g \cdot cm^{-2}$ of interstellar matter, ISM) of primaries. The separation of the acceleration region from that of the

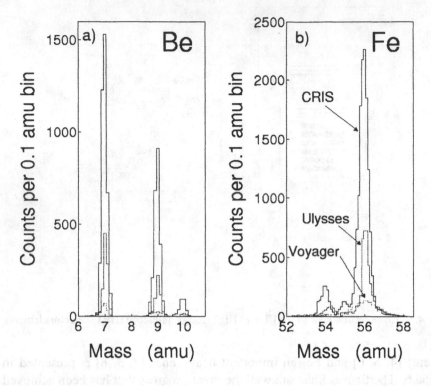

Figure 5. Comparison of mass distributions obtained from Voyager (dotted histogram), Ulysses (dashed) and ACE (solid) for a) beryllium and b) iron

propagation (secondary production) is necessary – if these two processes had occurred simultaneously (e.g. in the same region of the interstellar space), there would have been relatively more secondaries at higher energies, as it would have taken longer time (thus a longer path length) to accelerate particles to that energy. The secondary to primary ratios would have increased with energy (or finally flattened out due to secondary interactions with gas) [9]. One can not exclude, however, that *some* acceleration in the IMS does exist (so called "reacceleration"), the more so that such an assumption fits nicely with some other propagation predictions. We shall come to this issue in §4e.

4.2. Path length distribution and secondary to primary (sec/prim) ratios

In this paragraph we shall show that the measurable quantity, a secondary to primary ratio (e.g B/C, or subFe/Fe) at a given energy depends on the path length distribution $f(x)$ of primaries of that energy.

For any spatial distribution of CR sources and a particular position of the observer there is a distribution of path lengths that would have been traversed by primary nuclei arriving to the observer, had they not interact with the gas (the so called "vacuum distribution" [10]. As we believe that a nucleus trajectory in the ISM depends only on deflections in the Galactic magnetic fields, that is on its rigidity $R = pc/Ze$, $f(x)$ should be the same for all nuclei with the same R, providing the composition of all the sources is the same. (Historically, the first assumption was that there is a *unique* path length x to be traversed by nuclei (the slab model) but it has quickly been realised that no unique x fits *simultaneously* B/C and subFe/Fe data).

Let us assume that the source has emitted $dN_o(x)$ primaries that are to traverse paths between x and $x + dx$, that is $dN_o(x) = N_o f(x)dx$, where N_o is some total number of the emitted particles). Number of primaries that arrive to the observer $dN_p(x)$ equals

$$dN_p(x) = dN_o(x)\exp(-x/\lambda_p) \tag{1}$$

where λ_p is the mean path length for destruction of the primary (in the sens that it loses some of its nucleons in a collision with the gas and becomes another CR nucleus). Any such process we shall call a nucleus fragmentation. As a result of such a collision a secondary nucleus is produced, but to a good degree of approximation, its energy per nucleon remains the same as before the collision. If we assume further that the charge per nucleon does not change much after the collision (which on average is a good approximation as well) then the rigidity of the secondary will remain the same as that of the primary. Thus, the nucleus trajectory will be unchanged and the secondary will follow the path that would have been followed by the primary, had the collision not occured.

Adopting this picture it is easy to find the number of secondary nuclei dNs(x) arriving to the observer, produced by the primaries $dN_p(x)$:

$$dN_s(x) = dN_o(x) \int_0^x \exp(-y/\lambda_p) \cdot \frac{dy}{\lambda_{ps}} \cdot \exp\left[\frac{-(x-y)}{\lambda_s}\right] \tag{2}$$

where λ_s – is the mean path length for fragmentation of the secondary and λ_{ps} is the mean path length for the production of the given secondary by the primary. Integrating (1) and (2) over all possible path lengths x of the primaries one obtains the numbers of primaries N_p and secondaries N_s (respectively) arriving to the observer. Their ratio equals

$$\frac{N_s}{N_p} = \frac{1}{\lambda_{ps}} \cdot \frac{1}{\frac{1}{\lambda_p} + \frac{1}{\lambda_s}} \left[\frac{\int_0^\infty f(x) \exp(-x/\lambda_s) dx}{\int_0^\infty f(x) \exp(-x/\lambda_p) dx} - 1 \right] \qquad (3)$$

Of course, any λ is related to the cross-section σ for the corresponding process by $\lambda = m/\sigma$, where m is the average mass of the interstellar gas atom.

Formula (3) demonstrates how a sec/prim ratio depends on $f(x)$. Usually a given secondary isotope has several "parents", but a generalisation of (3) to this case is straightforward. It is necessary, however, to assume then the relative parent abundances *at the sources*. Another factor which has to be taken into account (particularly for large Z nuclei at low energy per nucleon) are the energy losses for ionisation occurring during particle propagation is the ISM. (Roughly, they are ~ 2 MeV/g $\cdot$ cm^{-2} $\cdot$ Z^2). The complication is that $f(x)$ depends on energy as, to some extent, do the fragmentation cross-sections. Thus, in practice calculations are being done numerically: assuming energy spectra of primaries and they relative abundances one propagates the nuclei through path lengths with an appropriate distribution. (An example of how to deal with the problem that during propagation $f(x)$ is changing because particle energy is changing, can be found in [11]).

The path length distribution $f(x)$, its shape and energy dependence, contain, in principle information about spatial distribution of the sources and particle propagation in the Galaxy. A long confinement time in the propagation region would correspond to an exponential shape: $f(x) \sim \exp(-x/\overline{x})$ (see §4.3 about the leaky box model). If the sources were surrounded by matter (and confined there for some time) an observer outside them would see a cut-off of short path lengths (a rough illustration of the so called "nested leaky box" model [12]). If particles were totally confined in the Galaxy and CR production was time independent then $f(x) \approx const$ ("closed Galaxy" model). We have already mentioned the "slab" model for which $f(x) = \delta(x - x_o)$.

As better and better σ's have been measured and better chemical and isotopic resolution of CR fluxes obtained, some of these models have been rejected. "Closed Galaxy" did not quite fit various sec/prim ratios and its prediction of a large positron flux was in striking disagreement with observations [13]. For some time it seemed that it is the nested leaky box $f(x)$ that was necessary to account for B/C and subFe/Fe ratios simultaneously. Now the situation is different and it looks that the exponential shape of $f(x)$, with no cut-off of small x, fits best. This corresponds to the "leaky box" model or (almost) to the diffusion model which will be described and analysed in the next two paragraphs.

Before that we would like to make some important comments about possibilities to determine $f(x)$. It can be determined only if it is unique for all primary nuclei (for given rigidity). This means that all CR sources emit the CR flux with the same composition, or the CR propagation time between different sources is much smaller than a typical CR age (quick mixing). We shall consider two extreme cases:

(a) Mean free paths λ_p and λ_s for fragmentation of a primary and a secondary are much larger than a typical vacuum path length x. Then

$$\int_0^\infty f(x) \cdot \exp\left(\frac{-x}{\lambda}\right) dx \cong \int_0^\infty f(x) \cdot \left(1 - \frac{x}{\lambda}\right) = 1 - \frac{\bar{x}}{\lambda}$$

for $\lambda = \lambda_p, \lambda_s$. From (3) we obtain that

$$\frac{N_s}{N_p} \cong \frac{\bar{x}}{\lambda_{ps}} \tag{4}$$

Thus, the ratio depends only on the average value $\bar{x}$. If this case applied to many sec/prim ratios, then the determination of *the shape* of $f(x)$ would not be possible.

(b) The second extreme case would be if $\lambda_p, \lambda_s \ll x$. Then

$$\int_0^\infty f(x) \cdot \exp\left(\frac{-x}{\lambda}\right) dx \cong f(0) \cdot \lambda$$

and from (3) we get that

$$\frac{N_s}{N_p} \cong \frac{\lambda_s}{\lambda_{ps}} \tag{5}$$

Now there is no dependence of the ratio even on $\bar{x}$. Had the ratios fit the above relation we would be able to tell only that the CR vacuum path lengths are much longer than the mean fragmentation paths λ_p, λ_s.

However, an information about the shape of $f(x)$ can be drawn if we have an intermediate situation, i.e λ_p, λ_s have values similar to various path lengths x. Then these ratios with shorter λ_p, λ_s would be sensitive to $f(x)$ for lower values of x, whereas those with larger λ_p, λ_s would also be sensitive to larger x. It seems that we are just in this lucky situation (although such a coincidence may seem a bit suspicious).

4.3. The leaky box model

The most frequently used and the simplest propagation model is the leaky box. This model assumes that:

- there is a volume (box) containing CR sources which emit particles with a constant rate;

– the particles propagate somehow within this volume but in such a way that in any small time interval dt there is a probability dp that a particle escapes ("leaks out") from the box:

$$dp = \frac{dt}{T_{esc}} \qquad (6)$$

Thus, the escape probability per unit time does not dependent on particle history and the escaping law is identical with e.g. that of the radioactive decay of a nucleus. In this circumstances the total number of particles in the box would be stabilised at a value for which particle production rate is equal to their escaping rate. It is easy to show that the "age" t distribution $g(t)$ of particles in the box would be exponential, with T_{esc} being its mean value, i.e.

$$g(t) = \frac{1}{T_{esc}} \exp(-t/T_{esc}) \qquad (7)$$

If the box is filled uniformly with gas, the time distribution $g(t)$ corresponds to the same distribution of path lengths so that

$$f(t) = \frac{1}{\lambda_{esc}} \exp(-x/\lambda_{esc}) \qquad (8)$$

with $\lambda_{esc} = \rho v T_{esc}$, where ρ is the gas density and v – particle velocity.

There is no dependence on spatial position in this model, so that we can speak about CR number density $N_i(E)$ where index i denotes a particular isotope, and its production rate density $Q_i(E)$ (e.g. in $m^{-3} \cdot s^{-1} \cdot \text{GeV}^{-1}$). The equilibrium value of $N_i(E)$ depends, of course, not only on the escape rate but on nuclear fragmentation rate, on the isotope "i" production rate by heavier "parents" in the ISM, on the energy losses and radioactive decay of the unstable isotopes. This is described by the equation:

$$Q_i(E) + \sum_j \frac{N_j(E)}{T_{ji}} = \left(\frac{1}{T_{esc}} + \frac{1}{T_i} + \frac{1}{T_i^r} \right) N_i(E) + \frac{\partial}{\partial E} \left(\frac{dE}{dt} N_i(E) \right) \qquad (9)$$

The left hand side describes production rates and the right hand side – the loss rates. T_{ji} is the mean time for production of isotope 'i' by isotope 'j' ; T_i and T_i^r are the mean times for fragmentation and decay of isotope 'i' respectively. The last term on the r.h.s. describes a change rate of $N(E)$, $\partial N/\partial t$, caused by particle energy loss rate dE/dt (mainly ionisation).

Dividing formula (9) by ρv we obtain an analogous formula, with all times substituted by corresponding path lengths and all rates per unit time substituted by rates per unit path lengths (per $g \cdot cm^{-2}$). Neglecting (for the time being) energy losses one gets very simple solutions. For a primary isotope (with a negligible contribution of its production during propagation of heavier nuclei) we get

$$N_p = \frac{Q_p}{1/\lambda_p + 1/\lambda_{esc}}$$

and for a secondary ($Q_s = 0$):

$$N_s = \frac{\sum_j N_{pj}/\lambda_{pj,s}}{1/\lambda_s + 1/\lambda_s^r + 1/\lambda_{esc}}$$

For one primary the secondary to primary ratio equals

$$\frac{N_s}{N_p} = \frac{1}{\lambda_{ps}} \cdot \frac{1}{1/\lambda_s + 1/\lambda_s^r + 1/\lambda_{esc}} \tag{10}$$

For a stable secondary ($\lambda_s^r = \infty$) with a long fragmentation path length ($\lambda_s \gg \lambda_{esc}$) we obtain a simple relation

$$\frac{N_s}{N_p} \cong \frac{\lambda_{esc}}{\lambda_{ps}} \tag{11}$$

which is the same as (4) applied to the leaky box ($\bar{x} = \lambda_{esc}$). The above formulae (10) and (11), although approximate, illustrate well by what main quantities a sec/prim ratio is determined. We see that a ratio decreasing with energy means a decreasing of $\lambda_{esc}(E)$.

There is, however, an exact solution to the leaky box equation (9) which can be expressed in an integral from:

$$N_i(E) = \int_0^\infty dx \left| \frac{h(E')}{h(E)} \right| q_i(E') \cdot$$

$$\cdot \exp \left\{ -\int_0^x \left[\frac{1}{\lambda_i(E'')} + \frac{1}{\lambda_i^r(E'')} + \frac{1}{\lambda_{esc}(E'')} \right] dy \right\} \tag{12}$$

where $h(E) = dE/dx$ and $E' = E - \int_0^x h[E''(y)]dy$. The exponential factor describes the probability that a nucleus, produced with energy E', will survive path length x. $E''(y)$ is the particle energy after traversing path length y. $q_i(E')$ is the total production rate (per $g \cdot cm^{-2}$), i.e. including production in the gas by heavier nuclei (corresponding to the l.h.s. of (9)). The formula describes both primary and secondary nuclei, which in the leaky box model differ only by the source term.

Coming back to Fig. 3a, where experimental results on B/C are presented, we can see that below ~ 1 GeV/n the ratio decreases with *decreasing* energy. A similar behavior, although less significant, can be seen in Fig. 3b for the (Sc+Ti+V)/Fe ratio. Therefore it is impossible to fit these observations by

$\lambda_{esc}(E)$ decreasing with energy in its whole range, at least, in this model. The usually adopted form of λ_{esc} to fit the data is the following

$$\lambda_{esc}(R) = \begin{cases} \lambda_o \beta & \text{for} \quad R \leq R_o \\ \lambda_o \beta \left(\frac{R}{R_o}\right)^{-\alpha} & \text{for} \quad R \geq R_o \end{cases} \tag{13}$$

where βc is particle velocity and its rigidity R rather than energy per nucleon is being used (for reasons explained earlier). Fits made by different authors give slightly different values for the parameters λ_o, R_o and α. Those given by Webber et al. (14) are: $\lambda_o = 14.5$ g $\cdot$ cm^{-2}, $R_o = 3.3$ GeV and $\alpha = 0.5$.

Jones et al. (15) find (strictly speaking –within a framwork of another, disk-halo diffusion model which, however, is equivalent to the leaky box for a large halo) that

$$\lambda_o = 11.8 \text{ g} \cdot \text{cm}^{-2}, \qquad R_o = 4.9 \text{ GeV} \qquad \text{and} \qquad \alpha = 0.54 \tag{14}$$

The differences are probably caused by differents sets of data used for fitting –Webber et al. include ^{3}He/^{4}He ratio as well as the other two. Solid lines in Fig. 3a,b show the leaky box prediction (as well as that of the diffusion model, but see later) for the parameters (14). We stress here that the same $\lambda_{esc}(R)$ fits both ratios very well. There is no wonder that it could fit one ratio, as the parameters could have been fitted just to describe it. The fact that the unique $\lambda_{esc}(R)$ describes well the two, quite different, sets of nuclei confirms our conjecture that CR propagation depends on particle rigidity and that the source composition can be treated as unique.

To be able to calculate accurately the nuclei fluxes it is necessary to know many fragmentation cross-sections and their behavior with energy. Above some 2 GeV/n most cross-sections depend on energy rather weakly. At lower energies, however, the dependencies may be significant and various (compare §5.1). Calculating a secondary flux it is necessary to include all possible parents, even themselves being secondaries because the total contribution of many "weak" parents may be significant. Thus, all these partial cross-sections have to be known.

Many important σ's have been measured in the last decade by Webber et al. [16] and by the Transport Collaboration [17]. To deal with the unmeasured ones, two groups have worked out phenomenological approaches predicting σ's below ~ 2 GeV [18, 19]. Although it happens that the earlier predicted σ's differ from their values measured later (sometimes by several standard deviations) [29] the work has proved to be very useful for propagation calculations.

4.4. The diffusion model

Although the leaky box model can reproduce quite well the sec/prim ratios below several tens of GeV/n, it is not obvious that it will do so at higher energies. Actually, there is a slight indication (see Fig.3a,b) that there might be too many secondaries above some 50 GeV/n. If, however, λ_{esc} continued to fall with rigidity as indicated by (13) and (14) it would become very short at high energies, meaning that those particles should not spend long times in a confinement region but should travel to us more and more from the directions of their sources. As we believe that the sources are in the Galactic disk rather than far away from it we should expect a significant anisotropy. This, however, is not observed causing a problem.

It is obvious that, although the leaky box model describes remarkably well the low energy data its simple assumptions are not satisfactory for our understanding of the actual CR propagation mechanism. We would like to be able to explain the data in a more realistic picture of the Galaxy and CR confinement region. A step in this direction is to assume that CR are produced in the Galactic disk and their propagation mechanism is diffusion in a region extending further up from the Galactic plane than does the gas. This is the disk-halo diffusion model.

Let us first justify why diffusion can be used as a propagation mechanism. The Galactic magnetic field contains a significant irregular component: $B_{irr} \sim$ a few μG with a decoherence distance of the order of ≤ 102 pc. These fields may be caused by many Alfven waves propagating along the regular field $\vec{B}_{reg}$, by chaotic gas motions with the field lines frozen in the gas and/or by larger scale turbulences produced by SN shocks passing though the interstellar medium. The Larmor radius of a nucleus with $E_{kin} = 1$ GeV/n and $Z/A = 0.5$ in a field 3μG is only $4 \cdot 10^{12}$ cm ($\sim 1/4$ a.u.), so particles are tied up to the field lines but can be scattered by Alfven waves. Whether it is this scattering or the field line motions that determine particle large scale propagation it is a question of debate. If it was the former mechanism then particle would be most effectively scattered by Alfven waves with wave lengths close to the particle Larmor radius (so called resonant scattering). For a given power spectrum $P(k)$ of the waves (energy density contained in waves of wave number k per unit k) it is possible to derive particle diffusion coefficient D. If $P(k) \sim k^{-\alpha}$ then

$$D \sim vR^{2-\alpha} \tag{15}$$

The most often considered $P(k)$ is that derived by Kolmogorov to describe a distribution of structure sizes in a turbulent motion. From very general considerations (energy is being supplied to the biggest structure and being fully transferred to the smaller and smaller ones, so that a dynamical equilibrium exists, with simple relations between structure size, its velocity and time to transfer energy) he obtained that $P(k) \sim k^{-5/3}$. Rigidity dependence of the diffusion coefficient would be then

$$D \sim v R^{1/3}$$

We shall come back to the implications of this result after discussing in more detail the diffusion model.

From the above considerations it follows that large scale gas motions (e.g. galactic wind) may cause CR large scale drifts, not described by diffusion, corresponding to a convection, which should also be considered.

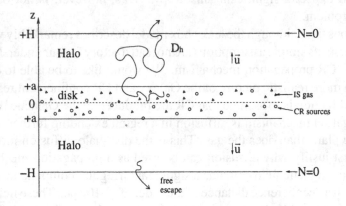

Figure 6. Cross section of the Galaxy, perpendicular to the Galactic disk. A picture in the CR diffusion model (see text) where any Galaxy and CR characteristic depends on z only

As the thickness of the ISM disk is small in comparison with its radial size, one can consider the Galaxy characteristics as functions of one spatial dimension z only, perpendicular to the Galactic disk. The situation is represented in Fig.6. CR originate uniformly in the disk of total thickness $2a$ (in length units) containing also uniform distribution of interstellar gas. CR diffuse in the disk with a diffusion coefficient D_d and to the halo (with size $H - a$ on both sides of the plane) where they diffuse with a different D_h. Outward convection with velocity u in the halo can also be taken into consideration. In the equilibrium that should be achieved, CR particle density $N_i(p, z)$ (per unit momentum) obeys the following general equation:

$$\frac{\partial}{\partial z} D \frac{\partial N}{\partial z} - u \frac{\partial N}{\partial z} + \frac{\partial u}{\partial z} \frac{p^3}{3} \frac{\partial(N/p^2)}{\partial p} - \frac{\partial}{\partial p}\left(\frac{dp}{dt}N\right) - \frac{N}{T_{nuc}} - \frac{N}{T_r} + Q = 0 \quad (16)$$

for any isotope i (for simplicity we have omitted index i). T_{nuc} and T_r are the nucleus mean times for fragmentation at z and for radioactive decay respectively. The source $Q(p, z)$ includes also a contribution from fragmentation of heavier nuclei. The first term describes $\partial N/\partial t$ caused by diffusion, where D may depend on z. The second term corresponds to particle convection. The third term describes $\partial N/\partial t$ that would be caused if the convection velocity depended on position. Then expansion (or compression) of the gas would occur and, as magnetic field is frozen in it, particle would lose (gain) energy. This is called an adiabatic expansion. The forth term is analogous to that in the equation (9) for leaky box, describing changes of the particle spectrum due to energy losses for gas ionisation, expressed in terms of momentum p. The boundary conditions for any p are:

$N(z = \pm H) = 0$ free escape at halo boundary

$N(a + \varepsilon) = N(a - \varepsilon)$ for $\varepsilon \to 0$ (continuity of particle density at disk-halo boundary)

$Flux|_{a+\varepsilon} = Flux|_{a-\varepsilon}$ for $\varepsilon \to 0$ (continuity of particle flux at disk-halo boundary)

For the particular case of the situation from Fig. 6 most of the terms in the halo will be equal to zero (all apart from the first two). In the disk the second and third term will vanish ($u = 0$, $\partial u/\partial z = 0$). Particularly simple solutions can be obtained in an approximation that $a \ll H$, $D_h(z) = const$ and $u = 0$ (see e.g.[12]). Introducing particle flux $J(E_k)$ per unit kinetic energy per nucleon so that $J(E_k)dE_k = v \cdot N(p)dp$, one obtains from (16) an equation for $J(E_k)$ observed in the disk ($z = 0$):

$$\left(\frac{1}{\lambda_{diff}} + \frac{1}{\lambda_i} \right) J + \frac{d}{dE_k} \left(\frac{dE_k}{dx} J \right) = Q' \tag{17}$$

where

$$\lambda_{diff} = \frac{v\rho a H}{D_h} \quad \text{and} \quad Q' = \frac{A}{2v\rho a} \cdot \int_{-a}^{+a} Q dz \tag{18}$$

A is the atomic number of the CR particle. Comparing it with equation (9) we see that (17) is the same as that for the leaky box model with $\lambda_{esc} = \lambda_{diff}$ and the source term being the average CR production rate per unit volume in the disk. The path length distribution must then be exponential with the mean equal λ_{diff}. Thus, it has been proved that the diffusion model is equivalent to the leaky box if the disk size is much smaller than that of the halo. It can be understood as follows: CR particles can quickly reach disk-halo boundary (small disk) getting a small chance to diffuse through the large halo and leak out of the Galaxy; it is much more probable than the particle returns to the disk, repeating this situation many times as it is in the leaky box model. If this is so, the solution of the diffusion model in the approximation $a \ll H$ is given by (12), where $\lambda_{esc} = \lambda_{diff}$ given by (18).

It turns out that the inclusion of the convection (keeping $a \ll H$ valid) results in a similar situation [15]: one obtains a leaky box equation, similar to (16), where the role of λ_{esc} is played by

$$\lambda_w = \frac{\upsilon \rho a}{u} \left[1 - \exp \left(-\frac{uH}{D_h} \right) \right] \tag{19}$$

for an observer at $z = 0$. The energy loss term includes now an additional dE_k/dx term, due to gas expansion at the disk – halo boundary (the mentioned adiabatic energy losses):

$$\frac{dE}{dx} \bigg|_{ad} = -\frac{up}{3c\rho a}$$

where p is particle momentum per nucleon. Relation (19) brings a new feature to the energy dependence of λ_w: as it is natural to assume that convection velocity does not depend on energy it follows from (19) that for energies such that $D_h < uH$, diffusion becomes unimportant and $\lambda_w \approx \frac{v}{u}\rho a$, becoming energy independent. Therefore, convection could explain the behaviour of λ_{esc} in the leaky box model (flattening at low R) which in this model had to be put "by hand" (see formula (13)).

We refer the reader to the instructive paper by Jones et al. [15] to learn about another considered case of turbulent diffusion (large scale turbulences cause energy independent particle diffusion (convection by turbulence) together with energy dependent diffusion in the gas). Also in this case one obtains a flattening of λ_{esc} at low rigidities for the same reason.

Before discussing the results of calculations of the measurable sec/prim ratios for the models described above, we shall consider the case where the disk thickness a is not necessarily much smaller than that of the halo, H. It is not our aim to find solutions to equation (16) with all possible terms included and the boundary conditions discussed above (which is not a simple task). We want, however, to illustrate by some simple examples how to find the lifetime and path length distributions.

Let us consider the simple case when there is no convection, diffusion is z-independent and energy losses can be neglected. For the steady state (equilibrium) we have

$$D_d \frac{\partial^2 N}{\partial z^2} + q = 0 \text{ in the disk}, \quad \text{and } D_h \frac{\partial^2 N}{\partial z^2} = 0 \text{ in the halo} \tag{20}$$

where q is the source rate per unit length along z (different for primaries and secondaries). The boundary conditions are as above. In particular the net particle flux on both sides of the disk-halo boundary must be same:

$$D_d \frac{\partial N}{\partial z} \bigg|_{a-\varepsilon} = D_h \frac{\partial N}{\partial z} \bigg|_{a+\varepsilon} \quad \text{for } \varepsilon \to 0 \tag{21}$$

and equal to $1/2$ of the total source production $= \int_0^a q(z)dz$. At each point particles have some vacuum distribution of lifetimes $n(t; z)$. If the equilibrium was reached long time ago, then

$$N(z) = \int_0^\infty n(t; z)dt \qquad (22)$$

and $n(t; z)$ fulfil the time dependent equation

$$\frac{\partial n}{\partial t} = D_d \frac{\partial^2 n}{\partial z^2} + Q \cdot \delta(t) \qquad (23)$$

and without the source term – in the halo. In other words, if the particles were produced only at an instant $t = 0$, their density at a given point z would change with time as $n(t; z)$. In the equilibrium we would see all of them simultaneously at any time, what is expressed by (22).

Note that we have omitted the nuclear fragmentation term (as well as that for radioactive decay), and that was because we wanted to find the *vacuum* lifetime distribution. Solving (23) by variable separation one obtains a solution of the following from:

$$n(t; z) = \sum_{i=1,3,5\ldots}^{\infty} A_i \cos \frac{i\pi z}{2H} \exp(-i^2 \alpha_i t) \qquad (24)$$

where constants A_i and α_i depend on boundary conditions and the source term. Thus, in general, the age distribution is a sum of many exponential terms, while for $a \ll H$ only the first term becomes important.

To find the actual density N' of a given stable isotope one has to find the solution of (20) with the nuclear fragmentation in the disk included. To find the same for a secondary is not easy by this way, as the source term depends on position. If we knew, however, the distribution of path lengths t_d traversed by the primaries in the disk, we could find the measurable sec/prim ratio using (3) (and allowing for more than one primary). To find it [20] let us note that

$$N'(z; T_i) = \int_0^\infty f(t_d; z) \exp(-t_d/T_i)dt \qquad (25)$$

where $\int_0^\infty f(t_d; z)dt = N(z)$ is the solution to (20) without the fragmentation term. The factor $\exp(-t_d/T_i)$ is just the probability that the nucleus survives time td in the disk against nuclear interaction. Let us note further that $N'(T_i)$ is just the Laplace transform of $f(t_d)$ (for any z). Thus, the inverse transform is:

$$f(t_d) = \frac{1}{2\pi i} \int_{x-i\infty}^{x+i\infty} N'(\nu) \exp(\nu t_d)d\nu \qquad (26)$$

where the integration line is to the right of all the poles of $N'(\nu)$, a function of the complex variable $\nu = 1/T_i$. The integral can be found (by closing this line on the left side in infinity) as a sum of the integrand residua

$$f(t_d; z) = \sum_m r_m(z) \exp(-\beta_m t_d) \qquad (27)$$

Thus, the distribution of lifetimes t_d in the disk has a similar form (a sum of exponential terms) as that of the total particle age t. And again only the first term becomes important for $a \ll H$.

If we are interested in the mean $\overline{t_d}$ only, we can calculate it using relation (25), *without* calculating the distribution $f(t_d)$. For $T_i \to \infty$ (vacuum situation) we can expand the exponential factor and keep the first two terms only:

$$N'(T_i) \approx_{T_i \to \infty} \int_0^\infty f(t_d) \left(1 - \frac{t_d}{T_i}\right) dt_d = N(z) \left(1 - \frac{\overline{t_d}}{T_i}\right)$$

so

$$\overline{t_d} = \lim_{T_i \to \infty} \left[\frac{N - N'(T_i)}{N} \cdot T_i\right] \qquad (28)$$

Thus, having $N(z)$ and $N'(z)$ we can determine $\overline{t_d}(z)$. Similar treatment can be applied successfuly (analytical solutions can be found) also if convection is taken into account [20].

4.5. Diffusion model with reacceleration

So far it has been assumed that CR particles can only lose energy while propagating through the ISM. However, if their diffusion is caused by scattering on Alfven waves (or on other moving magnetic irregularities) then, in general, each change of particle direction is associated with a change of its energy. The latter process can be described as a diffusion in particle momentum space with a diffusion coefficient $K(p)$; that is: $\overline{dp^2} \sim K(p)dt$. The two diffusion coefficients have to be related and for the scattering on the Alfven waves it can be proved that

$$K(p) \cdot D(p) = \frac{V_A^2}{9} p^2 \qquad (29)$$

where V_A is the Alfven velocity of the waves. A typical relative momentum change is then

$$\frac{1}{p} \sim \frac{1}{p^2} \frac{dp^2}{dt} \sim \frac{K(p)}{p^2} \sim \frac{1}{D(p)} \qquad (30)$$

This means that if $D(p)$ is an increasing function of p, then the relative acceleration rate decreases with p. It can be calculated that at $E_{kin} \leq 1$ GeV/n and the diffusion coefficient $7 \cdot 10^{29}$ cm$^2 \cdot$ s^{-1} [15] the r.m.s. of momentum change within CR lifetime $\sim 2 \cdot 10^7$ years (see later) would correspond to

$\geq 20\%$ increase in kinetic energy. At $E_{kin} \geq 20$ GeV/n the effect would be negligible.

An existence of such a reacceleration, with its role decreasing with energy, would have an important effect on the sec/prim ratios and on our deduction about the dependence of CR life time (diffussion coefficient) on energy. The reacceleration would steepen the sec/prim ratios: the secondaries produced at lower energies would be observed at an energy shifted to the right on the logarithmic scale by a factor larger than those produced at higher energies. (Note, that if the relative reacceleration rate $\frac{1}{p}\frac{dp}{dt}$ was constant, the sec/prim ratio would not change its shape, being only shifted paralelly towards higher energies). We have discussed in §4.4 that for the Kolmogorov power spectrum of Alfven waves one would expect that $D(p) \sim p^{1/3}$, whereas the observed sec/prim ratios fall as $p^{-(0.5 \div 0.6)}$. Now, this discrepancy could be explained because a part of this steep dependence would be attributed to the reacceleration.

5. Model predictions, comparison with observations and discussion

5.1. Stable isotopes

Predictions of all the described above models for the two main sec/prim ratios, B/C and (Sc+Ti+V)/Fe, have been calculated by Jones et al. and are presented in Fig.3a,b, together with observations from many experiments. To any model its free parameters have been adopted to fit best *simultaneously* the two ratios. It can be seen that the fit is practically equally good for any model ($\chi^2 = 1.3 \div 1.8$ for one degree of freedom). The turbulent diffusion model, however, gives the turbulence scale comparable to the halo size, which is unphysical. The fitted halo size is $H = 2.1$ kpc for the halo diffusion model. In the standard diffusion model (no convection) $D \sim R^{0.54}$ (above 4.9 GV), whereas for the reacceleration model the authors obtain $D \sim R^{0.3}$, compatible with the Kolmogorov spectrum of Alfven waves. Unfortunately, basing on these two sets of observations it is impossible to discerne between the models.

There are, however, some ways which, with accurate flux and cross section measurements, could tell us something about the possible role of reacceleration. One method [23] takes advantage of the energy dependence (usually an unwanted thing) of partial cross sections. Let us consider a primary producing two different secondaries such that the ratio of the corresponding partial cross sections depends on energy. The observed ratio of the two secondaries, depending rather weakly on the primary path length distribution, is determined by the ratio of the two partial cross sections at the *production* energy. If there was an acceleration it would correspond to the different cross section ratio at a lower energy. Such a situation concerns the secondaries of iron - σ's for production of heavier fragments fall with energy (between 0.3 and 2 GeV/n), whereas σ's for the lighter ones tend to increase. For example, the ratio $\sigma(\text{Fe} \rightarrow \text{Cr})/\sigma(\text{Fe} \rightarrow \text{K})$ at 0.3 GeV/n is almost 4 times larger than that at 2 GeV/n. Thus, a big observed ratio Cr/K at 2 GeV would indicate a reacceleration. Unfortunately, the uncertainties in the partial cross sections and their energy dependences as well as those of the interstellar ratios (solar modulation is effective at ≤ 1 GeV/n but still, to some extent, uncertain) are too large for such an analysis to be conclusive.

Another method, which has been recently revived with the new measurements of Ulysses and CRIS, is based on an analysis of the electron-capture secondary isotopes [24, 25]. The very existence of such isotopes in the CR flux proves that they must spend most of their propagation time at energies when e-capture is not probable (the relative electron velocity must be much larger than that for the inner electron orbit, $v/c \sim Z/137$), that is already at relativistic energies, because the half-lives for this process range from a few to several hundreds of days. If some part of the interstellar propagation time was spent at a low enough energy then the e-capture process could occur. Of particular interest is the ratio $^{49}\text{V}/^{51}\text{V}$. Both isotopes are the products mainly of the same primary (Fe). ^{49}V decays by e-capture whereas ^{51}V is a result of ^{51}Cr e-capture decay (^{51}Cr being also a secondary from Fe). Thus, if e-capture was possible ^{49}V flux would decrease whereas that of ^{51}V would increase. The $^{49}\text{V}/^{51}\text{V}$ ratio is practically unsensitive to path length distribution, as both isotopes have similar masses and thus similar fragmentation cross sections. It is, however, necessary to know well the production σ's of ^{49}V, ^{51}V, ^{51}Cr from iron and other contributing isotopes at different energies. The recent Ulysses results [25], where for the first time CR vanadium isotopes have been clearly separated, indicate a possible reacceleration (e.g. the measured $^{49}\text{V}/^{51}\text{V}$ ratio is about 20% lower (with an error of 10%) than that predicted without reacceleration. The predicted ratio, however, must also have an uncertainty and it is difficult to imagine that it is $\sim$10% only.

A corresponding analysis from CRIS would be much welcomed. So far the question, whether there is some CR reacceleration in the interstellar space, remains open.

5.2.   Radioactive isotopes

Of great interest for CR propagation studies are the radioactive secondaries with half-lifes comparable to the propagation time in the Galaxy. An estimation of the latter may be obtained from CR mean path length: $10 \text{ g} \cdot \text{cm}^{-2}$ of ISM would be traversed by a particle with velocity c, in the gas of the number density 1 atom cm^{-3}, in about 5 mln years. But an independent determination of the CR lifetime can be obtained by measuring arriving (relative) fluxes of the radioactive isotopes and comparing them to those expected if no decay had occured. There are four suitable isotopes: ^{10}Be ($\tau_{1/2} = (1.5 \pm 0.3) \cdot 10^6$ yr), ^{26}Al ($7.1 \cdot 10^5$ yr), ^{36}Cl ($3 \cdot 10^5$ yr) and ^{54}Mn ($\sim 6.3 \cdot 10^5$ yr, when stripped of electrons). ^{10}Be was the first radioactive isotope studied, from which CR age was estimated for the first time about 20 years ago, as $\sim 2 \cdot 10^7$ years (in the leaky box model) being ~ 4 times larger than that for the mean gas density in the Galactic disk of 1 atom cm^{-3}.

This lifetime can be easily derived in the framework of the leaky box model (neglecting energy losses), when the ratio of ^{10}Be (or ^{7}Be), the stable isotopes, is measured:

$$\frac{^{10}\text{Be}}{^9\text{Be}} = \frac{\lambda_9}{\lambda_{10}} \cdot \frac{1/\lambda_9 + 1/\lambda_{esc}}{1/\lambda_{10} + 1/\lambda_{esc} + 1/\lambda_{10}^r} \qquad (31)$$

where λ_9, λ_{10} are the mean path lengths for ^{9}Be and ^{10}Be production by carbon (for simplicity, we omit other contributions) and $\lambda_{10}^r = \rho v T_{10}^r$ is the mean path for ^{10}Be decay with the mean time T_{10}^r. The only unknown in (31) is the density ρ of ISM, which can be calculated once ^{10}Be/^{9}Be is measured. Fig.7, taken from Simon and Molnar [26], with the CRIS point [6] added, shows a compilation of the ^{10}Be/^{9}Be measurements (see also Fig. 5a to compare accuracies of isotope separation), together with predictions of the leaky box and diffusion models. The latest and apparently most acurate measurement by CRIS corresponds to a number density of gas slightly larger than 0.3 H atoms cm^{-3}. It can be calculated that the surviving fraction of ^{10}Be (the ratio of the observed ^{10}Be flux to that expected, had ^{10}Be been stable) is only about 0.2.

CRIS group have also measured ratios of other radioactive "clocks" to their neighboring stable isotopes. Their results are confronted with the previous experiments on Fig. 8 from [6]. The horizontal "error bars" for all points are the energy intervals for which the ratios have been measured. We can see that the accuracy in the ratios determination is by far best in CRIS. CR escape life times determined by CRIS from each ratio coincide (apart from that from ^{54}Mn), giving a value $T_{esc} = 22$ mln years. For the diffusion model this would need a ~ 3.5 kpc halo.

It should be noted, however, that the various experiments have been performed at different periods of the solar cycle through which the solar modulation effect is varying. A ratio measured at an energy within the Solar System corresponds to a higher interstellar energy and it should be compared with the calculated curves drawn at that energy. For example, for the period of CRIS measurement (1997, around solar minimum) the modulation parameter Φ=555 MV, which corresponds to a typical change of nucleus energy per nucleon by $\Delta E_k/n = \frac{Ze}{A}\Phi = 222$ MeV/n for ^{10}Be. As the predictions are rather flat below several hundreds MeV/n, allowing for solar modulation would not change much the conclusions about mean gas density and CR lifetime.

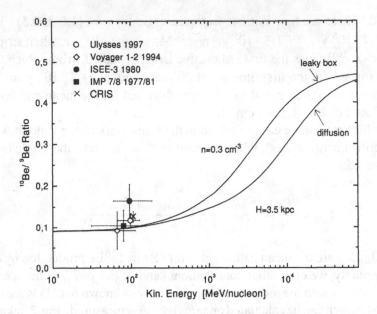

Figure 7. ^{10}Be/^{9}Be flux ratio measured by various experiments versus measured energy. Lines are calculations [26] for the leaky box model with gas number density $n = 0.3$ H-atom cm^{-3} and diffusion in the $H = 3.5$ kpc halo and $a = 0.1$ kpc disk with $n = 1$ H-atom cm^{-3}

Measurements of the surviving fractions of the radioactive nuclei in a larger energy range would be a test for propagation models. Simple assumptions that halo diffusion coefficient does not depend on the height above the Galactic plane, or that the halo size does not depend on particle energy may turn out not to be true. The role of convection and/or reacceleration awaits also its verification.

6. Source composition

In the previous chapters we have shown that having several sec/prim ratios, with a secondary having only one parent, it is possible in principle to determine the path length distribution of the primaries. Having this it is possible to calculate the necessary CR composition (chemical or even isotopic) such that after propagating with the obtained earlier path length distribution, the resulting fluxes of various isotopes would match those observed. The determination of the Galactic CR source (GCRS) composition is, of course, of crucial importance for their origin. It is then natural to compare the GCRS composition with the solar one. The history of interpretations of the GCRS composition is complicated and here we shall describe only the recent, most seriously founded picture, proposed by J.P.Meyer, L.O'C.Drury and D.C.Ellison [27]. Their main ideas are illustrated in Fig.9 where the GCRS abundances, relative to solar, are

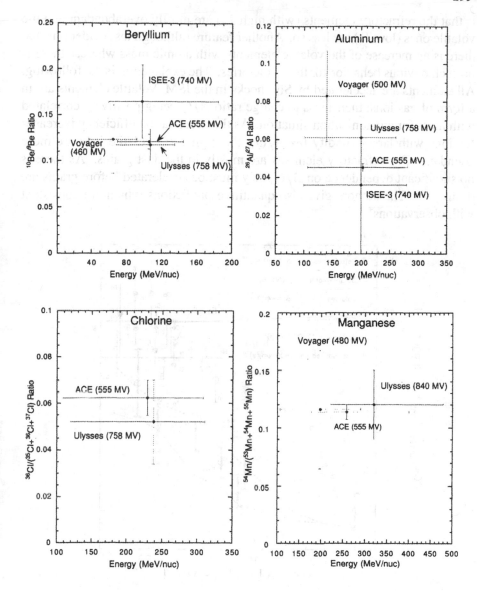

Figure 8. Comparison of four radioactive isotope abundance ratios measured by various experiments (taken from [6]). Horizontal bars are energy intervals for which a ratio has been determined

presented as a function of the element atomic mass A (with hydrogen as the normalization point). The authors devide elements according to their ability to condense into solid compounds, that is according to condensation temperature T_c (at a very low pressure), as marked on the figure. The conspicuous feature is that the refractory elements (with high T_c) are mostly overabundant and the volatile ones (low T_c) are less so. Another feature (although less evident) is that there is an increase of the volatile elements with atomic mass whereas there is no such obvious behavior for the refractories. The explanation is the following. All elements are accelerated by SN shocks in the ISM. Volatile elements are in a form of gas ions, their mass to charge ratio A/Q being positively correlated with mass for any ionization situation. As the acceleration efficiency increases for ions with larger rigidity ($\sim A/Q$), the high A/Q volatile ions are more abundant. The refractory elements are mostly in the dust grains. As there is no significant dependence on A/Q, they must be accelerated before grains are destroyed. The authors give also quantitive predictions which are consistent with observations.

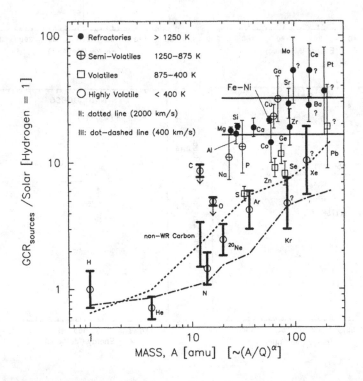

Figure 9. Galactic CR source abundance relative to solar abundance vs. atomic mass number A (taken from [27]), normalized to hydrogen. Dotted and dot-dashed lines are predictions for SN shock acceleration with quoted shock velocities. Two thick horizontal lines contain most of the refractory elements

The interesting thing is that it is not the freshly synthesized SN material which undergoes acceleration but it is the interstellar matter and that it is the *atomic* properties of its constituents which determine the composition of the produced high energy particles. This picture is in accordance with another recent interesting discovery by CRIS, about absence of the isotope ^{59}Ni in the CR flux [28]. This isotope decays by e-capture with $T_{1/2} = 7.6 \cdot 10^4$ years. It should be synthesized in the SN explosion and if this material was to be accelerated by the blast wave of the same explosion, ^{59}Ni would be stripped of its electrons and would be present in the CR flux. Its absence strongly suggests that there is at least 10^5 years time delay between nucleosynthesis and CR acceleration – a result of great importance for models of CR origin.

7.   Conclusions

There is a wealth of information about CR propagation and chemical isotopic composition of their sources. Production of the secondary nuclei in collisions with the ISM can be studied quite well thanks to the fact that some nuclei are absent in the CR sources. The derived path length distribution is consistent with the exponential one, meaning that a halo large in comparison with the disk (gas and CR source region) is the necessary CR propagation region. Measurements of radioactive secondaries confirm this picture, indicating that the mean gas density of the CR confinement volume is several times lower than that in the disk. An increasing role of a convection in the halo of the lower energy particle would explain in a natural way the behaviour of the mean escape path length with energy: flat up to $\sim$ 2 GeV/n, then decreasing with energy. This rather quick decrease, derived in the framework of the leaky box model, is a little worrying, as it is difficult to obtain it from any reasonable power spectrum of Alfven waves and, when extrapolated to higher energies, would be in conflict with the small CR anisotropy observed. An existence of some CR reacceleration in the ISM would solve these problems. Studies of some secondary to secondary ratios as well as those of the e- capture decaying isotopes can, in principle, verify the reacceleration hypothesis. Further studies of secondary to primary ratios at higher energies, as well as the radioactive isotope measurements as a function of energy would allow to distinguish better between the propagation models.

For a given path length distribution, CR source composition can be obtained. When compared to that of the Solar System it shows a large overabundance of the refractory elements present mostly in the interstellar grains and, to a lesser extent, a mass dependent overabundance of the volatile ones, present in the gas. This picture confirms nicely the SNR CR acceleration mechanism.

The fluxes of the ultra heavy elements ($Z > 28$) are still rather badly separated, not mentioning that their isotopic composition is not known. Thus, a lot of precious information that could be drawn from their studies (not only about CR propagation along short path lengths but also about the nucleosynthesis of the accelerated material) is yet unavailable. An example may be the non-observation of the ^{59}Ni isotope by the CRIS experiment - meaning that the accelerated material must be least 10^5 years old.

New ambitious experiments planning to cover the non-measured charge and energy regions with a better precision would be much welcomed. One of them, the ACCESS (Advanced Cosmic-ray Composition Experiment for Space Station) has a goal to measure the rare ultra-high energy and ultra-heavy nuclei [30] and its installation on the International Space Station would enable to obtain the necessary big collecting factor.

Acknowledgments

The author thanks the Organisers of the School for the invitation to give the lectures and for their all support. This work has been also sponsored by the Polish State Committee for Scientific Research under grant no 2 P03C 006 18.

References

[1] Engelmann, J.J. et al. (1990) Charge composition and energy spectra of cosmic-ray nuclei for elements from Be to Ni. Results from HEAO-3-C2, *Astron.Astrophys.* **233**, 96-111.

[2] Müller, D. et al. (1991) Energy spectra and composition of primary cosmic rays, *ApJ* **374**, 356-365.

[3] Connell, J.J. and Simpson, J.A. (1997) Isotopic Abundances of Fe and Ni in Galactic Cosmic-Ray Source, *ApJ* **475**, L61-L64.

[4] Lukasiak, A. et al. (1999) Voyager Measurements of the Charge and Isotopic Composition of Cosmic Ray Li, Be and B Nuclei and Implications for Their Production in the Galaxy, *Proc. 26th ICRC* (Salt Lake City) **3**, 41-44.

[5] Connell, J.J. (1998) Galactic Cosmic-Ray Confinement Time: Ulysses High Energy Telescope. Measurements of the Secondary Radionuclide ^{10}Be, *ApJ* **501**, L59-L62.

[6] Binns, W.R. (1999) Radioactive Clock Abundance Measurements from the CRIS Experiment aboard the ACE Spacecraft, *Proc. 26th ICRC* (Salt Lake City) **3**, 21-24.

[7] Lukasiak, A. et al. (1995) Voyager Measurements of the Isotopic Composition of Sc, Ti, V, Cr, Mn and Fe Nuclei, *Proc. 24th ICRC* (Rome) **2**, 576-579.

[8] Wiedenbeck, M.E. et al. (1999) The Isotopic Composition of Iron, Cobalt and Nickel in Cosmic-Ray Source Material, *Proc. 26th ICRC* (Salt Lake City) **3**, 1-4.

[9] Giler, M. et al. (1987) On the continuous acceleration of cosmic rays in the ISM, *Proc. 20th ICRC* (Moscow) **2**, 214-217.

[10] Cowsik, R. et al. (1967) Steady State of Cosmic-Ray Nuclei – Their Spectral Shape and Path Length at Low Energies, *Phys.Rev.* **158**, 1238-1242.

[11] Giler, M. and Wibig, T. (1990) Reacceleration and the path length distribution, *Proc. 21st ICRC* (Adelaide) **3**, 365-368.

[12] Cowsik, R. and Wilson, L.W. (1975) The nested leaky-box model for Galactic cosmic rays, *Proc. 14th ICRC* (Munich) **2**, 659-664.

[13] French, D.K. and Osborne, J.L. (1976) Comments on a closed galaxy model for cosmic-ray propagation, *Nature* **260**, 372.

[14] Webber, W.R. (1999) A Study of the Propagation of Cosmic Rays in the Galaxy Using a Monte Carlo Diffusion Model, *Proc. 26th ICRC* (Salt Lake City) **4**, 222-224.

[15] Jones, F.C. et al. (2001) The modified weighted slab technique: models and results, *ApJ* **547**, 264-271.

[16] Webber, W.R. et al. (1998) Production cross section of fragments from beams of 400–650 MeV per nucleon ^{9}Be, ^{11}B, ^{12}C, ^{14}N, ^{16}N, ^{16}O, ^{20}Ne, ^{22}Ne, ^{56}Fe, and ^{58}Ni nuclei interacting in a liquid hydrogen target, I. Charge changing and total cross sections, *ApJ* **508**, 940-948.

Webber, W.R. et al. (1998) Production cross section of fragments from beams of 400–650 MeV per nucleon ^{9}Be, ^{11}B, ^{12}C, ^{14}N, ^{16}N, ^{16}O, ^{20}Ne, ^{22}Ne, ^{56}Fe, and ^{58}Ni nuclei interacting in a liquid hydrogen target, II. Isotopic cross section of fragments, *ApJ* **508**, 949-958.

Webber, W.R. et al. (1990) Total charge and mass changing cross sections of relativistic nuclei in hydrogen, helium, and carbon targets, *Phys.Rev.C* **41**, 520-532.

Webber, W.R. et al. (1990) Individual charge changing fragmentation cross sections of relativistic nuclei in hydrogen, helium, and carbon targets, *Phys.Rev.C* **41**, 533-546.

Webber, W.R. et al. (1990) Individual isotopic fragmentation cross sections of relativistic nuclei in hydrogen, helium, and carbon targets, *Phys.Rev.C* **41**, 547-565.

[17] Chen, C.X. et al. (1997) Relativistic Interaction of ^{22}Ne and ^{26}Mg in Hydrogen and the Cosmic – Ray Implications (the Transport Collaboration), *ApJ* **497**, 504-521.

Chen, C.X. et al. (1994) Interactions in hydrogen of relativistic neon to nickel projectiles: Total charge-changing cross sections, *Phys.Rev.C* **49**, 3200-3210.

Knott, C.N. et al. (1996) Interaction of relativistic neon to nickel projectiles in hydrogen, elemental production cross sections, *Phys.Rev.C* **53**, 347-357.

[18] Webber, W.R. et al. (1990) Formula for calculating partical cross sections for nuclear reactions of nuclei with $E \geq 200$ MeV/nucleon in hydrogen targets, *Phys.Rev.C* **41**, 566-571.

[19] Silberberg, R. et al. (1998) Updated Partial Cross Sections of Proton-Nucleus Reactions, *ApJ* **501**, **911-919**, (.and references therein)

Tsao, C.H. et al. (1999) Updated Semiempirical Cross Sections for Cosmic Rays Propagation, *Proc. 26th ICRC* (Salt Lake City) **1**, 13-16.

[20] Freedman, I. et al. (1980) Derivation of the Age Distributions of Cosmic Rays in a Galaxy, *Astron.Astrophys.* **82**, 110-122.

[21] Simon, M. et al. (1986) Propagation of injected cosmic rays under distributed reacceleration, *ApJ* **300**, 32-40.

Heinbach, U. and Simon, M. (1995) Propagation of galactic cosmic rays under diffusive reacceleration, *ApJ* **411**, 209.

[22] Giler. M. et al. (1989) Distribution processes as contributions to the acceleration of cosmic rays, *Astron.Astrophys.* **217**, 311-318.

[23] Giler, M. et al. (1998) A method for determining the relative importance of the continuous accelaration process for cosmic rays, *Astron.Astrophys.* **196**, 44-48.

[24] Mahan, S.E. et al. (1999) Secondary Electron-Capture Clock Isotopes as a Probe of Reacceleration, *Proc. 26th ICRC* (Salt Lake City) **3**, 17-20.

[25] Connell, J.J. and Simpson, J.A. (1999) Ulysses HET Measurements of Electron-Capture Secondary Isotopes: Testing the Role of Cosmic Ray Reacceleration, *Proc. 26th ICRC* (Salt Lake City) **3**, 33-36.

[26] Simon, M. and Molnar, A. (1999) Test of Diffusion Halo and the Leaky Box Model by means of secondary radioactive Cosmic Ray Nuclei with different lifetimes, *Proc. 26th ICRC* (Salt Lake City) **4**, 211-214.

[27] Meyer, J.P. et al. (1997) Galactic cosmic rays from supernova remnants. I.A cosmic-ray composition controlled by volatility and mass – to – charge ratio, *ApJ* **487**, 182-196.

Ellison, D.C. et al. (1997) Galactic cosmic rays from supernova remnants. II. Shock acceleration of gas and dust, *ApJ* **487**, 197-217.

[28] Wiedenbeck, M.E. et al. (1999) Constraints on the time delay between nucleosyntesis and cosmic-ray acceleration from observations of ^{59}Ni and ^{59}Co, *ApJ* **523**, L61-L64.

[29] Guzik, G. for the Transport Collaboration (1997) The systematics of isotopic production cross sections and implications for cosmic ray astrophysics, *Proc. 25th ICRC* (Durban) **4**, 317-320.

[30] Wefel, J. and Wilson, T. (1999) The ACCESS Mission: ISS Accommodation Study, *Proc. 26th ICRC* (Salt Lake City) **5**, 84-87.

REACCELERATION OF GALACTIC COSMIC RAYS: SECONDARY ELECTRON CAPTURE ISOTOPES MEASURED BY THE COSMIC RAY ISOTOPE SPECTROMETER

SUSAN M. NIEBUR
Washington University
Campus Box 1105
One Brookings Drive
St. Louis, MO 63130

Abstract: Isotopes that decay only by electron capture decay preferentially at the lower cosmic-ray energies. Data from the Cosmic Ray Isotope Spectrometer on ACE show energy dependence in the abundances of these isotopes. We discuss these results and their application to reacceleration of galactic cosmic rays in the interstellar medium.

1. Introduction

It is generally agreed that galactic cosmic rays are initially accelerated by a supernova explosion, the only known mechanism that can impart the energy necessary to accelerate nuclei to cosmic ray energies. Since supernova shocks propagate through the interstellar medium, it is also likely that the cosmic rays encounter diffuse supernova remnant shocks later during propagation. Cosmic rays that encounter such supernova remnants would gain energy in a first-order Fermi process termed distributed acceleration [1] or continuous acceleration [2]. Cosmic rays could also gain energy in a second-order Fermi process in encounters with the associated Alfven waves or other turbulent magnetic fields in the interstellar medium [3, 4]. There are a number of propagation models that incorporate acceleration during propagation, referred to here as reacceleration.

All of these models can be brought into agreement with the elemental ratios measured to date, including secondary/primary ratios such as B/C and subFe/Fe, which are a measure of the amount of material that the cosmic rays traversed during propagation. However, there is now new isotopic data from the Cosmic Ray Isotope Spectrometer (CRIS) with such a large number of particles collected over its energy range that it can be used to examine the energy dependence of isotopes affected by energy-dependent processes. One such method is the exploitation of the energy dependence of the fragmentation of primary cosmic rays into secondary cosmic rays [5]. Another is the energy dependence of abundances of isotopes that decay only by electron capture, originally proposed by Raisbeck [6], expanded by Letaw et al. [7] and Silberberg and Tsao [8], and implemented by Soutoul et al. using a combination of Voyager and ISEE-3 data to obtain a $^{51}V/^{49}V$ ratio [9].

M. M. Shapiro et al. (eds.), Astrophysical Sources of High Energy Particles and Radiation, 305–309.
© 2001 *Kluwer Academic Publishers. Printed in the Netherlands.*

The isotopes discussed in this paper were created at cosmic ray energies by fragmentation during propagation. These secondary isotopes decay only by electron capture, which occurs primarily at low cosmic-ray energies where the kinetic energy of ambient electrons in the frame of the cosmic rays is comparable to the binding energy of the k-shell electrons. These energies are approximately equal for ^{49}V and ^{51}Cr at the lowest energies detected by CRIS; at the highest kinetic energies, very little electron attachment, and therefore decay, can occur. The lower-energy nuclei can decay and the higher-energy nuclei are effectively stable; variations with energy can reveal the amount of electron-capture decay that has occurred, and therefore the energy of propagation. Secondary electron-capture decay isotopes that can be studied include ^{7}Be, ^{37}Ar, ^{44}Ti, ^{49}V, ^{51}Cr, ^{55}Fe, and ^{57}Co and their decay products ^{7}Li, ^{37}Cl, ^{44}Ca, ^{49}Ti, ^{51}V, ^{55}Mn, and ^{57}Co. These isotopes all have half-lives of less than 2 years in the laboratory and approximately twice that with a single electron attached, as would usually be the case for a low-energy cosmic ray that picks up an electron [10]. We know that the measured abundances are purely secondary, as any primary abundance initially accelerated would have decayed during the 10^5 year time delay prior to initial acceleration indicated by recent measurements of Ni-Co isotopes [11].

2. Data Analysis

The Advanced Composition Explorer (ACE) was launched August 25, 1997, and began transmitting data from outside the magnetosphere a few days later. CRIS detects He-Zn nuclei in the energy range 50-550 MeV/nucleon with a large collecting power (geometrical factor of 250 cm^{2}sr) as described in Stone et al. [12]. The abundances of Ti, V, and Cr isotopes discussed in this paper were measured at energies of 150-500 MeV/nucleon.

During the first three years of operation, CRIS has collected 15,200 Ti, 7,300 V, and 15,000 Cr events. This large number of events allows division of the data into several energy bins and subsequent examination of possible energy dependence of isotopic abundance ratios. The mass resolution of these binned data are ≤ 0.25 amu; the peaks are clearly separated, as shown in the mass histograms in Figure 1. These peaks can be fit with Gaussian curves to derive abundances.

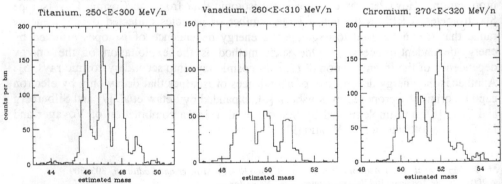

Figure 1. Titanium, Vanadium, and Chromium data collected by CRIS September 1997-October 2000 in the energy interval specified (bin size = 0.1 amu).

3. Results

In order to discuss the amount of electron-capture decay that has occurred, we must compare the measured isotopic abundances with those predicted by propagation models that account for fragmentation of primary isotopes, ionization energy losses, diffusion, convection, and other processes that affect the cosmic rays during their propagation through the interstellar medium and the heliosphere. The leaky-box propagation model [13] that we use incorporates all of these effects, but does not include reacceleration. The model uses nuclear fragmentation cross sections measured and calculated in Webber et al. [14] and electron attachment cross sections calculated using a modified OBK calculation as detailed in Wilson [10] and Crawford [15]. Solar modulation effects are incorporated using the spherically symmetric model of Fisk [16]; the detected energies of 150 – 550 MeV/nucleon correspond to interstellar energies up to 1000 MeV/nucleon.

The data shown in Figures 2, 3, and 4 [17] include the parent isotope abundance relative to the abundance of uninvolved stable reference isotopes, the daughter isotope relative to the same stable isotopes, and the sum of these quantities, so that energy dependences can be isolated and examined. Comparison of ^{51}Cr with the stable isotope ^{52}Cr (Figure 2) shows that ^{51}Cr is depleted at the lower energies and ^{51}V is enriched at the lower energies in qualitative agreement with the propagation model that allows electron capture. An additional calculation with electron-capture cross sections set to zero (no electron capture allowed at any energy) is shown in each figure as the dashed line.

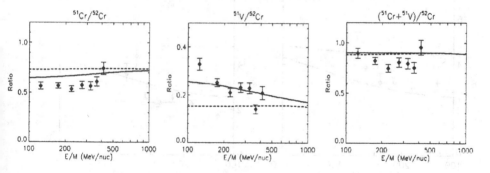

FIGURE 2. Parent and daughter abundances for the ^{51}Cr $\rightarrow$ ^{51}V decay, relative to stable isotope ^{52}Cr. Plotted curves are leaky-box propagation model results, with (solid) and without (dashed) electron-capture decay. Reacceleration is not included in this modeling calculation.

Comparison of ^{49}Ti with the sum of stable, uninvolved isotopes ^{46}Ti, ^{47}Ti, and ^{48}Ti shows that ^{49}Ti is enriched at the lower energies, in qualitative agreement with the propagation model that allows electron capture (Figure 3).

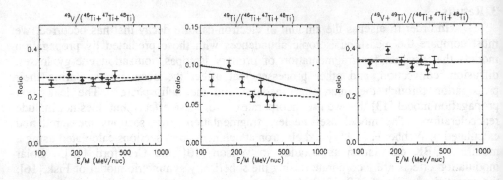

FIGURE 3. Parent and daughter abundances for the ^{49}V $\rightarrow$ ^{49}Ti decay. Plotted curves are leaky-box propagation model results, with (solid) and without (dashed) electron-capture decay. Reacceleration is not included in this modeling calculation.

A comparison of the parent/daughter ratios such as ^{49}V/^{49}Ti and ^{51}Cr/^{51}V shows steeper energy dependence, as the effects of electron-capture decay are doubled (Figure 4). Both pairs show a depletion of the parent isotope and/or enrichment of the daughter isotope at lower energies, in qualitative agreement with the model that allows electron capture.

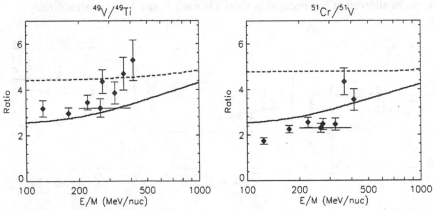

FIGURE 4. Parent/daughter ratios for ^{51}Cr $\rightarrow$ ^{51}V and ^{49}V $\rightarrow$ ^{49}Ti. Plotted curves are the same as figures 2 and 3. The points with horizontal bars show Ulysses results over the energy interval indicated, as in Connell and Simpson [18].

Although the data in Figures 2-4 indicate an energy dependence due to electron-capture decay, the data are in qualitative agreement with our leaky-box model which does not include reacceleration. To address the question of reacceleration, other isotope pairs must be analyzed and all compared with models with and without reacceleration. Other isotope pairs with short halflives ($\tau_{1/2} < 2$ years) that decay purely by electron capture are ^{37}Ar $\rightarrow$ ^{37}Cl, ^{44}Ti $\rightarrow$ ^{44}Sc $\rightarrow$ ^{44}Ca, ^{55}Fe $\rightarrow$ ^{55}Mn, and ^{57}Co $\rightarrow$ ^{57}Fe. These isotope pairs are currently being investigated using CRIS data.

4. Conclusions

The CRIS data presented here show energy dependence consistent with electron-capture decay at low energies, in qualitative agreement with results from our leaky-box model. Comparison of model predictions with other measured isotopic ratios not reported here, such as $^{37}Ar/^{37}Cl$, $^{44}Ti/^{44}Ca$, $^{55}Fe/^{55}Mn$, $^{57}Co/^{57}Fe$ and their abundances relative to stable reference isotopes, will also help determine the amount of electron-capture decay that has occurred. Additional comparison with propagation models that incorporate reacceleration, such as Webber [19], will allow limits to be set on the amount of reacceleration that may have occurred during cosmic ray propagation through the Galaxy.

Acknowledgements

This research was supported by the National Aeronautics and Space Administration at the California Institute of Technology (under grant NAG5-6912), the Goddard Space Flight Center, the Jet Propulsion Laboratory, and Washington University. This author also wishes to acknowledge the generous support of the Mr. and Mrs. Spencer T. Olin Foundation, which supported her during the initial phase of this research.

References

1. Silberberg, R. et al. (1983) *Phys. Rev. Lett.* **51**, 1217-1220.
2. Cowsik, R. (1986) *Astron. Astrophys.* **155**, 344-346.
3. Seo, E.S. and Ptuskin, V.S. (1994) *Astrophys. J.* **431**, 705-714.
4. Heinbach, U., and Simon, M. (1995) *Astrophys. J.* **441**, 209-221.
5. Giler, M., Wdowczyk, J., Wofendale, A. W. (1988) *Astron. Astrophys.* **196**, 44-48.
6. Raisbeck, G.M. et al. (1975) *Proc. 14th Internat. Cosmic Ray Conf.* **2**, 560-563.
7. Letaw, J. R., Silberberg, R., and Tsao, C. H. (1984) *Astrophys. J. Supp.* **56**, 369-391.
8. Silberberg, R., and Tsao, C. H. (1990) *Phys. Reports* **191**, 351-408.
9. Soutoul, A. et al. (1998) *Astron. Astrophys.* **336**, L61-L64.
10. Wilson, L.W. (1978) Ph.D. thesis. University of California at Berkeley. LBL-7723.
11. Wiedenbeck, M. E. et al. (1999) *Astrophys. J.* **523**, L61-L64.
12. Stone, E.C. et al. (1998) *Space Sci. Rev.* **88**, 285-356.
13. Leske, R.A. (1993) *Astrophys. J.* **405**, 567-583.
14. Webber, W. R. et al. (1998) *Phys. Rev. C.* **58**, 3539-3552.
15. Crawford, H.J. (1979) Ph.D. Thesis. University of California at Berkeley. LBL-8807.
16. Goldstein, M.L., Fisk, L. A., and Ramaty, R. (1970) *Phys. Rev. Lett.* **25**, 832-835.
17. Niebur, S.M. et al. (2000) In R.A. Mewaldt et al. (eds.), Acceleration and Transport of Energetic Particles Observed in the Heliosphere, New York, pp. 406-409.
18. Connell, J.J., and Simpson, J.A. (1999) *Proc. 26th Internat. Cosmic Ray Conf.* **3**, 33-36.
19. Webber, W.R. (2000) In R.A. Mewaldt et al. (eds.), Acceleration and Transport of Energetic Particles Observed in the Heliosphere, New York, pp. 396-401.

THE ASTROPHYSICS OF ULTRA-HEAVY GALACTIC COSMIC RAYS (AND HOW WE CAN DETECT THEM)

J. T. Link

*Department of Physic, and the McDonnell Center for the Space Sciences,
Washington University
One Brookings Drive, St. Louis, Missouri, 63105*

1. Introduction

Ultra-heavy galactic cosmic rays (UH GCRs) are those nuclei in the cosmic rays with atomic number greater than 30. Knowledge of the relative abundances of elements in these cosmic rays allows us to probe several outstanding astrophysical questions.

The best current measurements of UH GCRs come from Ariel 6 and HEAO-3 data and the TREK experiment. Ariel 6 and HEAO –3 measured the elemental abundances of even Z elements with $30 \leq Z \leq 60$ and charge groups with $58 \leq Z \leq 82$ [1,2]. The TREK experiment provided measurements of even elemental abundances with $70 \leq Z \leq 82$ [3]. These experiments lacked the resolution and exposure needed to separate elemental abundances of odd-Z elements.

In this paper I will outline some of the outstanding astrophysical questions that can be addressed by studying the UH GCRs. I will also discuss two future experiments -- TIGER and HNX, which might provide us with new measurements of the UH GCRs - including the currently unmeasured odd-Z UH GCR elements.

2. Galactic Cosmic Ray Material Source

There is a general consensus that cosmic-ray nuclei with energies up to about $\sim 10^{15}$ eV are accelerated by supernova shocks. What is not known however is the source of material for the cosmic-ray nuclei. If one infers the elemental abundances of the cosmic-ray source(s) by looking at the galactic cosmic rays and correcting those measurements for nuclear interaction during interstellar propagation, and compares them to the abundances of elements in the solar system, some interesting results emerge. We find that he two sources are remarkably similar, however some of the elements seem to be more abundant. We assume this is because they are preferentially accelerated, most likely due to their atomic properties. Two models have emerged to explain the reason certain nuclei are preferentially accelerated over others. Each points to a particular galactic cosmic-ray material source.

The first of these models notes that one can order the elements based on their first ionization potential (FIP). As seen in Figure 1 those elements with low FIP are more abundant that those elements with high FIP. The decrease in the GCR source to

M. M. Shapiro et al. (eds.), Astrophysical Sources of High Energy Particles and Radiation, 311–315.
© 2001 *Kluwer Academic Publishers. Printed in the Netherlands.*

solar-system composition ratio occurs at a FIP of about 10 eV which corrosponds to a temperature of about 10^4 K. One model suggested to explain this is that the GCR source is a stellar type atmosphere (which has a temperature on the order of 10^4 K) and that there is a preferential acceleration of those elements that are ionized.

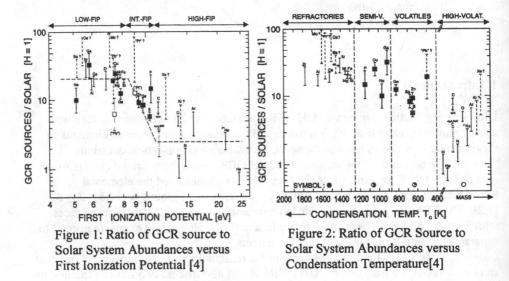

Figure 1: Ratio of GCR source to Solar System Abundances versus First Ionization Potential [4]

Figure 2: Ratio of GCR Source to Solar System Abundances versus Condensation Temperature[4]

The second model notes that one can also order the elements based on their condensation temperature and that those elements with a high condensation temperature (refractory elements) are more abundant than those with a low condensation temperature (volatile elements). The model suggested here is that the GCR source is ambient dust and gas in the ISM. The enhancement in the refractory elements is due to the fact that dust grains are more easily accelerated than individual particles due to their high rigidity. Once accelerated, particles would be sputtered off these grains leading to an enhancement in the material forming these grains, which are primarily refractory elements, in the GCRs.

Differentiation between these two models is difficult because most low FIP elements are refractory and hence one expects to see an enhancement of these elements no matter which model is used. There are a few elements that break this degeneracy and would allow for a differentiation between these two models. Among the best elements for this differentiation are the UH GCRs Ge, Rb, Sn, Cs, Pb, Bi. Among these elements, individual abundances have only been measured for Ge, Sn and Pb. These measurements suggest that acceleration is governed by volatility rather than FIP [1,2,3] but the data are not yet conclusive. Further measurements of these elements are needed to determine which process governs the GCR acceleration process.

3. R-Process Enhancement in the GCRs

Data from HEAO-3 and Ariel-6 [1,2] showed a GCR composition very similar to solar system abundances for Z < 60. However there was an enhancement in r-process cosmic ray nuclei relative to solar system abundances for $Z > 60$. R-process nuclei are formed by rapid neutron capture, which occurs during a supernova explosion. HEAO-3 and Ariel-6 had limited charge resolution and low statistics in the $Z > 60$ region. It was therefore not possible to cleanly separate primary GCRs from secondaries or to confirm that individual elements had relative abundances expected of an r-process source. If there is an enhancement in r-process elements compared to solar abundances it suggests GCR material is accelerated in a region that is dominated by r-process material. Such an enhancement could be seen for material synthesized and directly accelerated within a supernova explosion

4. Cosmic Ray Age

Recent measurements from the ACE spacecraft of the $^{59}Co/^{59}Ni$ ratio indicate that there a time of $> 10^5$ years passes between synthesis and acceleration of GCRs.[5] These measurements suggest that GCRs are not produced and accelerated by the same supernova. A recent model set forth by Higdon, Lingenfelter and Ramaty [6] argues that the source of material from which the GCR are accelerated must include freshly synthesized nuclear matter. This model suggests that GCR are accelerated in superbubble regions. There the freshly synthesized material from supernovae does not mix with the ambient ISM material and can be accelerated by later supernova shocks.

To determine if GCRs are from ambient galactic material or freshly synthesized material, we need to determine their age. This age is the time from the synthesis of a galactic cosmic ray nucleus to it's detection. If GCRs are accelerated from superbubble regions they cannot be older than 10^6 years. There are not yet any measurements of the age of the primary galactic cosmic rays.

To determine this age we need to measure the relative abundances of nuclei with a radioactive half-life between 10^7 and 10^9 years, the timescale for galactic chemical evolution. The radioactive actinides are such elements Figure 3 shows the predicted abundances of the individual actinides as a function of mean GCR age.

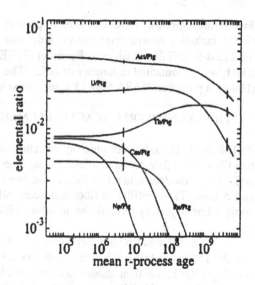

Figure 3: Predicted abundances of the individual actinides with respect to the PT-group, as a function of mean GCR age, assuming uniform synthesis. [7]

5. Cosmic Ray Pathlength Distribution

Another aspect of GCR astrophysics that can be explored by looking at UH GCRs is propagation. The amount of interstellar material traversed by GCRs between acceleration and observation is measured by the abundance of galactic cosmic ray secondaries that are rare relative to heavier elements. Current measurements of light secondaries (Li, B, Be) and Iron secondaries (elements with $21 \leq Z \leq 25$) can be explained by an exponential distribution of pathlengths in the interstellar medium with a mean of about 8 g/cm^2 and a truncation in the distribution at about 1 g/cm^2. To test the validity of such a distribution we need to understand better what is happening at the shorter pathlengths. The light and Iron secondaries are not the best probes at short pathlengths because their interaction mean free path in the ISM is too long (> 3 g/cm^2). Secondaries of the UH GCRs are much better probes of short pathlengths as their interaction mean free paths in the ISM are between about 1.2 and 1.6 g/cm^2.

6. Future Experiments to Measure UH GCRs

Two future experiments hold promise to provide additional measurements of the UH GCRs, including measurements of the individual odd-Z elements. The first of these is the Trans-Iron Galactic Element Recorder (TIGER), a balloon born cosmic ray detector which will be launched in January of 2002. The other is the Heavy Nuclei Explorer (HNX), a proposed NASA Small Explorer mission, accepted for phase A study.

6.1. THE TRANS-IRON GALACTIC ELEMENT RECORDER (TIGER)

TIGER is a balloon-born cosmic-ray detector that consists of four scintillation detectors, two Cherenkov detectors and a scintillating fiber hodoscope. The scintillation detectors and Cherenkov detectors allow for an unambiguous measurement of a particle's charge and energy. The scintillating fiber hodoscope allows for a determination of a particle's straight line trajectory through the detector. TIGER has an active detector area of about $1m^2$.

TIGER was flown in September of 1997. This 23-hour flight demonstrated that the TIGER instrument had the necessary resolution to measure GCRs heavier than iron.[8] TIGER was then chosen to be the first instrument to fly in the new Ultra Long Duration Balloon (ULDB) program. The ULDB will be the first balloon capable of flights approaching 100 days in duration. The duration of flight will allow TIGER to measure the individual elemental abundances of GCRs with $20 \leq Z \leq 40$. TIGER is currently scheduled to fly in December 2001 from the McMurdo station in Antartica.

6.2. THE HEAVY NUCLEI EXPLORER (HNX)

HNX is a proposed NASA Small Explorer (SMEX) mission. HNX is composed of two instruments, the Energetic Trans-Iron Composition Experiment (ENTICE) and the Extremely-heavy Cosmic-ray Composition Observer (ECCO). HNX would measure the abundances of elements with $10 \leq Z \leq 100$. Should HNX be selected it would launch in 2005 or 2006 for a mission of about three years.

ENTICE is based on the TIGER instrument and consists of two silicon detector arrays, two Cherenkov detectors and a scintillating fiber hodoscope. The silicon arrays and Cherenkov detectorss provide a measurement of the particles charge and energy. The scintillating fiber hodoscope measures a particle's straight line trajectory through the detector. ENTICE would have a detector area four times that of the TIGER instrument and measure elements with $10 \leq Z \leq 83$.

ECCO is a large array of glass track-etch detectors based on the TREK experiment. A cosmic-ray nucleus incident on the ECCO detector would pass through the glass plate and leave an ionization damage trail. Later the glass plates are retrieved and chemically etched. The size of etched pits where particles passed through the glass is then measured and based on that measurement the composition of the particle can be determined. ECCO will be able to measure GCRs with $Z > 70$ and will have sufficient collecting power to detect $Z > 100$ actinides.

7. Conclusions

Ultra-heavy galactic cosmic rays hold a wealth of information about the source, acceleration process and propagation of cosmic rays. Their low fluxes make them difficult to observe, requiring instruments with large areas, long exposure times and good resolution. Currently our measurements of ultra-heavy galactic cosmic ray composition are restricted to even-Z elements and, for the heaviest GCRs, charge groups. Future experiments like TIGER and HNX will be able to provide more complete measurements of the elemental abundances of all the ultra-heavy elements and allow us to have a better understanding of the source, acceleration and propagation of galactic cosmic rays.

8. References

1. Binns, W. R., *et al.*, (1989) Abundances of Ultraheavy Elements in the Cosmic Radiation – Results from HEAO 3, *Astrophysical Journal* **346**, 997-1009.
2. Fowler *et. al.*, (1987) Ariel 6 Measurements of the Fluxes of Ultraheavy Cosmic Rays *Astrophysical Journal* **314**, 739-746.
3. Westphal, A.J., *et al.*, (1998) Evidence against stellar chromospheric origin of Galactic cosmic rays *Nature* **396**, 50-52.
4. Meyer, J.P, Drury, L. Ellison, D.C. (1998) Galactic Cosmic Rays from Supernova Remnants. I. A Cosmic-Ray Composition Controlled by Volatility and Mass-to-Charge Ratio *Astrophysical Journal* **487**, 182-196.
5. Wiedenbeck, M.E. ., et al., (1999) Constraints on the Time Delay Between Nucleosynthesis and Cosmic-Ray Acceleration From Observations of ^{59}Ni and ^{59}Co *Astrophysical Journal* **523**, L61-L64
6. Higdon, J.C., Lingenfelter,R.E., Ramaty,R. (1998) Cosmic-Ray Acceleration from Supernova Ejecta in Superbubbles *Astrophysical Journal* **509**, L33-6.
7. Westphal, A.J. based on r-process yields of Pfeiffer et. al., private communication (2000).
8. Sposato, S.H. et al., (1999) The Trans-Iron Galactic Element Recorder (TIGER): A Balloon-borne Cosmic Ray Experiment, *Proc. 26th ICRC (Salt Lake City)* **5**, 29-32.

EAS: N_μ - N_e measurements with the EAS-TOP array

Silvia Valchierotti [†]

University of Torino (Italy) - INFN Torino

Abstract. One of the puzzles about the cosmic rays in the knee region ($E_0 \approx 3 \cdot 10^{15} eV$) is the chemical composition. A useful tool of investigation is the relation between the total number of muons (N_μ) and the electro-magnetic size (N_e) of the Extensive Air Showers (EAS). The experimental data taken by the EAS-TOP array will be presented.

The N_μ - N_e relation measured at different zenith angles are shown and the experimental data taken at vertical incidence has been compared with simulation results in order to obtain preliminary conclusions on the primary composition.

1. Introduction

At the "knee" energy ($E_0 \approx 3 \cdot 10^{15} eV$) it's impossible to make direct measurements of the cosmic rays. The study with ground based detectors of Extensive Air Shower (EAS) produced by primary interactions in the atmosphere is needed.

The theories about the knee origin give predictions on the primary chemical composition and the spectrum of the primary nuclei, so it's necessary to find out a way to recognize the kinds of nuclei producing the showers.

An investigation tool to study the primary composition is the relation between N_μ and N_e.

To illustrate the principle let's consider a simple superposition model. For showers produced by a proton the relations between the primary energy E_0 and the total number of particles are:

$$N_e = K_e \cdot E_0^\alpha \qquad (1)$$

for the electromagnetic component (N_e = electromagnetic size)

$$N_\mu = K_\mu \cdot E_0^\beta \qquad (2)$$

for the muon component (N_μ = number of muon in a shower)

[†] on behalf of the EAS-TOP collaboration

M. M. Shapiro et al. (eds.), Astrophysical Sources of High Energy Particles and Radiation, 317–323.
© 2001 *Kluwer Academic Publishers. Printed in the Netherlands.*

$$\alpha \neq \beta \tag{3}$$

Where K_e and K_μ are constant and α and β are the spectral indexes. For a nucleus with mass number A we have:

$$N_e \propto A \left(\frac{E_0}{A}\right)^\alpha \tag{4}$$

$$N_\mu \propto A \left(\frac{E_0}{A}\right)^\beta \tag{5}$$

So the relation between N_e and N_μ is:

$$N_\mu \propto A^{(\alpha-\beta)/\alpha} N_e^{\beta/\alpha} \tag{6}$$

This relation shows a clear dependance of N_μ and N_e produced in the shower with the mass number of the primary. The experimental data can be compared with simulations to obtain information on the primary composition.

2. The EAS-TOP experiment

The EAS-TOP (INFN - National Gran Sasso Laboratories) array was a multi-component detector ((Aglietta et al., 1993) , (Aglietta et al., 1998)). It was located at Campo Imperatore near l'Aquila in central Italy at 2005 m a.s.l., at the atmospheric depth $x_0 = 810 \ g \cdot cm^{-2}$. The apparatus included an electromagnetic detector (EMD), a muon hadron detector (MHD) and telescopes for Cherenkov light measurements. For this analysis only the EMD (to measure the e.m. componenent) and the muon tracker MHD (to measure the muon component) have been considered.

2.1. EMD

The electro-magnetic detector was made of 35 scintillator modules organized in 15 subarrays for trigger requirements and data taking organization. Event selection for our analysis requires at least 6 (or 7) modules fired with the highest particle density recorded by the inner detector. The arrival direction of the shower is obtained through

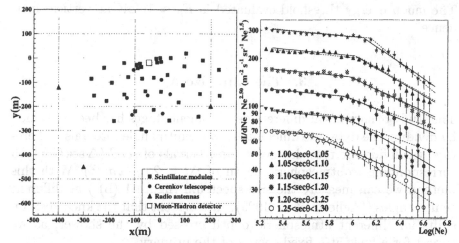

Figure 1. (a) the EAS-TOP array plan and (b) the measured size spectrum

the times of flight technique among the different detectors (0.5 ns of sensitivity).

The shower size N_e and the core location are computed through the χ^2 method fitting the e.m. lateral distribution with the NKG ((Kamata K. et al., 1958)) formula:

$$\rho(r) = \frac{N_e \cdot C(s)}{r_M^2} \left(\frac{r}{r_M}\right)^{s-2} \left(\frac{r}{r_M}+1\right)^{s-4.5} \tag{7}$$

where N_e is the e.m. size, $C(s)$ is a normalization constant depending from the slope parameter s, r_M ($\simeq 100m$) is the Moliere radius and r is the distance from the shower axis.

The accurancy of the measurements, at $\log N_e > 5.2$, are: $\Delta r < 10\ m$ for the core location and $\frac{\Delta N_e}{N_e} \approx 10\%$ for the shower size.

2.2. MHD

The Muon Hadron Detector was a calorimeter of 9 planes, $3\,m$ height with a total surface of $144\,m^2$. Each plane was composed of 2 layers of streamer tubes, 1 layer of quasi proportional tubes and $13\,cm$ thick iron absorber. The quasi-proportional tubes were used for the hadronic calorimetry purpose. The muon tracking is made with the streamer tubes of each layer. The signals were collected directly from the anode wires for the X-view, for the Y-view we used the induced signals on strips placed orthogonally to the wires. For the reconstruction a minimum of 6 hits in different streamer layers is required.

The muon energy threshold evaluated is $E_\mu \approx 1 GeV$ at vertical incidence.

3. The e.m. size N_e

The study of showers at different development stages has been obtained by dividing the data into different arrival zenith angle intervals . Therefore, we divide the recorded events on intervals of $\sec\theta$ ($\Delta\sec\theta = 0.05$ corresponding to atmospheric depth of $\Delta t = 20\,g \cdot cm^{-2}$). With this method we can measure the size spectrum (figure 1 (b)) at different zenith angles ((Aglietta et al., 1999)). It's evident that the knee position shifts for different atmospheric depth crossed by the shower, as we expect for a knee at a fixed energy of the primary.

4. The muon density ρ_μ

As a measurement of the muon number we use the muon density ρ_μ. The relation between ρ_μ at the distance r and the total number of muons in a shower is the lateral distribution:

$$\rho_\mu = \frac{N_\mu}{3.717} r^{-0.75} \cdot R_0^{-1.25} \cdot \left(\frac{r}{R_0} + 1 \right)^{-2.5} \tag{8}$$

where $R_0 = 300\,m$ is the natural unit of spread of the muon component like the Moliere radius for the e.m. component.
As one can see from equation [8] the relation between ρ_μ and N_μ is a proportional one so the dependance of the density from the e.m. size is like the [6]:

$$\rho_\mu = K'_\mu N_e^{\alpha'} \tag{9}$$

The events have been grouped by means of the e.m. size ($\Delta\log N_e = 0.05$) and the zenith angle θ ($\Delta\sec\theta = 0.05$) .
By this the mean value of the number of tracks $< n_T >$ in MHD has been obtained and so the muon density $\rho_\mu = \frac{<n_T>}{A_{eff}(\theta)}$. $A_{eff}(\theta)$ is the effective area of detection depending on the zenith angle θ.
The figure 2 (a) shows the muon lateral distribution obtained for a fixed size interval by the experimental data, the line is the NKG fit.

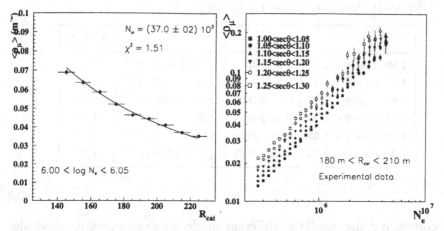

Figure 2. (a) muon lateral distribution (the line is the best fit with the [8] lateral distribution) and (b) ρ_μ-N_e relation at different zenithal angle

5. ρ_μ vs N_e

Figure 2 (b) shows the experimental relation between ρ_μ and N_e for 6 different zenithal angle intervals.

A best fit of these curves by the following expression:

$$\rho_\mu = K\left(\frac{N_e}{N}\right)^\alpha \tag{10}$$

has been performed both under and above the "knee".N is a normalization. It's value is $N = 3 \cdot 10^5$ for the fit under the "knee" and over all the range and $N = 2 \cdot 10^6$ above the "knee".

The results are shown in the tables I and II.

Table I. Under the "knee" $\log N_e < 6.2$

	α	K	$\chi^2/d.f.$
$1.00 < \sec\theta < 1.05$	0.99 ± 0.01	$(1.246 \pm 0.009)10^{-2}$	0.81
$1.05 < \sec\theta < 1.10$	0.97 ± 0.01	$(1.37 \pm 0.01)10^{-2}$	0.88
$1.10 < \sec\theta < 1.15$	0.99 ± 0.02	$(1.53 \pm 0.01)10^{-2}$	0.64
$1.15 < \sec\theta < 1.20$	0.97 ± 0.02	$(1.72 \pm 0.02)10^{-2}$	1.06
$1.20 < \sec\theta < 1.25$	0.97 ± 0.03	$(1.90 \pm 0.03)10^{-2}$	1.71
$1.25 < \sec\theta < 1.30$	0.86 ± 0.08	$(2.19 \pm 0.03)10^{-2}$	0.73

Table II. Above the "knee" $\log N_e > 6.2$

	α	K	$\chi^2/d.f.$
$1.00 < \sec\theta < 1.05$	0.93 ± 0.03	$(7.74 \pm 0.08)10^{-2}$	1.15
$1.05 < \sec\theta < 1.10$	0.95 ± 0.03	$(8.6 \pm 0.01)10^{-2}$	1.05
$1.10 < \sec\theta < 1.15$	0.90 ± 0.03	$(9.2 \pm 0.01)10^{-2}$	1.07
$1.15 < \sec\theta < 1.20$	0.96 ± 0.04	$(10.3 \pm 0.2)10^{-2}$	1.31
$1.20 < \sec\theta < 1.25$	0.94 ± 0.05	$(11.0 \pm 0.3)10^{-2}$	0.60
$1.25 < \sec\theta < 1.30$	0.89 ± 0.03	$(11.9 \pm 0.4)10^{-2}$	1.23

By comparing the results at different angle intervals one can conclude that the α parameter does not change within experimental errors. Therefore there is an indication of a constant absorption of the e.m. size at different primary energies. The K parameter, as we expect, grows with the angle because the member of muon at the same size is greater in a more inclined shower. For vertical events the best fit from below to above the "knee" give the parameter shown in table III.

Table III. Fit over all the size range for vertical events

α	K	$\chi^2/d.f.$
0.969 ± 0.007	$(1.258 \pm .008)10^{-2}$	23.15

The relation

$$\rho_\mu = K N_e^\alpha \tag{11}$$

seems to be good over all the size range. The results are compatible with ones found under and above the "knee". These means that the slope α is almost the same all over the range showing no change in the knee region.

For the vertical events we also made a comparison with simulation. In figure 3 experimental data have been compared with simulated ones (pure protons, pure iron and a uniform composition of p, He, N, Mg, Fe assuming the same spectrum, i. e. a constant composition for all the elements). The experimental data seems to be closer to pure iron than to pure proton composition for higher N_e. This likely give us an indication of a heavier composition above the "knee" region.

The constant slope under and above $\log N_e = 6.2$ indicates that if something changes this happens in a smooth way and not drastically.

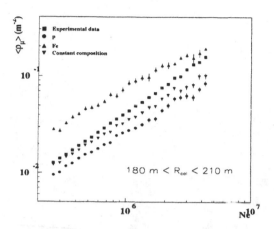

Figure 3. Comparison between vertical experimental data and simulation

6. Conclusions

The data analyzed are homogeneus at different zenith angles and there isn't indication of different absorbtion of the e.m. component for different energy. The [11] relation describes well the data over the whole size range giving no indication of some kind of change at the knee. So we can say that the N_e spectrum and the ρ_μ spectrum change their slope at $E_0 = 3 \cdot 10^{15}$ eV in the same way. Moreover the preliminary comparisons with simulation seems to indicate a primary composition heavier for greater energy.

References

Kamata K. et al., Suppl. Progr. Theor. Phys. 6, (1958), 93
Aglietta et al., Nucl.Instr. and Method A 336, (1993) 310
Aglietta et al., Nucl.Instr. and Method A 420, (1998) 117
Aglietta et al., Proc. XXV ICRC, Durban HE 1.2.23 (1997)
Aglietta et al., Astr. Part. Phys. 10 (1999) 1-9

L3+Cosmics: an atmospheric muon experiment at CERN

Tommaso Chiarusi
University of Bologna, University of Padova

L3 Collaboration
CERN

Abstract. The L3+C experiment studies the muon component of the atmospheric showers induced by primary cosmic rays. It combines the high precision spectrometer of the L3 detector at LEP, CERN, with a small air shower array. The momenta of the cosmic ray induced muons can be measured from 20 to 2000 GeV/c. Up to now, almost 12 billion of muon events have been recorded on tape, as well as over 33 million air shower events. Here the first results on the muon momentum spectrum and charge ratio will be presented.

1. Introduction

Nowadays many arguments, particularly the new evidence for neutrino oscillations, make the attention focus on the neutrinos'related muon flux (originating in the atmospheric showers mostly from charged pion decays $\pi^{\pm} \to \mu^{\pm} + \nu_{\mu}(\overline{\nu}_{\mu})$), its energy spectrum and the ratio between the abundances of the positive and negative muon components. These are the main targets of the L3+Cosmics experiment (fig. 1), which is an extension of the already existing L3 detector.

L3 (Adeva et al, 1990) is one of the four experiments at LEP, the $e^+ e^-$ accelerator at CERN Laboratories, located near Geneva (6.02° E, 46.25° N) at an altitude of 450 m. L3 is underneath 30 m of molasse ($\sim$ 72 m.e.w), with a 15 GeV minimum energy cutoff for downgoing muons. The high precision muon spectrometer is composed of layers of drift chambers, occupying a volume of $\sim$1000 m^3, and is completely inside an uniform magnetic field $\vec{B}$ of 0.5 T.

L3 has operated very successfully since 1989 up to the first days of November 2000 when the LEP machine was shut down. Since April 1998, the muon spectrometer has been converted into a powerful detector of atmospheric muons, without affecting the normal L3 activity. Such conversion has been possible thanks to additional equipment, the t_0 system, composed of plastic scintillators, and by means of an independent trigger and DAQ electronics. The wide energy range under inspection (20 - 2000 GeV), *ad hoc* ground-filtering of unwanted non-muon secondaries, a precise momentum resolution and huge statistics characterize the measurement of the muon spectrometer of L3. A continuously monitored detector, active during more than 6 months per

M. M. Shapiro et al. (eds.), Astrophysical Sources of High Energy Particles and Radiation, 325–330.
© *2001 Kluwer Academic Publishers. Printed in the Netherlands.*

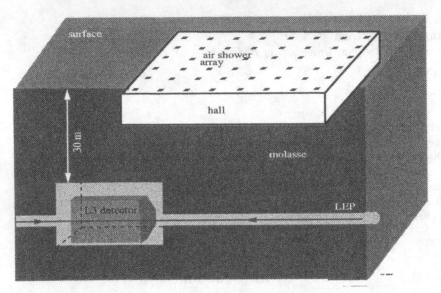

Figure 1. Schematic view of the whole L3+Cosmics experiment.

year, the spectrometer provides as well the conditions to study short time dependent phenomena such as muon bursts, or to determine fine effects, such as sideral asymmetries.

Since the beginning of 2000, a small air shower array has been added on the roof of the L3's pit access hangar, located at the surface level above L3. The trigger signal from both the spectrometer and the shower array are exchanged, to enable an offline merging of the two data streams, in order to reconstruct a possible primary energy and composition, and to put constraints to models describing the development of a shower in the atmosphere.

2. The L3+C experiment

The muon spectrometer, magnet and scintillators are shown in fig. 2.a. as part of the L3 detector. Fig. 2.b. shows a simplified cross section. The muon spectrometer consists of 80 precision drift chambers (P-chambers) to measure positions in the XY bending plane of the magnet, transverse with respect to $\vec{B}$, and 96 drift Z-chambers to measure the coordinate along the magnetic field. The P-chambers are arranged in 2 groups of 8 *octants*, in 3 layers per octant. The Z-chamber are arranged in 2+2 layers *sandwiching* the inner and the outer P-chambers layers. Chamber alignments are continuously monitored and are known to 20 μm. Each chamber's single wire resolution is 200 μm. The transverse muon momentum $P_\perp$ is determined in one octant by measuring the sagitta S of the muon trajectory, according to the formula

$$P_\perp = \frac{c\,B\,L^2}{8 \times 10^9\,S},$$

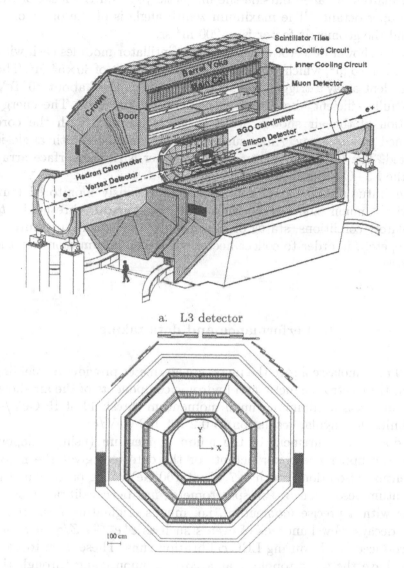

a. L3 detector

b. Section of L3 muon spectrometer.

Figure 2. The muon spectrometer inside L3 detector. The outer octagons correspond to coil and yoke of the magnet. Scintillators are installed on the top

where L is the extension of the octant and c the speed of light (expressing P in GeV/c and the other quantity according to SI). The information along Z is then used to perform a helix-fit to the trajectory, refining the momentum measurement.

The detector is enclosed in a 12 m diameter magnet (coil and yoke). The arrival time of the muons is needed to determine the drift time in the chambers. It is provided by the t_0 detector, consisting of 202 m^2 of scintillators installed outside the magnetic yoke, on the faces of the three upper octants. The maximum zenith angle is of the order of $\sim$ 50°, and the geometric factor is $\sim$ 200 m^2 sr.

The air shower detector consists of 50 scintillator modules each with a surface of 0.5 m^2, which are distributed over an area of 30×54 m^2. The independent array's trigger has an energy threshold of about 10 TeV and is fully efficient above a shower energy of 100 TeV. The energy resolution of the air shower array is 30% for events with the core contained in the array. An accuracy of 1°-2° on the zenith angle is expected. The L3 muon spectrometer together with the surface array form the L3+C experiment.

The status of the L3+C detector is continuously monitored (run conditions, muon chamber voltages and discriminator thresholds, t_0 scintillators conditions, status of the magnet, atmospheric pressure at surface, etc.), in order to collect necessary on-line informations for the data base.

3. Performance and data taking

The 30 m of molasse above the muon spectrometer provides a shielding against the electro-magnetic and hadronic components of the air showers. It also sets a minimum muon momentum threshold of 15 GeV/c, and limits the angular resolution to 0.2° at 100 GeV/c.

A double measurement of the muon momentum (using independently the upper and lower octants of the detector) gives the muon momentum resolution, which is 7.4% at 100 GeV/c. In particular, the momentum resolution of the spectrometer has been calibrated using muons with a precise momentum, i.e. muons originating from the Z boson decays. Few hundreds of events such as $e^+e^- \rightarrow Z/\gamma \rightarrow \mu^+\mu^-$ are produced yearly during LEP calibration runs. These back to back muons have the same topology as a cosmic muon going through the spectrometer by its centre, thus it has been possible to determine a momentum resolution of 5.1% at 45.6 GeV/c.

The spectrometer data taking officially started in 1999. From May to November a total 5 billion events were recorded, with a live-time of $\sim$ 124 days. From April to November 2000 a total of 6.8 billion events have been collected with a live-time of 188 days. The air shower array has been fully operational since April 2000 and by November 12th $\sim$ 33 million air shower events have been recorded. Almost one third of these shower events are accompanied by muon(s) down in the spectrometer.

4. First results: the muon spectrum and charge ratio

A first muon momentum spectrum from 50 to 500 GeV/c has been determined by using data from September to November 1999. The related total live-time was slightly more than 30 days. Strict quality cuts have been applied, requiring in particular a good measurement of the tracks in both upper and lower octants and a good matching of them, to get the entire muon track. The zenith angle has been restricted to a range of 0° to 10° to measure the vertical flux. In fig 3 is shown the measured flux weighted by P^3 together with the results of other experiments (Allkofer et al., 1975) (Baschiera et al., 1979) (Rastin et al., 1984) (De Pascale et al., 1993) (Kremer et al., 1999). The systematic error of $\sim 9\%$ dominates the total error bars. In the future we hope to reduce the total error less than 3%, and to extend the momentum range from 20 to 2000 GeV/c.

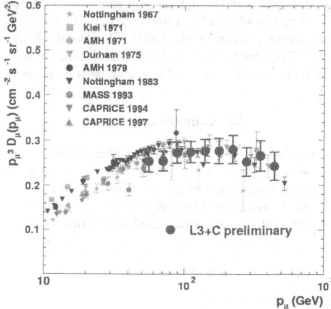

Figure 3. Spectrum of down-going vertical muons multiplied by p_μ^3 within the energy range 10-500 GeV

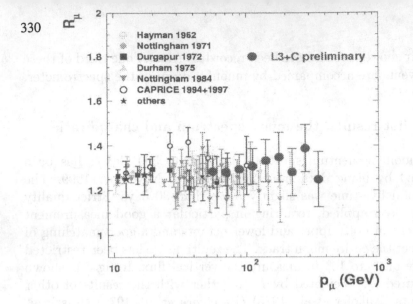

Figure 4. Charge ratio of down-going muons within the range of 10-500 GeV

The charge ratio was extracted from the same event sample. The preliminary result can be seen in fig. 4 together with the results of other published experiments (Rastin et al., 1984) (De Pascale et al., 1993) (Kremer et al., 1999). Due to the small sample of data considered for the analysis, at high momenta the statistical error is dominating.

5. Conclusion

L3+C is a new type of cosmic rays detector, which combines air shower data with precise muon momentum measurements. First preliminary results on the muon spectrum and charge ratio in the range from 50 to 500 GeV/c have been presented. A substantial reduction of the statistical and systematic errors is expected in the future.

References

Adeva, B. et al. *NIM*, A289:35, 1990.

Allkofer, O. C. et al. Lett. Nuovo Cim. 12: 107, 1975.

Baschiera, B. et al. Il Nuovo Cim. 2C: 473, 1979.

De Pascale, M. P. et al. J. Geophys. Res. 98: 3501, 1993.

Kremer, J. et al. Phys. Rev. Lett. 83: 4241, 1999.

L3 Collaboration http://l3www.cern.ch/l3_cosmics/

Rastin, B. C. et al. J. Phys. G.: Nucl. Phys. 10: 1629, 1984.

NUCLEAR PALEOASTROPHYSICS: PROSPECTS AND PERSPECTIVES

G. KOCHAROV[1], P. DAMON[2], H. JUNGNER[3], I. KOUDRIAVTSEV[1],
M. OGURTSOV[1]

[1] *A. F. Ioffe Physico-Technical Institute of Russian Academy of Sciences,
194021 St. Petersburg, Russia*

[2] *Department of Geosciences, University of Arizona,
Tucson, Arizona 85721, U.S.A.*

[3] *Dating Laboratory, University of Helsinki
P. O. Box 64, FIN-00014 Helsinki, Finland*

1. Introduction

Palaeoastrophysics, as a branch of science, could be defined in the following way: it studies astrophysical phenomena whose signals reached the solar system before the advent of instrumental astronomy. Instrumental astronomy was borne in the early 17th century through the systematic studies of the sky by Galileo Galilei. Using only a small optical telescope built with his own hands he discovered Jupiter's satellites and the Moon's phases. He reported officially on the discovery of spots on the Sun in Padua (Italy) in 1610 B. P.

Konstantinov and Kocharov [1] were the first to formulate the basic ideas underlying the new branch of science, Paleoastrophysics. It was shown that by studying natural archives one can quantitatively explore a number of astrophysical phenomena which occurred in the distant past. Among these are: long-period variations of the intensity of galactic and solar cosmic rays, the flare and modulation activity of the Sun, catastrophic events in the past, the amplitude and temporal characteristics of Supernova explosions, and gamma ray bursts.

The problem of the future of the Sun and of the solar system has been receiving increasing attention in recent years. To construct a concrete theoretical model of the present and future Sun, one should know the dynamics of the processes occurring in the solar material during a long time period extending over millions and even billions of years. The only source of such information can be the natural archives, which are capable of fixing the time of arrival of a solar signal, its type, and amplitude.

2. Natural Archives of Astrophysical Phenomena

The Earth's crust is a permanent detector of cosmic-ray particles and radiation. Only the crust offers a fundamental possibility of establishing the dynamics of generation of

M. M. Shapiro et al. (eds.), Astrophysical Sources of High Energy Particles and Radiation, 331–343.
© 2001 *Kluwer Academic Publishers. Printed in the Netherlands.*

thermonuclear energy in the deep solar interior for the last tens of millions of years. This possibility is connected with high-precision measurements of the content of lead and technetium isotopes in the Earth's crust [2].

Annual tree rings have become a traditional source of quantitative data on the time variations of the galactic cosmic ray intensity on time scales from the present to 10,000 years back. The width of the tree rings and their isotope content contain information on the solar activity and climatic effects over large time scales in the past. It has recently been established that tree rings can even be used to reconstruct the dynamics of solar activity over a huge time scale, up to 25 Myr [3].

Polar ice is also an important source of information on solar flare activity, Supernova explosions, and climatic effects. The frequency and amplitude of solar proton flares are determined from the concentration of nitrates in dated polar-ice layers. The amplitude and time characteristics of supernova explosions are derived by measuring the evolution of the cosmogenic isotopes ^{14}C, ^{10}B and ^{36}Cl in independently dated samples of polar ice. These isotopes form in nuclear reactions in the Earth's atmosphere initiated by galactic cosmic rays, which are believed to be produced in Supernova explosions. These isotopes are produced also by gamma quanta of cosmic origin. According to present concepts, Supernova explosions generate both high-energy protons and hard gamma quanta. Polar ice is a unique detector of the amplitude and temporal characteristics of the cosmic gamma radiation and high-energy protons. The time scale over which astrophysical information can be obtained using polar ice covers, presently, hundreds of thousands of years.

Thus the archives already revealed, and explored, permit one to study the nature of the physical processes throughout the solar volume over a huge time scale in the past. An analysis of the results of present high-precision measurements of the solar radiation and particles, combined with decoded data from the natural archives, may provide unique information on solar-terrestrial relations in the distant past. This information is needed, not only for reconstructing the history of the Sun but also for predicting the future of our Solar System.

3. What Do Solar Neutrinos Tell Us About the Thermonuclear History of the Sun

Neutrino paleoastrophysics, a fairly new branch of science, has not yet been experimentally established. We shall consider here briefly only the principal potential of this area of science as applied to the thermonuclear history of the Sun. The idea underlying it is based on creation in the Earth's crust of characteristic isotopes by neutrinos of various energies [2].

The $^{205}Tl + \nu_e \rightarrow {}^{205}Pb + e^-$ reaction is sensitive to low-energy neutrinos because of a low reaction threshold (43 keV) and the large pp-neutrino flux. This group of neutrinos are generated in the first reaction of proton-proton cycle, $p + p \rightarrow {}^2D + e^+ + \nu_e$. Because this reaction is rate limiting in the proton-proton cycle, it is a direct indicator of the hydrogen burning power in the solar interior. A comparison of the results obtained in the Tl-Pb experiment with the present studies made with the gallium detector may provide an answer to the questions of whether hydrogen is the only fuel, and whether the hydrogen burning power varies in time, both questions of fundamental importance.

Experiments based on the reactions:

$$^{98}Mo + \nu_e \rightarrow {}^{98}Tc + e^-,$$
$$^{97}Mo + \nu_e \rightarrow {}^{97}Tc + e^-,$$
$$^{97}Mo + \nu_e \rightarrow {}^{96}Tc + n + e^-,$$

are sensitive to high-energy (^{8}B) neutrinos. Because the ^{8}B neutrino flux depends very strongly on the solar core temperature, the results obtained in such experiments, combined with the data of present studies, will hopefully permit one to find an answer to the question of the dynamics of, and physical conditions in, the solar core.

It was established that the best minerals for the Thallium experiment are the lorendites from Yugoslavia. Ore from Colorado satisfies the requirements imposed by the Molybdenum detector. The separation from the huge detector mass of a small amount of Lead and Technetium atoms, and their radioactive counting, present a formidable problem. The experiments on neutrino paleoastrophysics are large-scale and expensive, and they require cooperation of many countries. Taking into account the fundamental importance of the problem, we have only to hope that in the new century experimental neutrino paleoastrophysics will become one of the key fields of science.

4. Annual Tree Rings Report on Deep Minima in the Solar Activity

Fedor Nikiforovich Shvedov, Professor at the Novorossiisk University, was the first to grasp the unique possibilities of annual tree rings for environmental studies. His paper "The tree as a record of droughts" was published in 1892 in the "Meteorological Vestnik" magazine. Studies on dendrochronology, dendroclimatology, and the isotope content in annual tree rings are presently pursued in many countries. There are grounds to assume Shvedov to have been the founder of this complex branch of science.

In 1965, B. P. Konstantinov and G. E. Kocharov formulated a coordinated research task "Astrophysical Phenomena and Radiocarbon", which involved high-precision measurement of the content of radiocarbon in the annual tree rings. During the past 30 years, a large series of dendroclimatochronological and radiocarbon studies were carried out by scientists from Russia, Lithuania, Ukraine, Georgia, the U.S.A., Finland, and Japan.. We shall consider, in what follows, the most remarkable results obtained within the framework of this problem.

Major attention in these studies was focused on investigating the long-period variations of the galactic cosmic ray intensity by measuring the radiocarbon concentration in annual tree rings. We may recall that the radiocarbon, ^{14}C, is produced in the Earth's atmosphere in nuclear reactions initiated by cosmic rays. Next, radiocarbon enters the annual tree rings together with the stable isotopes ^{12}C and ^{13}C. The information introduced into the rings remains there for hundreds and thousands of years. Therefore, by determining the radiocarbon content in precisely dated tree rings, one can reconstruct the temporal variation of galactic cosmic rays over large time scales in the past. It is known that the cosmic ray flux decreases with increasing energy. However the higher the energy, the easier it is to overcome the magnetic shielding and to reach the Earth's atmosphere. As a result of the interplay between these factors, the most efficient cosmic ray energy region was found to extend from 200 to 50,000 MeV. The dynamics of the solar magnetic field manifest themselves particularly clearly during the deep and extended minima of solar activity.

Figure 1 presents experimental data on the temporal variation of galactic cosmic-ray intensity during the last 8,000 y [4], which was reconstructed from radiocarbon measurements made on dated tree rings [5]. It is seen that the cosmic ray intensity underwent regular and long minima, coordinated with solar activity, similar to the Maunder minimum (1645—1715). The theory of such minima is still lacking.

Tree rings, the unique witnesses of the past, keep in memory the year-by-year variation not only of the radiocarbon content in the Earth's atmosphere, but of the concentration of stable isotopes, ^{13}C and deuterium, which are indicators of paleotemperature. Besides, the width of the tree rings contains information not only on solar activity but the local conditions as well (temperature, precipitation etc.). Our analysis of growth records of the pine tree abundant in dry mountain regions of the U.S.A., revealed deep depressions which are in phase with the maxima in galactic cosmic ray intensity (see Fig. 1). Clearly enough, the principal conductor of the 8,000-year global cosmic ray performance on the Earth is the Sun. Note that the characteristic scale of the regular decrease in the depression amplitude is 5,000 y. We witness here, possibly, a modulating effect of auto-oscillations in the atmosphere--deep-ocean--glaciers climatic system. During the last two thousand years, one observed a decrease of the average ring growth, against which one can, nevertheless, clearly see minima, which occur in phase with those of the solar activity.

The evolution of the Sun during the last one thousand years was studied, most comprehensively, from indirect data. Both the variations of the radiocarbon concentration in tree rings and the content of ^{10}Be in Antarctic and Greenland ice cores revealed, reliably, three extremal periods in solar activity, specifically, the Maunder, Spoerer, and Wolf minima. An analysis of five chronologies of pine trees growing in mountain and northern regions of Eastern Europe and Western Siberia led to the conclusion that one observes not only a general in-phase decrease of tree growth during these periods, but also a coincidence of the characteristic details in the variations, both in the ring width and in the ring radiocarbon content.

Thus, one may conclude that periods of the type of the Maunder minimum are typical phenomena in solar evolution, and that they result in global changes on the Earth, which are fixed clearly by annual tree rings.

5. Galactic Cosmic Ray Modulation by Solar Activity during the Past 400 Years

Historical records state the existence of long periods during which no spots were observed on the Sun. The most recent deep and long solar minimum lasted from 1645 to 1715. This minimum is called the Maunder minimum, by the name of the English scientist who published in 1912 a paper on the existence of this phenomenon. The method of cosmogenic isotopes permits obtaining quantitative data on cosmic ray modulation in the past. Therefore, high-precision measurements of radiocarbon content in annual tree rings during the past 400 years, including periods before, during, and after the Maunder minimum, were considered to be of considerable interest.

The first series of measurements was carried out in the 1970s at the Ioffe Physicotechnical Institute [6]. The second series was made at the Tbilissi State University in the first half of the 1980s [7]. The third series of measurements was performed in the U.S.A. in the late 1980s [8] at the laboratory of Stuiver, a well-known specialist in radiocarbon methods.

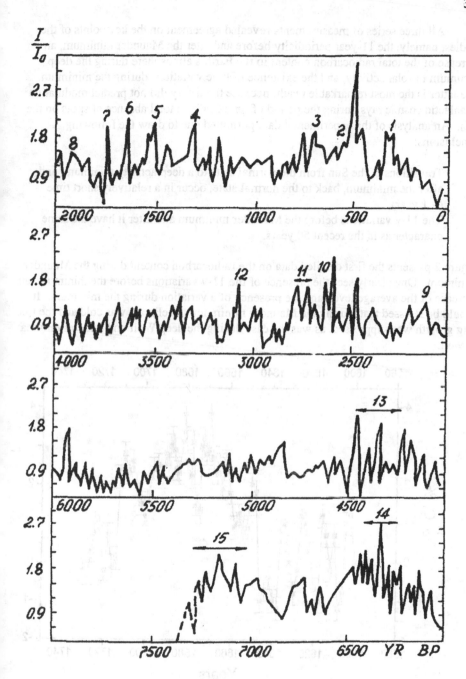

Figure 1. GCR flux over the past 8000 years.

All three series of measurements revealed agreement on the key points of the studies, namely, the 11-year periodicity before and after the Maunder minimum, an increase of the total radiocarbon content in the Earth's atmosphere during the deep minimum in solar activity, and the existence of time variations during the minimum. The latter is the most remarkable result, because the theory did not predict modulation of galactic cosmic rays during the period of, practically, a total absence of spots on the Sun. An analysis of the experimental data permitted one to draw the following conclusions:

1. Transitions of the Sun from the normal state to a deep activity minimum, and, after the minimum, back to the normal state, occur in a relatively short time (~ 1 year).
2. The 11-y variations before the Maunder minimum and after it have the same character as in the recent 50 years.

Figure 2 presents the first detailed data on the radiocarbon content during the Maunder minimum. One clearly sees the presence of the 11-y variations before the minimum, an increase of the average level, and the presence of a variation during the minimum. It should be stressed that, during the Maunder minimum, the climate was cold and the tree ring growth was suppressed, as was the case during the deep Wolf and Spoerer minima as well.

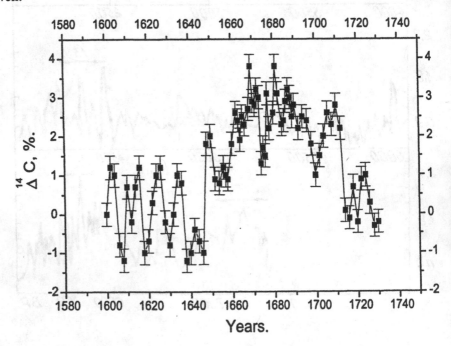

Figure 2. Concentration of radiocarbon in tree rings for A.D. 1600-1728.

The available experimental data permitted us to construct, for the first time, a year-by-year profile of the intensity of galactic cosmic rays during the last 400 years. Of particular interest are the variations of the intensity of galactic cosmic rays (I_p) and of the Wolf numbers (W) before the deep minimum and after it. One readily sees that the number of solar spots during the Maunder minimum was small. At the same time one observed a variation of the galactic cosmic rays in intensity; note that the amplitude of this variation is larger that before and after the minimum, and the characteristic variation period was not 11 years, but rather close to 22 years, which corresponds to the period of reversal of the general magnetic field of the Sun.

Let us formulate the main properties of the Maunder minimum:

1.   A practically complete absence of magnetic activity during 70 years;
2.   A fairly fast transition to the state of deep minimum and a fast recovery of the solar activity at the end of the minimum.
3.   The existence of galactic cosmic-ray modulation during the deep minimum, which was not predicted by theory.

It appears that the latter feature is extremely important; it was confirmed in experiments of other authors bearing on the variation of the content of ^{14}C in tree rings, of the ^{10}Be isotope in polar ice, and of the solar diameter during the Maunder minimum. It would be of paramount importance to measure now the content of the cosmogenic isotopes ^{14}C and ^{10}Be in dated tree rings and polar ice, respectively, for the more ancient minimum in the solar activity, namely, the Spoerer minimum, (1450—1550) which lasted longer than the Maunder one. The Spoerer minimum extended over five 22-y cycles, so that one could determine with a higher reliability the characteristics of the variations. In the time since our discovery of the galactic cosmic-ray modulation during the minimum, considerable progress has been reached in the theory of solar modulation of the galactic cosmic-ray intensity. Concrete physical processes have been proposed to account for the 11- and 22-y cycles.

6. Nitrates in Polar Ice: a New Window in the Investigation of Astrophysical Phenomena in Real Time and the Distant past

Systematic measurements of the nitrate concentration in Antarctic ices, carried out during many years by Dreschoff and Zeller [9,10], have led to development of a unique technique for studying astrophysical and terrestrial phenomena. The basis of this technique lies in the fact that ice contains a chemical record of the processes of ionization in the polar atmosphere by charged particles, x-rays and gamma rays. The Antarctic acts as a cold trap capable of freezing astrophysical signals and storing them for a long time. A substantial part of the nitrates, precipitating at the Earth's surface and concentrating in ice, is generated by ionizing cosmic rays. The model of nitrate generation due to ionization of the atmosphere by cosmic rays was developed in the work of Dreschoff et al. [11]. It is possible to separate two steps in the process of generation of a NO_3^- ion. At the first step, the capture of an electron by oxygen molecules, O_3 and O_2, and origination of ion O_3^- takes place. The ozone molecule immediately creates O_3^- ion. The molecule O_2 initially dissociates to ion O^- and atom O. Then the O^- ion reacts with the molecule O_2 and forms O_3^- ion.

In the second step, the interaction between O_3^- ion and molecules NO_x takes place. This interaction leads to the formation of the ion NO_3^- and molecules CO_2. Calculations of these reactions [11] show that:

1. Generation of nitrates in the Earth's atmosphere by cosmic rays takes place mainly at altitudes 20-50 km. The lower limit is determined by the depth of cosmic rays penetration in the atmosphere. When the altitude increases, the rates of reactions decrease due to the decrease of atmospheric density.

2. The increase of input of nitrogen oxides due to industrial activity leads to an increase in the rate of generation of NO_3^- ion in the atmosphere and to an increase of the abundance of nitrates in polar ice.

So the nitrate content in polar ice really carries information about solar flares, the intensity of cosmic rays, variations of the nitrate oxides concentration in the atmosphere, and, probably, about climatic changes.

It is universally accepted that the solar magnetic field is the energy source of the solar cosmic rays. This follows primarily from the fact that the experimentally measured magnetic fields of the Sun and the geometric size of a flare region can account for the flare energy. At the same time, we just have no other choice; indeed, we do not know any other source of energy. As our experimental possibilities continue to grow, one succeeds in determining ever more accurately the main characteristics of the flare-accelerated protons and electrons (e.g. the total energy of all particles, their total number, the particle generation power etc.). However, interpretation of the data in terms of the hypothesis of the magnetic nature of the energy source grows ever more difficult. Because the magnetic field strength is finite, there should be an upper limit E_{max} on the total energy imparted to particles. Therefore experimental determination of this energy E_{max} is very important. The problem is very complex, because the higher the total energy of accelerated particles, the lower is the probability of this event. Obviously enough, a long series of studies is needed. The nitrate method permitted reliable detection of solar-flare protons from the flares of 1859, 1946, 1972, and others. This makes detection of the largest flare by studying the nitrate content in polar ices a realistic problem.

The problem of possible terrestrial manifestations of cosmological gamma ray bursts is presently being actively discussed. It is pointed out, in particular, that gamma-ray bursts can leave traces in the Earth's atmosphere in the form of nitrates and cosmogenic isotopes. Assuming the total energy of a gamma-ray burst to be 10^{57}erg, and the distance from the Earth to be 1 kpc, this yields 2.7 erg $cm^{-2}s^{-1}$ for the total flux of energy at the Earth from such a source. The radiocarbon generation rate would be 34 atoms $cm^{-2}s^{-1}$, which is 16 times lower than the generation rate by galactic cosmic rays. The generation rate of another important cosmogenic isotope, ^{10}Be, would be negligible, because the energies of burst gamma rays lie primarily from 30 keV to 2 MeV, which is less than the threshold of ^{10}Be generation from the nitrogen nucleus. This makes burst detection from radiocarbon possible, in principle. However the radiocarbon half-life is relatively short (5740 y), so that it can cover a time interval only of a few tens of thousands of years. The probability for a close gamma-ray burst to occur during this time period is very low. Estimates show that for a total energy of 10^{52} erg and a distance of 10 pc, the total number of the nitrate molecules would be 10^{34}, which makes detection of such a gamma-ray burst possible.

7. The Problem of the Origin of the Cosmic Rays

The origin of cosmic rays has been a key issue of high-energy astrophysics during several decades. This issue can be formulated in four questions:

1. What is the origin, galactic or metagalactic, of the cosmic rays detected in satellite, balloon, and ground-based experiments?
2. What is the source of the cosmic rays with energies extending to levels inaccessible for ground-based particle accelerators?
3. What physical mechanism accounts for the particle acceleration?
4. What is the explanation of the experimental observation that the intensity of cosmic rays practically does not vary in time?

Many years ago, Ginzburg [12-14] formulated clearly an experimental possibility of answering the first question, which lies in comparing the cosmic ray intensity in our Galaxy to the closest to us extragalactic object, the Magellanic Clouds. He proposed to detect the gamma radiation produced in the Magellanic Clouds from their interaction with cosmic rays. The gamma-ray intensity yields, directly, information on the extragalactic gamma rays. The idea is brilliant and straightforward; indeed, if the cosmic rays are of metagalactic nature, the gamma-ray flux should coincide with the value calculated from the measured flux of cosmic rays in our Galaxy. If, however, the gamma-ray flux is lower, then the cosmic rays detected on the Earth are of galactic origin (i. e., they are "ours"). The preparation and realization of this satellite experiment took about 20 years. In 1993, the experiment showed that the flux of gamma radiation is less by approximately 100 than if they were of metagalactic origin. Thus, the old idea of Ginzburg that cosmic rays are of galactic origin obtained experimental validation.

The possibility of cosmic ray generation in Supernova explosions was suggested in 1934 by Baade and Zwicki [15]. The validity of this hypothesis received support in the early 1950s, when it became clear from astronomical observations that the remnants of supernovae contain large amounts of relativistic electrons. The test of this fundamental hypothesis required, naturally, measurement of cosmic rays from an exploded Supernova. This is, however, a formidable problem. In order for the average cosmic-ray density (1 eV/cm^3) to change by a measurable amount, the Supernova has to explode close enough to the solar system, more specifically, closer than 100 pc. It is known that in a sphere 100 pc in radius an explosion should occur once every 100,000 years. Obviously enough, the probability of an explosion on a real time scale is extremely low. Moreover, even if such an explosion did occur, in order to establish the nature of the source (Supernova or not), continuous measurements covering tens of thousands of years would be required. B. P. Konstantinov and the present author formulated and developed a method of detection of cosmic rays from a Supernova that exploded in the distant past, by use of cosmogenic isotopes.

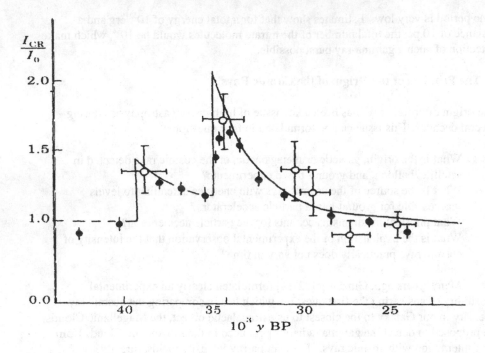

$$\frac{I_{CR}}{I_0}$$

Figure 3 . GCR flux over the past 30 000 yr from ^{10}Be content in polar ice cores from Antarctica and Greenland.

The first experimental measurements of the ^{10}Be content in the last 40,000 years were published in the early 1980s. One detected a considerable increase of the cosmic ray intensity around 10,000—40,000 y ago (Figure 3) [16,17]. Because the author has been long waiting for these results, a paper was published already in 1982, that showed that the temporal course obtained, provides qualitative and quantitative evidence for the explosion of a Supernova close to the solar system, with a total energy released in cosmic rays of 10^{50} erg. Experimental data on ^{14}C, ^{10}Be, and ^{36}Cl obtained later supported fully the conclusion of a Supernova explosion about 35,000 years ago.

An analysis of the data permits the following conclusions:

1. The existence of a maximum in the generation of cosmogenic ^{10}Be in ice cores 35,000 years ago was reliably established;
2. Experimental data reveal a simultaneous increase of the ^{10}Be content in the southern and northern hemispheres;
3. The available experimental data on the content of cosmogenic radiocarbon in the Earth's archives (corals and stalactites) suggest that the radiocarbon content in the atmosphere in the interval from 30 to 40 thousand years ago was twice the present level, with the maximum of the concentration also being found 35,000 years ago;

4.   The ^{10}Be and ^{14}C cosmogenic isotopes differ strongly both in the mechanism of
        their generation in the Earth's atmosphere and in their geophysical and
        geochemical behavior. The three isotopes, ^{10}Be, ^{14}C, and ^{36}Cl have a common
        parent, the cosmic rays.

Consider now the experimental time history of the cosmic ray intensity. The reliability
of the maximum at 35,000 years ago does not arouse any doubt. Somewhat earlier
(about 40,000 y back), the cosmic ray intensity underwent an increase. Let us accept as
a working hypothesis that within the interval from 35 to 40 thousand years ago there is
also a maximum. How could one explain such a premaximum? Particle acceleration in
any source, including Supernovas, takes time, and this time is spent by particles not in
vacuum but in the same "material" which supplies particles to the accelerator. This
entails nuclear interactions, generating gamma quanta which escape from the source.
Gamma quanta also create cosmogenic isotopes in the Earth's atmosphere. One cannot
therefore rule out the possibility that the prepulse is due to gamma quanta. At the same
time, a two-peak structure can be obtained within the shock-wave acceleration
mechanism. It would be very important to obtain more detailed experimental data on
the time period extending from 35 to 50 thousand years back.
        Figure 4 presents the results of high-precision measurements of the radiocarbon
content in annual tree rings for the period covering the Supernova explosions 1006 [18].
These unique studies were performed in cooperation between the U.S.A. (P. Damon)
and Russia (G. Kocharov). One established for the first time the profile of generation of
accelerated particles in stellar explosions and the total energetics of these phenomena.

8. The Main Results and Prospects

The most important result has been the founding of a promising new branch of science
-- Experimental Paleoastrophysics -- an achievement which has been made possible
only due to efforts of researchers in many countries. The most fundamental among the
accomplishments are:

1.   An experimental discovery of cosmic-ray modulation by the Sun during the deep
        minimum in activity, a phenomenon not predicted by theory;
2.   Cosmic rays from a supernova explosion have been detected for the first time,
        and the time history of the phenomenon has been reconstructed.

One could formulate the following most important issues to work on in the future.

1.   For the Supernova that flared up 35,000 y ago, it would be extremely important
        to reconstruct in detail the time history of the prepulse and, by measuring the
        concentrations of the ^{14}C, ^{10}Be, and ^{36}Cl isotopes, to establish the energy spectrum
        of the cosmic rays;
2.   It appears important to experimentally detect at least one more Supernova,
        establish the Supernova explosion frequency in our Galaxy. The most
        appropriate object appears, presently, to be GEMINGA;

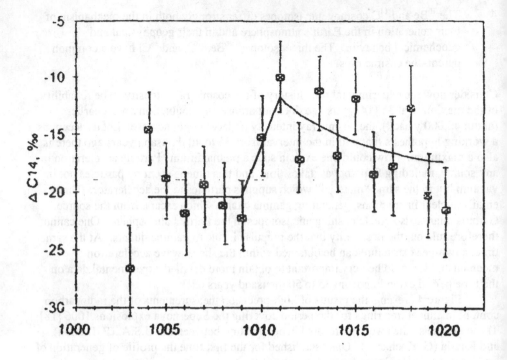

Figure 4. Concentration of radiocarbon in tree rings for A.D. 1003-1020 and time profile of $\Delta^{14}C$, calculated for a 90% increase of rate of radiocarbon generation in A.D. 1009.

3. It would be of paramount importance to obtain year-by-year data on cosmogenic isotopes for at least one more deep solar minimum, the Spoerer minimum of 1450—1550. If the cosmic rays are modulated with a period of 22 y in this case, as well, the problem of modulation will certainly become one of the key issues in cosmic ray astrophysics and the physics of the Heliosphere;

4. Correlated nitrate and radiocarbon studies promise to provide an answer to one of the key problems in solar physics, namely, the mechanism responsible for the energetics of solar flares;

5. We have considered above only ground-based cosmic-ray detectors. There are, however, other cosmic archives of radiation history. Meteorites are constantly irradiated by cosmic rays, which inevitably produce in them cosmogenic isotopes. Finding a cosmogenic trace of the Supernova that exploded 35,000 y ago would be of paramount importance, because all the information available thus far was obtained from the Earth's archives. The meteorite needed for this investigation should be not too old, otherwise the accumulated background of the isotope of interest to us will be too high to permit detection of the Supernova explosion signature. The best object is the Fargminton meteorite. Its space age is acceptable (~40,000 y). To detect the effect, one should measure the content of

the cosmogenic isotope ^{81}Kr. Unfortunately, the sensitivity of detection of this isotope needs to be increased by approximately an order of magnitude.

It should be noted in conclusion that the potential of Experimental Paleoastrophysics is very large, both for studying unique phenomena, and for investigation of the radiation history of the Solar System and of our Galaxy as a whole.

Acknowlegement: The work was made possible with the financial support of the Russian Foundation for Basic Research (Grant No. 99 02-18398) and Finnish Academy of Sciences (project No.16).
Authors are thankful to Prof. M.M. Shapiro and to Prof. J.P.Wefel for the possibility to participate in the International School of Cosmic Ray Astrophysics.

References

1. Konstantinov B.P, Kocharov G.E.: 1965, *Dokl. AN SSSR*, v. 165, p. 63 (in Russian).
2. Wolfsberg K., Kocharov G.E.: 1991, In: *The Sun in time*, eds. C.P. Sonett, M.S. Giampapa, M.S. Mattheus, Tuscon, pp. 288-313.
3. Cecchini S., Galli V., Nanni T., Ruggiero L.: 1996, Nuovo Cimento, v. 19C, _4, pp. 527-532.
4. Kocharov G.E., Vasiliev V.A., Dergachev V.A., Ostryakov V.A.: 1983, *Pis'ma A.Zh.*, v.9, p. 206 (in russian).
5. Suess H.E.: 1978, *Radiocarbon*, v. 20, p.1.
6. Vasiliev V.A., Kocharov G.E.: 1983, In: *Proceedings of XIII Leningrad cosmophysical seminar*, p.101 (in russian).
7. Zhorzholiani I.V., Kereselidze P.G.,Kocharov G.E., Lomtatidze Z.V., Masrchilashvili N.M., Metzhvarishvili R.Ya., Tagauri Z.A., Zereteli S.L., Chesnokov V.I.: 1988, In: *Experimental methods of investigation of astrophysical and geophysical phenomena*, p.92 (in russian).
8. Stuiver M., Braziunas T.: 1993, *The Holocene*, v.3, p.1.
9. Dreschoff G.A.M., Zeller E.J.: 1994, In: D.Wakeffield (ed.), *TER-QUA Symposium Series 2*, , Nebraska Academy Sciences, pp.1-24.
10. Dreschoff G.A.M., Zeller E.J: 1997, *U.S. Air Force Final Technical Report*, AFOSR F496020-95-0003, Nov. 30.
11. Dreschoff G.A.M., Boyarchuk K.A., Jungner H. Et al.: 1999, In: *Proceedings of the 26th International Cosmic Ray Conference*, v. 4, pp.318-321.
12. Ginzburg V.L.: 1953, *UFN*, v. 51, p.343.
13. Ginzburg V.L: 1972, *Nature Phys. Sci.*, v. 238, p.8.
14. Ginzburg V.L., Syrovatskij S.I.: 1964, Origin of Cosmic Rays (Oxford; Pergamon Press).
15. Baade W., Zwicky F.: 1934, *Phys. Rev.*, v. 46, p.76.
16. Konstantinov A.N., Kocharov G.E.: 1982, Preprint PTI-801.
17. Konstantinov A.N., Kocharov G.E.: 1984, *Pis'ma A.Zh*, v. 10, p.92.
18. Damon P.E., Kocharov G.E., Peristykh A.N., Mikheeva I.B., Dai K.M.: 1995, In: Proceedings of 24th International Cosmoc Ray Conference, v.2, pp. 311-314.

H.E.S.S. - THE HIGH ENERGY STEREOSCOPIC SYSTEM

Ira Jung

Max-Planck-Institut für Kernphysik, Heidelberg, Germany

Ira.Jung@mpi-hd-mpg.de

Abstract   The HESS project is one of the next generation instruments for VHE gamma-ray astronomy. In its first phase a system of 4 Imaging Atmospheric Cherenkov Telescopes (IACTs) with a mirror area of about $100 \, m^2$ will be built in the Komhas Highlands of Namibia. In the next phase, it will be extended to a maximum of 16 telescopes. This paper concentrates on the first phase of the project.

Keywords: Cherencov Telescopes, gammay-ray astronomy, telescope design

Introduction

High energy gamma rays are an important diagnostic tool for the investigation of the nonthermal universe. In the past, various techniques have been applied for the detection of very energetic γ-rays but, at present the IACTs have proven to be the most efficient ones. They have mainly detected and studied the four established VHE γ-ray sources: two galactic sources, the Crab Nebula and PSR 1706-44, and two extragalactic AGNs, Mrk 421 and Mrk 501. Nonetheless the origin and the acceleration mechanism of the nonthermal components are not known yet.

An increase in sensitivity and a lower energy threshold should increase the number of detected sources and the photon statistics of known sources. Detailed energy spectra measurements allow the investigation of the nonthermal particle acceleration mechanisms and the study of cutoff due to intrinsic absorption or absorption by the Diffuse Extragalactic Background Radiation (DEBRA) fields.

1.   The HESS performances

The simultaneous observation of air showers with several telescopes under widely different viewing angles is the principle of the stereoscopic technique used by the HESS project. A full three-dimensional reconstruction of the shower geometry and an unambiguous determination of the shower core, direction and

M. M. Shapiro et al. (eds.), Astrophysical Sources of High Energy Particles and Radiation, 345–348.
© 2001 *Kluwer Academic Publishers. Printed in the Netherlands.*

the primary energy is possible on an event to event basis. Therefore a good angular resolution will be achieved. The multiple images and the determination of the shower core lead to good energy resolution and effective gamma/hadron separation. Triggers generated by night sky noise or by local muons will be strongly suppressed by the use of a multi-telescope trigger. This allows a lower energy threshold. The power of the stereoscopic approach has been for the first time successfully demonstrated by the HEGRA-experiment Daum97. Detailed simulations have been carried out to determine the performance of the HESS system. More details can be found in Aharonian97 and Konopelko99.

The angular resolution will be less than 0.1 degree and the energy resolution less than 20 %. The energy threshold is supposed to be 40 GeV for the detection and 100 GeV for spectroscopy. The integral flux sensitivity above 100 GeV for 100 hours of observation will be 10^{-12} ph/(cm^2 s).

2. Technical Concept

2.1. Mirror and Telescope Structure

The HESS telescopes will use an alt-azimuth mount. Each reflector will consist of 380 individual mirror tiles with a total mirror area of about 100 m^2 based on Davies-Cotton optics with 15 m focal length. An uniform image over the field of view is achieved by a large ratio focal length to diameter of the telescope dish of 1.2.

The individual mirrors are quartz-coated aluminized glass mirrors with 60 cm diameter from two producers COMPAS (Czech Republic) and GALAKTICA (Armenia). The specification requires 80% reflectivity between 300 and 600 nm and a focus spot of 1 mrad diameter containing 80% of the light, which corresponds to a rms spot width of 0.28 mrad. The deformation of the dish under gravity and wind should be comparable to this value and is small compared to the pixel size of 2.8 mrad. Therefore the mirror support points are specified over the entire altitude range to be stable to 0.14 mrad rms.

The large number of mirrors requires an automatic alignment system. Each individual mirror is motorized and supported at three points at the edge of an equilateral triangle with side length of 30.3 cm. The accuracy of the mirror movement is 6 μm. Therefore a resolution of 0.05 mrad is achievable. A CCD camera will be used to monitor the mirror position.

2.2. Camera Electronics

A large field of view and a flat acceptance over the camera is essential for the observation of extended sources. Therefore the size of the pixels will be uniform within the camera (0.16 degree) and the field of view will be 4.3 degrees extendable up to 5.0 degrees.

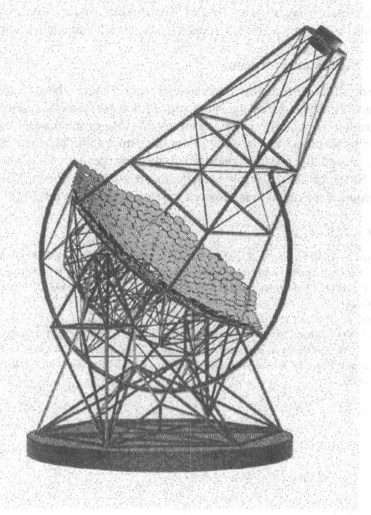

Figure 1. A single HESS telescope

Conventional 30 mm PMTs with bialkali cathodes and 8 dynodes, operated at a gain of $2 \cdot 10^5$ will be used as photodetectors. The camera will contain almost the entire trigger and readout electronics. Therefore wide bandwidth signal transmission are not needed and the number of conductors and interfaces is reduced.

The readout system will be based on the ARS (Analog Ring Sampler) ASIC to sample the PMT signal at 1 GHz rate. 128 samples are stored by the ARS leaving sufficient time for a trigger decision. The trigger requires a coincidence of a fixed number of pixels (typically 3-5) out of overlapping 8×8 pixel groups. This reduces the rate of the night sky induced noise trigger dramatically while keeping most of γ-ray triggers. The PMTs and the electronics are packed into 16 PMT units containing active HV supplies, current and temperature monitoring.

3.   Status and Outlook

The details of the technical implementation of HESS have been under intense study and the design parameters are fixed. Prototypes of the actuators and the specifications of the mirrors have been tested. 13% of the mirrors have been shipped to Namibia. The building of the infrastructure on the site was started in September 2000, and the foundation of the telescopes will be finished by the end of November 2000. First light on the first telescope will be in 2001 and the completion of the initial four telescope system is scheduled for 2002.

References

Aharonian, F. A., Hofmann, W., Konopelko, A., & V"olk (1997), *The potential of ground based arrays of imaging arhmospheric Cherenvoc telescopes* Asropart. Phys. 6, 343 and 369

Daum, A. et al.(1997) ,First Results on the Performance of the HEGRA IACT Array, Astropart. Phys.8 ,1

Hofmann, W. (1999), *The High Energy Stereoscopic System (HESS) Project* , Proc. Towards a Major Atmospheric Cherenkov Detector VI, ed. Kieda, D. , Salomon, M. , Dingus, B.,400

Konopelko, A. (1999),*On design studies of future 50 GeV arrays of imaging air Cherenvoc telescopes*, Proc. the Veritas Workshop, ed. Weekes, T.C. , Catanese, M. , Astropart. Phys. 11 (1999), 263

Development of Atmospheric Cherenkov Detectors at Milagro

Robert Atkins for the WACT collaboration
(atkins@titus.physics.wisc.edu)*
Department of Physics, University of Wisconsin, Madison, WI 53706 USA

1. Introduction

In 1912 Victor Hess discovered that ionizing radiation was coming from outer space. From this discovery the field of cosmic-ray physics was born. The detection of cosmic-rays is usually done in one of two ways. Cosmic rays are directly measured by placing particle detectors high in the earths atmosphere or in outer space. This is most often accomplished with balloons and satellites. The other way to detect cosmic-rays is to build large ground based detectors that observe the secondary particles created in extensive air showers (EAS). These methods have poor resolution but can have huge areas. The wide angle Cherenkov telescopes (WACT) is an array of telescopes in northern New Mexico.

The primary energy spectrum of cosmic-rays is a power law that falls like $E^{-2.7}$. For energies above 10^{15} to 10^{16} the power law changes to $E^{-3.0}$ (Gaisser, T.K., 1990). This spectral change, known as the knee, was first observed in 1958 by the Moscow State University (MSU) group (Kulikov, G.V. & Khristiansen, G.B., 1958) and was thought to be a sudden change, however it has since been observed to be a more gradual change (Amenomori, M. et al., 1996). The primary goal of WACT is to measure cosmic-ray composition from low energies (around 50 TeV), where is has been directly measured, up to the region of the knee.

The knee has been a region of great interest for quite some time, and in light of this it is remarkable how little we know about this region. What is the source of cosmic-rays in this region? What is the composition in this region and why do we see this spectral change? These are fundamental questions to which we still do not know the answer. There have been several theories to explain the presence of the knee. Some of these models predict a composition becoming heavier across the knee, while others predict a lighter composition. One possible explanation for a heavier composition is the magnetic confinement of cosmic rays in the

---

* This e-mail address is available for all problems and questions.

M. M. Shapiro et al. (eds.), Astrophysical Sources of High Energy Particles and Radiation, 349–354.
© 2001 *Kluwer Academic Publishers. Printed in the Netherlands.*

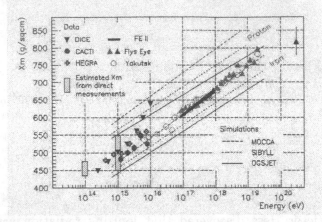

Figure 1. The above plot shows measured and inferred shower max. from different experiments. The solid and dashed lines represent expected shower max . . for an iron or proton composition for different simulation packages. It should be noted that the DICE points are from an older analysis. Also the last two DICE points have large error bars that are not shown.

galaxy. As the rigidity (pc/Z) of a cosmic ray increases the less likely it is to be contained in the galaxy. This means if the knee in the proton spectrum is observed at a certain $E_{knee} = E_o$, then the knee in the iron spectrum would occur at $E_{knee} = 26E_o$. This model is consistent with a smooth knee with a composition becoming more iron like at higher energies. In 1979 Hillas proposed that the knee, if the composition is becoming lighter, could be attributed to photodisintegration of more massive cosmic rays by optical and soft UV photons present around the source (Hillas, A.M., 1979). This theory has been explored more recently by Candia (Candia, J. et al., 2000).

Composition through the region of the knee has been measured by several different experiments and is summarized in figure 1. Some experiments find a composition becoming more massive through the knee (Fowler, J.W. et al., 2000). A more iron like composition is consistent with higher energy multicomponent measurements made by the High Resolution Fly's Eye (HiRes) and the Michigan muon array (MIA) (Abu-Zayyad, T. et al., 2000). The Dual Imaging Cherenkov Experiment (DICE) has reported composition becoming more proton like across the knee (Swordy, S.P. & Kieda, D.B., 2000). Still other experiments have seen evidence of an unchanging composition (Bernlöhr, K. et al., 1998).

The question of composition in the region of the knee could be solved by placing a large direct measurement detector in space. Direct measurements have very good resolution and can determine composition on

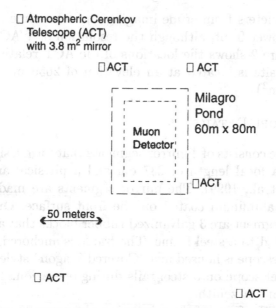

Figure 2. Locations of WACT telescopes relative to Milagro.

event-by-event basis. But due to the low flux at the knee (on the order of 1 particle per m^2-year) the required size would be to large to be practical. Therefore, the only way to make composition measurements at knee energies and above is to build EAS arrays and atmospheric Cherenkov telescopes (ACT). These types of detectors can have very large effective effective areas but poor resolution when compared to direct measurements.

2. Detector Description and Experimental Technique

2.1. GENERAL OVERVIEW

The WACT experiment is located in the Jemez mountains, about 40 miles west of Los Alamos New Mexico. The array consists of 6 ACT surrounding the Milagro observatory. Placement of the 6 ACT is chosen to make measurements of the Cherenkov light density at various distances from the shower core over the effective area of Milagro. The location of the shower core and the angle of the primary is reconstructed from the Milagro data. Milagro can reconstruct core and angle to about 5 meters and 0.7° respectively for hadronic showers whose core is on the Milagro pond. Both core, angular resolution, and total area will be significantly improved with the placement of outriggers (water tanks with a pmt in them) around Milagro. WACT will also be able to determine both of these parameters from crude imaging and from time measurements across the shower front, although the resolution of WACT will not be as good. Figure 2 shows the locations of the ACT relative to Milagro. The Milagro site is located at an elevation of 2650 meters above sea level ($750g/cm^2$).

2.2. TELESCOPE DETAILS

Each telescope consists of 4 mirror segments that form a single spherical mirror with a focal length of 237 cm and a physical area of 3.8 m^2 (Atkins, R. et al., 1999). The mirror segments are made of slumped glass that is aluminum coated on the front surface. On the back of each mirror segment are 3 galvanized rubber blocks that allow the glass to be connected to a steel frame. The frame is anchored to a concrete pad. Each telescope is housed in a "Covered Wagon" style building that rolls off the telescope onto steel rails during operation. The individual telescopes point at zenith.

Located at the focus of the mirror is a cluster of 27 pmts. The pmts are Amperex 2262 linear focusing tubes. Each tube has a diameter of 2 inches and is arranged in a hexagonal close pack. Hexagonal light cones (compound elliptical cone) are used to fill in the dead spots between the pmts. The cones give each pmt a 2.289^o field of view. The total field of view is 13.9^o on the long side and 9.9^o on the short side. An important goal of any cone design is to have a uniform response across the camera face. This is not possible if the off axis cones have the same angular acceptance as the on axis cones. To solve this problem cones have been designed to have an angular acceptance that changes with its position in the WACT camera. By doing this we can have an off axis cone that can see the entire mirror without picking up stray light from off the mirror. All 27 cones for a camera will be machined out of a single block of acrylic and then coated with aluminum (Mohanty, G., 2000).

2.3. ELECTRONICS

Each pmt uses a passive, positive high voltage base that is designed to have good collection efficiency and a gain of $\sim 10^5$ at $\sim$ 1200 volts. The pmts operate at a low gain so as to minimize the standing currant in the dynode chain from night sky background ($\sim$0.5 pes/ns at the Milagro site). Having a large anode current will cause saturation of the pmt and over time can cause damage to the anode (Atkins, R., 2000).

WACT makes use of the Milagro DAQ and hence the front end electronics are very similar to Milagro's electronics. Both WACT and Milagro use a time over threshold (TOT) method to measure pmt pulses. The TOT method has a charge resolution of about 10% over a large dynamic range and does not require the use of ADCs. The TOT is readout with a Lecroy 1887 FASTBUS TDC (Atkins, R. et al., 2000).

2.4. TECHNIQUE

The WACT experimental technique is based on the sensitivity of the Cherenkov lateral distribution (CLD) to the depth of shower max (Patterson, J.R. & Hillas, 1983). Most of the Cherenkov light in an EAS is generated at or near shower max. The amount of light generated scales well with the total energy of the primary. For a given energy, one would expect to see roughly the same amount of Cherenkov light generated by a iron primary as you would for a proton primary. But due to the larger cross section, the iron primary will develop higher in the atmosphere and thus produce a broader and flatter CLD. In general measurements of the Cherenkov density away the shower core reveal information about the primary energy, while measurements of the shape of the CLD near the core reveal information about shower max. As stated above information about the core location will be provided by the Milagro data.

There is also useful information contained in the muon content of an EAS. Muons are primarily produced by the decay of charged pions. Neutral pions decay into gammas and produce electromagnetic subshowers. The bottom layer of Milagro has a physical area of 2300 m^2 and can serve as a muon counter. Muons in Milagro manifest themselves as clusters in the bottom layer of the pond.

The main problem with using the CLD method to determine composition is that it depends on the details of EAS simulations to relate the CLD to the depth of shower max. With its coarse pixels, WACT will be able to make stereo measurements of shower max. This measurement can be used in conjunction with the CLD information and the muon information. WACT is not an imaging detector in the traditional sense, but the distribution of Cherenkov light across the face of the camera can be fit to some function. From this fit the center of the "image" be found and its location will correspond to the the direction of shower max. By having multiple measurements of this, the location of shower max in the sky can be geometrically constructed.

3. Status and Future

The six telescopes are completed and are in position around Milagro. The electronics and the cabling have been finished and are ready for data. Pmts are being tested for linearity characteristics and the light cones and camera box are being constructed. We expect to have a fully operational detector in the late spring of 2001. The outrigger upgrade to Milagro is currently under-way and the first stages should be finished by the summer of 2001.

4. Acknowledgments

The WACT collaboration would like to thank the Hi-Res experiment for the mirrors used with WACT. We would also like to thank the Cygnus collaboration for the pmts. We would also like to thank Scott Delay, Galen Gisler, David Kieda, Stan Thomas, Joe Herrera, and Gary McDonough for their technical support and assistance. This work is supported in part by the National Science Foundation, the U.S. Department of Energy Offices of High Energy Physics and Nuclear Physics, the University of California, Los Alamos National Laboratory, the University of Utah, the University of Wisconsin at Madison, and the U.C. Institute of Geophysics and Planetary Physics

References

Gaisser, T.K. Cosmic Rays and Particle Physics (New York: Cambridge University Press, 1990)

Kulikov, G.V. and Khristiansen, G.B., JETP, 35 635 1958

Amenomori, M. et al., ApJ 461, 408 1996

Paling, S. M, Ph.D. Thesis, University of Leeds, Dept of Physics and Astronomy 1997

Hillas, A.M., Proc. 16th Int. Cosmic Ray Conf., Kyoto 1979 Vol.8 p.7

Candia, J. et al., astro-ph/0011010

Fowler, J.W. et al., astro-ph/0003190

Abu-Zayyad, T. et al., Phys. Rev. Lett. Vol. 84 p. 4276. 2000

Swordy, S.P. & Kieda, D.B., Astropart. Phys. 13 (2000) 137

Bernlöhr, K. et al., Astropart. Phys. 8 (1998) 253

Atkins, R. et al., 26th Int. Cosmic Ray Conf., Salt Lake City 1999, OG.4.3.34

Mohanty, G., Internal Collaboration Memo. Oct. 24 2000

Atkins, R., Internal Collaboration Memo. Feb. 2 2000

Atkins, R. et al., Nucl. Instrum. Methods Phys. Res., Sect. A 449 (2000) 478

Patterson, J.R. & Hillas, A.M., 1983, J. PHYS. G: Nucl. Phys. 9 1433

NEMO: NEutrino Mediterranean Observatory

G. Riccobene

Laboratori Nazionali del Sud INFN, Catania, Italy
riccobene@lns.infn.it

Abstract. The NEMO (NEutrino Mediterranean Observatory) Collaboration aims at R&D for the construction of an underwater Čerenkov neutrino detector. Great attention is dedicated to projects for the apparatus electronics and mechanical structures. At this stage the main effort of the collaboration is devoted to the selection and characterization of a marine site suitable for the detector deployment. Several sites near the Italian coast have been investigated. In addition, during year 2001, the collaboration will install a test site at 2000 m depth near the Sicilian coast. The facility will be fundamental to study the deep sea environment and to test the reliability of submersed structures.

1. Underwater Telescopes for Neutrino Astronomy

In the last decade, the observation of cosmic rays of Very High Energy, Ultra High Energy, and even with energy greater than 10^{20} eV, has attracted the attention of the scientific community (Cronin et al., 1997). The sources of such events are supposed to be the most luminous and energetic objects observed in the universe such as Gamma Ray Bursters and Active Galactic Nuclei. The detection of intense extra galactic gamma ray sources with energy ~ 10 TeV seems to confirm this hypothesis. If these high energy photons are generated through the production and decay of neutral pions, it is reasonable to expect, from the same sources, an associated flux of high energy neutrinos, generated through the production and decay of charged pions.

It has been demonstrated that the GZK mechanism (Greisen, 1966) does not allow the observation of photons with energy > 10 TeV and protons with energy $> 10^{18}$ eV from sources located at cosmological distance. On the contrary, weakly interacting neutrinos are not significantly absorbed in the universe and are not deflected by the intergalactic magnetic fields: the identification of neutrino events will allow trace back to the source. This is the goal of the new exciting field of neutrino astronomy (for a complete review, see (Gaisser, Halzen, Stanev, 1995)). Already in 1960, Markov proposed to use seawater and the rocks of seabed as a huge target to detect UHE neutrinos, and to look at charged current (CC) weak interactions between neutrinos and water or rock nuclei. If one looks at outgoing muons (and anti muons), those particles carry $\sim 50\% \div 60\%$ of the neutrino energy and

M. M. Shapiro et al. (eds.), Astrophysical Sources of High Energy Particles and Radiation, 355–361.
© 2001 *Kluwer Academic Publishers. Printed in the Netherlands.*

preserve the neutrino direction. The neutrino-induced muon propagates in seawater for several km ($\sim$ 10 km at 10 TeV) and generates, along its path, Čerenkov light that can be detected by a lattice of optical sensors (e.g. large area photo multiplier tubes, PMTs). In order to identify faint neutrino fluxes from astrophysical sources the detector should have an effective area $\geq 10^6$ m^2. Several calculations show that the reconstruction of neutrino energy and direction, is affordable with a detector instrumented with limited number ($\sim$ 7000) of PMTs displaced over a volume of ~ 1 km^3. Such detectors are usually named km^3 neutrino telescopes. Underwater telescopes may allow the reconstruction of the neutrino direction within a few tenths of degree and the reconstruction of neutrino energy, within an order of magnitude. The expected angular resolution will allow a catalogue of the expected neutrino sources in the sky to produce a map and search for correlations with the known gamma sources to be made. The previous statements make the construction of the km^3 neutrino telescope one of the main goals of the astro-particle physics today.

2. The NEMO Project

Several collaborations are involved in neutrino astronomy. The major effort in this field has been conducted up to now by AMANDA-ICECUBE at the South Pole (Spiering, 2000) and BAIKAL-NT, located in lake Baikal (Russia) (Spiering, 2000), the first undewater-ice detectors which observed neutrino events. In the Mediterranean Sea, a region which offers optimal conditions over a worldwide scale for locating the detector, three collaboration are active: NESTOR (Resvanis, 1993) which, since 1990, aims for the construction of a 10^4 m^2 demonstrator detector near the Greek coast at 3800m depth; ANTARES collaboration[1] which is working to build a 0.1 km^2 demonstrator in the vicinity of Toulon (France) at 2500 m depth; and NEMO. The realisation of the ANTARES demonstrator will be a fundamnetal step to acquire experience in submarine technology and to start the activity of neutrino astronomy in the Mediterranean area. ANTARES and NEMO collaborations are also developing new technologies for the realization of the km^3 detector in the Mediterranean Sea. In particular NEMO is an R&D is project funded by INFN which concerns: i) a complete characterisation of several deep sea sites in the Mediterranean area; ii) Monte Carlo simulation studies of the detector capabilities; iii) design of low power consumption, high rate and high reliability electronics for

---

[1] see ANTARES Collaboration web page: *http:antares2.in2p3.fr*

data acquisition and transmission to shore; *iv*) design of mechanical and connection layout of the detector.

Concerning the detector design, the collaboration is simulating the performances of various arrays of phototubes. Several geometrical configurations, instrumented with different number of large area PMTs have been tested. The simulations show that a detector instrumented with ~ 7000 PMTs arranged in a lattice of square towers (side =20 m, height= 300 m) placed at a distance ~ 200 m one from the others, may achieve an effective volume greater than 3 km^3 for E>100 TeV muons, and an angular reconstruction resolution < 0.5°. At this stage the needed CPU time to run the simulation is a very important parameter to allow freqeuent change of the detector configuration. For this reason parametrisations have been used to describe muon propagation and Čerenkov light production (DeMarzo et al., 1999). INFN scientists are also collaborating with industrial partners, leaders in telecommunications (*ALCATEL*) and deep-sea operation (*SONSUB*) to design a mechanical layout and connection net which may allow easy and fast deployment (within few years) and the manteniance of the detector. Marine tests and R&D are also carried out with the helpful collaboration of Marina Militare Italiana and NATO-Saclantcen.

3. Deep Sea Site Selection

The choice of the km^3 scale neutrino telescope location is such an important task that careful studies in candidate sites must be performed in order to identify the most suitable one. The NEMO Collaboration is going to conclude a two year program to characterise selected deep-sea sites along the Italian Coast. The Collaboration has identified four areas corresponding approximately to the coordinates (figure 1):

- 35° 50 N 16° 10 E in the Jonian Sea, South-East of Capo Passero

- 39° 05 N 13° 20 E in the Tyrrhenian Sea, North-East of Ustica island

- 39° 05 N 14° 20 E in the Tyrrhenian Sea, North of Alicudi island

- 40° 40 N 12° 45 E in the Tyrrhenian Sea, South of Ponza island.

The characterisation programme includes measurements of deep sea water optical properties: absorption and diffusion; measurements of optical background: bioluminescence and Čerenkov light produced by β-radioisotopes dissolved in seawater (e.g. ^{40}K); measurements of site oceanographic properties: water temperature, water salinity, dissolved

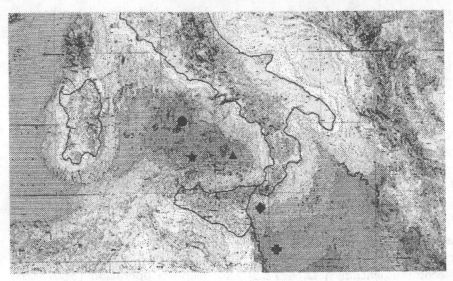

Figure 1. Location of the four sites selected by the NEMO collaboration as candidates for the km^3 deplyment. Alicudi (triangle), Ustica (star), Capo Passero (cross), Ponza (circle). The "Catania test site" location is also shown (diamond).

compounds and particulate; measuremnts of deep sea currents, sedimentation rate, fouling. This work is fundamental since the site must satisfy several requirements.

The site has to be deep enough to filter out the low energy downgoing atmospheric muons. At 3500 m depth, the atmospheric muon flux is reduced by 5 order of magnitude; this would dramatically reduce the number of downgoing muons reconstructed as *fake* upgoing events by the track reconstruction algorithms. At this depth, there will be also an enhancement of the capabilities of the telescope to detect horizontal tracks, which are expected to overcome upgoing events at energy ≥ 10 PeV. The site has to be close enough to the coast. The data and power transmission to/from shore, are obtained via electro-optical multi-fiber cables. At distances closer than 100 km from the coast, commercial systems allow data and power transmission without particular requirements (amplifiers) which would increase costs and reduce the reliability of the project. Moreover, the proximity of the detector location to the coast and to existing infrastructures (well instrumented ports and laboratories) will reduce the time and the difficulties of sea operations. On the other hand, one should also consider the possible occurrence of catastrophic submarine events such as *turbidity* and *density* currents that may occur near the continental shelf boundary, and locate the detector far (≥ 30 km) from shelf breaks and canyons.

3.1. WATER OPTICAL PROPERTIES

The site has to show good optical underwater properties. The detector effective area is indeed not only directly determined by the extension of the instrumented volume but is strongly affected by the light transmission in the water. A muon track crossing the water at a certain distance from the detector can be easily observed since the emitted Čerenkov photons have a non vanishing probability to reach the PMTs. Mainly two microscopic processes affect the propagation of light in the water: absorption and scattering. Light absorption directly reduces the effective area of the detector; the scattering has a negative effect on track reconstruction (based on the measurement of the photon arrival time on the Photon Detectors). To characterise the deep-sea water optical properties, we measure the absorption and attenuation coefficients, down to 3500 m under the sea surface, over 9 different wavelengths in the range 412÷715 nm. The basic device for our optical measurements is an *AC9 transmissometer* by *WetLabs*. During several cruises aboard the Research Vessel *URANIA*, we carried out measurements in all the above mentioned sites and in the vicinity of Matapan Abyss, near the Greek coast, a few miles from the site selected by the NESTOR collaboration. A preliminary analysis of collected data shows that the optical properties of deep-sea water in the selcted sites are not very different from pure water (Smith and Baker, 1981). A preliminary estimation of the measured absorption coefficient for blue light ($\lambda = 440$ nm) close to Capo Passero site at 3300 m depth is equal to 0.014 ± 0.003 m^{-1}, while the value of attenuation coefficient at the same wavelength is 0.025 ± 0.003 m^{-1}. These values will allow us to evaluate the light transmission length in water when the diffused light angular distribution will be known. In deep-sea water, light transmission suffers both from Rayleigh and Mie scattering. Assuming for Mie scattering a conservative value of the average diffusion angle ($< cos\vartheta > \sim 0.9$) we can evaluate a transmission length for blue light close to 60 m.

3.2. SEDIMENTATION AND FOULING

Another requirement is that the sedimentation rate in the selected region must have very low values. The presence of sediments in the water can affect, seriously, the performances of the detector. Sediments increase the light scattering and so worsen the track reconstruction angular resolution. Moreover, a deposit on optical surfaces which host photon detectors reduces the global detector efficiency. Micro-organisms in submarine environment could also produce fouling, a thin organic film formed in submersed surfaces. Fouling can trap deposited sedimets and quickly reduce the transparency of optical surfaces. Starting from

August 1999, sedimentation rate was measured in Capo Passro. The first data (August-December 1999) show that sedimentation rate is extremly low when compared to other coastal regions. Data confirms the suspected low biological activity in the central regions of Mediterranean Sea. In order to measure the effect of sediments and fouling on optical surfaces, NEMO has constructed a deep sea station that was deployed in Capo Passero from December 1999 to March 2000. The station, moored at 3300 m depth, was composed of two blue LEDs that illuminated an array of 14 photodiodes (PDs). The photodiodes were positioned at different angles inside a pressure resistant transparent sphere (of the same type that will be used to contain the PMTs). As a function of mooring time, fouling formed on the surface of the sphere would reduce the light collected by the PDs. Data analysis of the first 40 days shows a constant value of the light collected for all the PD. Therefore, biofouling appears to be negligible over this time scale.

3.3. DEEP SEA CURRENTS

The site has to be *quiet*, i.e. the water current has to show low intensity and stable direction. This is important because it does not imply special requirements for the mechanical structure; the detector deployment and positioning is easier if the water current is limited; the *optical noise* due to bioluminescence, mainly excited by variation of the water currents, is reduced. Current metre chains have been moored in the region of Capo Passero since July 1998. The used lines carry current metres located within few hundreds meters from the seabed (in the range that can be covered by the km^3 structures). The deep-sea water current measured over 18 months is quite stable in direction and intensity: the average value is about 2.8 cm/s, and the maximum value is lower than 20 cm/s.

4. NEMO Deep Sea test Site

The collaboration is also installing a test site in proximity of Catania (Sicily). A 25 km long electro-optical cable will connect the structures moored at 2000m with a laboratory located in the port of Catania, and with the Laboratori Nazionali del Sud of INFN located in the town of Catania. The structure will be devoted to implement deployment, connection and recovery techniques in deepsea, and to perform long-term tests for the electronics. A branch of the electro-optical cable is assigned to other experiments: GEOSTAR and CREEP. The first is devoted to the survey of geoseismic and volcanic phenomena occurring in the Etna region; the second will perform long-term measurements to study the creeping of rocks under extreme pressure.

5. Conclusion

The construction of an underwater neutrino telescope seems to be feasible within the next ten years. However a strong technological effort is still needed. Careful and continuous survey of oceanographic and optical properties in few selected sites must be performed in order to choose the best location for the detector. A neutrino telecope located in a selected region of the Mediterranean Sea could have excellent capabilities to detect events with angular resolution of the order of $\sim 0.4°$. This detector could also look at sources located in the opposite hemisphere with respect to ICECUBE. Moreover, a detector moored at depth greater than 3000m could successfully search for E>10PeV horizontal events. It must be mentioned also that the installation of the km^3 underwater telescope will give other scientists a unique opportunity to study the mysterious world of deep-sea abysses.

References

Smith, R.C., Baker, K.S.: 1981, *Applied Optics* **Vol. 20.**, pp. 177.

Gaisser, T.K., Halzen, F., Stanev, T.: 1995, *Physics Reports* **Vol. 258.**, pp. 173.

Cronin, W., Gaisser, T.K., Swordy, S.P.: 1997, 'Cosmic Rays at the Energy Frontiers', *Scientific American* **January, 1997**.

Greisen, K.,: 1966, *Phys. Rev. Lett.* **Vol.16, 1966** pp. 784.

DeMarzo, C., et al.: 1999, *Nucl. Phys. B P.S.* **Vol.87, 2000** pp. 433.

Spiering, C.: 2000, *Proceedings of Int. Conf. Neutrino Physics (Neutrino 2000). Astro-ph/0012532*

Resvanis, L.K.: 1993, '3rd NESTOR International Workshop', *October 19-21, 1993* , Pylos, Greece.

SUBJECT INDEX

PARTICIPANTS

Dr. Mahmoud Abbas
INFN-Sezione di Genova
Via Dodecaneso 33
I-16146 Genova
Italy
Abbas@ge.infn.it

Mr. Torsten Antoni
Institut fuer Kernphysik
Forschungszentrum Karlsruhe
Postfach 3640
76021 Karlsruhe
Germany
antoni@ik3.fzk.de

Mr. Robert Atkins
University of Wisconsin-Madison
941 Calle Mejia Apt. 503
Santa Fe, NM 87501
USA
atkins@milagro.lanl.gov

Mr. Alexey V. Bakaldin
INCOS, Dep. #7
Moscow State Eng. Physics Institute
Kashirskoe shosse, 31
115409, Moscow
Russia
bakaldin@space.mephi.ru

Dr. Evegny G. Berezhko
Institute of Cosmophysical
    Research and Aeronomy
Lenin Avenue 31
677891 Yakutsk
Russia
berezhko@ikfia.ysn.ru

Mr. Nilay Bhatt
Nuclear Research Laboratory
Bhabha Atomic Research Center
Mumbai 400 085
India
nilayb@magnum.barc.ernet.in

Prof. Peter L. Biermann
Max Planck Inst. fur
Radioastronomie
Auf dem Hugel 69
D-53121 Bonn
Germany
plbiermann@mpifr-bonn.mpg.de

Mr. Oscar Blanch Bigas
IFAE. Edifici Cn. Campus
Universitat Autonoma de Barcelona
08193 Cerdanyola del Valles
Barcelona
Spain
blanch@ifae.es

Dr. Ciro Bigongiari
Department of Physics
University of Padova
Via F. Marzolo, 8
35131 Padova
Italy
ciro.bigongirai@pd.infn.it

Ms. Carla Bleve
Department of Physics
University of Lecce
via per Arnesano
73100 Lecce
Italy
bleve@le.infn.it

Mr. Oliver Bolz
Max Planck Institut fur Kernphysik
Postfach 103980
69029 Heidelberg
Germany
oliver.bolz@mpi-hd.mpg.de

Ms. Laura Brocco
Via San Mamolo, 71
40136 Bologna
Italy
Laura.Brocco@bo.infn.it
Laura.Brocco@cern.ch

Dr. Andrew Bykov
Ioffe Institute for Physics
and Technology
Polytekhnicheskaya 26
194021, St. Petersburg
Russia
byk@astro.ioffe.rssi.ru

Mr. David Paneque Camarero
IFAE Exp. Edf. Cn.
Fac. de Ciences UAB
E-08193 Bellaterra (Barcelona)
Spain
dpaneque@ifae.es

Mr. Diego Casadei
Bologna University
Dipartimento di Fisica
via Irnerio 46
I-40126 Bologna
Italy
Diego.Casadei@bo.infn.it
Diego.Casadei@cern.ch

Dr. Igor Cherednikov
Joint Institute of Nuclear Research
141980 BLTP JINR
Dubna, Moscow Region
Russia
igorch@thsun1.jinr.ru

Mr. Vitaliy A. Cherkaskiy
Electro-Physical Scientific and
Technology Centre
National Academy of Sciences
of Ukraine
Chernyshevsky st., 28
61002, Kharkiv
Ukraine
cherkaskiy@yahoo.com

Dr. Dmitry V. Chernov
Moscow State University
NIIYAF MGU, Vorob'evi Gory
119899 Moscow
Russia
chr@dec1.npi.msu.su

Mr. Tommaso Chiarusi
Via dell'Olivuzzo n 67
50143 Firenze
Italy
Tommaso.Chiarusi@cern.ch

Ms. Mihaela Chirvasa
Department Theoretical Physics
University of Bucharest
P.O. Box MG-11 Bucharest magurele
76900 Bucharest
Romania
chirvasa@personal.ro
mchirvasa@hotmail.com

Dr. Eric Christian
NASA/GSFC Code 661
Greenbelt, MD 20771
USA
erc@cosmicra.gsfc.nasa.gov

Prof. Ramanath Cowsik
Indian Institute of Astrophysics
Sarjapur Road
Koramangala
Bangalore, 560034
India
cowsik@iiap.ernet.in

Dr. Lev I. Dorman
Israel Cosmic Ray Center and
Emilio Segre' Observatory
P.O. Box 2217
QAZRIN 12900
Israel
lid@physics.technion.ac.il

Ms. Tatyana Dorokhova
Astronomical Observatory
Odessa State University
Park Shevchenko
270014 Odessa
Ukraine
tnd@pulse.tenet.odessa.ua

Dr. Anne-Marie Elo
University of Oulu
Dept. of Physical Sciences
P.O. Box 3000
FIN-90014, Oulu
Finland
anne-marie.elo@oulu.fi

Ms. Tulun Ergin
Physics Department
Middle East Technical University
06531 Ankara
Turkey
tulune@newton.physics.metu.edu.tr

Mr. Anant Eungwanichayapant
MPI fur Kernphysik
Saupfercheckweg 1
D-69117 Heidelberg
Germany
anant@mickey.mpi-hd.mpg.de

Mr. Abe Falcone
University of New Hampshire
Department of Physics
Space Science Center, Morse Hall
Durham, NH 03824
USA
afalcone@comptel.sr.unh.edu

Dr. Maria Cristina Falvella
Agenzia Spaziale Italianna
Viale Liegi 26
I-00198 Roma
Italy
falvella@asi.it

Dr. Daniele Fargion
Physics Department
Rome University-1
Ple. A. Moro 2
00185 Rome
Italy
daniele.fargion@roma1.infn.it

372

Mr. Stephen J. Fegan
Smithsonian Astrophysical Observatory
Steward Observatory
933 N. Cherry Avenue
Tucson, AZ 85719
USA
sfegan@egret.sao.arizona.edu

Mr. Lazar Fleysher
Physics Department
New York University
4 Washington Place
New York, NY 70003
USA
Lazar.Fleysher@physics.nyu.edu

Mr. Roman Fleysher
Physics Department
New York University
4 Washington Place
New York, NY 70003
USA
Roman.Fleysher@physics.nyu.edu

Prof. Victoria Fonseca
Facultad Ciencias Fisicas
Universidad Complutense
Ciudad Universitaria
E-28040 Madrid
Spain
fonseca@gae.ucm.es

Ms. Cristina Florina Galea
Dept. of Atomic & Nuclear
Physics-Astrophysics
University of Bucharest
P.O. Box MG-11 Mucharest Magurele
76900 Bucharest
Romania
gal@phobos.cs.unibuc.ro

Prof. Piero Galeotti
Dipartmento di Fisica Generale
Universita di Torino
Via P. Giuria, 1
110 126 Torino
Italy
galeotti@to.infn.it

Mr. Nerses Gevorgyan
Alikhanian Brothers 2
Yerevan Physics Institute
Cosmic Ray Division
Yerevan 375036
Armenia
nerses@crdlx5.yerphi.am

Dr. Mauro Giavalisco
Space Telescope Science Institute
Johns Hopkins University
Baltimore, MD
USA
mauro@stsci.edu

Prof. Maria Giller
Division of Experimental Physics
University of Lodz
Pomorska 149/153
90-236 Lodz
Poland
mgiller@kfd2.fic.uni.lodz.pl

Mr. Stefan Gillessen
Max-Planck-Institut fuer Kernphysik
P.O. Box 103980
D-69029 Heidelberg
Germany
Stefan.Gillessen@mpi-hd.mpg.de

Mr. Holger Göebel
University of Siegen
Walter-Flex-Str. 3
D-57068 Siegen
Germany
goebel@idal.physik.uni-siegen.de

Mr. Niels Goetting
University of Hamburg
II. Institute for Experimental Physics
Luruper Chaussee 149
D-22761 Hamburg
Germany
Niels.Goetting@desy.de

Dr. Anna (Michelazzi) Gregorio
Astronomy Department
University of Trieste
c/o CARSO, Area Science Park
Padriciano 99
34012 Trieste (TS)
Italy
gregorio@sci.area.trieste.it

Mr. Thomas Hams
ENC-B 128
University of Siegen
Walter-Flex-Str. 3
D-57068 Siegen
Germany
hams@idal.pysik.uni-siegen.de

Ms. Rellen Hardtke
Department of Physics
1150 University Ave.
University of Wisconsin-Madison
Madison, WI 53706
USA
rellen@alizarin.physics.wisc.edu

Dr. Subhon Ibadov
Institute of Astrophysics
Tajik Academy of Sciences
Dushanbe 734042
Tajikistan
subhon@ac.tajik.net

Mr. Alaa I. Ibrahim
Laboratory for High Energy Astrophysics
NASA Goddard Space Flight Center
Mail Code 662
Greenbelt, MD 20771
USA
alaa@milkyway.gsfc.nasa.gov

Mr. Alexei Illarionov
Laboratory of High Energy
Joint Institute for Nuclear Research
141980 Dubna, Moscow Region
Russia
illar@thsun1.jinr.ru

Mr. Ira Jung
Max-Planck-Institut fur Kernphysik
Saupfercheckweg 1
69117 Heidelberg
Germany
Ira.Jung@mpi-hd.mpg.de

Ms. Bianca Keilhauer
Forschungszentrum Karlsruhe
Postfach 3640
76021 Karlsruhe
Germany
Bianca.Keilhauer@hik.fzk.de

Prof. Grant E. Kocharov
Ioffe Physico Technical Institute
Polytekhnicheskaya 26
194021 St. Petersburg
Russia
grant.kocharov@pop.ioffe.rssi.ru

Ms. Katarina Kovac
University of Belgrade
Dept. of Physics
Studentski trg 16, Box 550
11000 Belgrade
Yugoslavia
kkatarina@sezampro.yu

Ms. Aleksandra Kozyreva
Sternberg Astronomical Insitute
University av., 13
119899 Moscow
Russia
sasha@sai.msu.su

Dr. Leonid T. Ksenofontov
Institute of Cosmophysical Research
and Aeronomy SB RAS
Lenin Ave. 31
677891 Yakutsk,
Russia
ksenofon@sci.yakutia.ru

Ms. Izabela Kurp
A. Soltan Inst. Nuclear Studies
Cosmic Ray Laboratory
P.O. Box 447
ul. Uniwersytecka 5
90-950 Lodz 1
Poland
kurp@zpk.u.lodz.pl

Mr. Michel Leclerc
University of Gottingen
Gutenbergstr.40
37075 Gottingen
Germany
michel.leclerc@stud.uni-goettingen.de

Mr. Christian Lendvai
Physik Department E15
Technische Universität München
James-Franck-Strasse
D-85748 Garching
Germany
clendvai@physik.tu-muenchen.de

Mr. Jason Link
Laboratory of Experimental Astrophysics
Washington University
One Brookings Drive CB 1105
St. Louis, MO 63105
USA
jason@cosray2.wustl.edu

Mr. Fabrizio Lucarelli
Departamento de Fisica Atomica
Facultad de Ciencias Fisicas
Universidad Complutense
Avda. Complutense s/n
28040 Madrid
Spain
lucarel@gae.ucm.es

Mr. Paolo Maestro
Istituto Nazionale di Fisica Nuclear
INFN
via Vecchia Livornese, 1291
56010 S. Piero a Grado (Pisa)
Italy
paolo.maestro@pi.infn.it

Mr. Pratik Majumdar
High Energy Cosmic Rays
Tata Institute of Fundamental Research
Homi Bhaba Road
Mumbai - 400005
India
pratik@mailhost.tifr.res.in

Mr. Oliver Mang
Institut fuer Experimentelle und
Angewandte Physik
CAU Kiel
Germany
mang@ifkhep.uni-kiel.de

Dr. Sera Markoff
Max-Planck-Institut fuer
Radioastronomie
Auf dem Huegel 69
53121 Bonn
Germany
smarkoff@mpifr-bonn.mpg.de

Mr. Josep Flix Molina
IFAE Exp. Edf. Cn. Fac. de Ciences
UAB.
08193 Bellaterra
Spain
jflix@ifae.es

Dr. Alexey Murashov
Institute of Space Physics
Moscow State Eng. Phys. Institute
Kashirskoe shosse 31
115409 Moscow
Russia
murashov@space.mephi.ru

Ms. Marianne Goeger-Neff
Physik Department E15
Technische Universität München
James-Franck-Strasse
D 85748 Garching
Germany
marianne.goeger@ph.tum.de

Ms. Susan M. Niebur
Washington University
Campus Box 1105
One Brookings Drive
St. Louis, MO 63130
USA
smahan@hbar.wustl.edu

Mr. Ludwig Niedermeier
Physik Department E15
Technische Universität München
James-Franck-Strasse
D-85748 Garching
Germany
neidermeier@paule.e15.physik.
tu-muenchen.de

Dr. Dmitry S. Oshuev
Nuclear Physics Institute
Moscow State University
119899 Moscow
Russia
dima@dec1.npi.msu.su

Prof. Mikhail I. Panasyuk
Skobeltsyn Institute of Nuclear Phys.
Moscow State University
119899 Moscow
Russia
panasyuk@srdlan.npi.msu.su

Mr. Manuel Pavon
Zimmer 126
Max Planck Institut fuer Physik
(Werner Heisenberg Institut)
Foehringer Ring 6
80805 Muenchen
Germany
mpavon@mppmu.mpg.de

Dr. Oleg V. Pavlovsky
Physical Department
Quantum Theory & H.E. Physics
Moscow State University
119899 Moscow
Russia
ovp@goa.bog.msu.su

Mr. Jason Peterson
Department of Astronomy
Box 30001/MSC 4500
New Mexico State University
Las Cruces, NM 88003
USA
jaspeter@nmsu.edu

Dr. Oleh Petruk
IAPMM NAS of Ukraine
3-b Naukova St.
79053 Lviv
Ukraine
petruk@astro.franko.lviv.ua

Mr. Alessio Piccioli
I.N.F.N.
I-56010 S. Piero - Pisa
Italy
alessio.piccioli@pi.infn.it

Prof. Tsvi Piran
Racah Institute of Physics
Hebrew University
Jerusalem 91904
Israel
tsvi@nikki.fiz.huji.ac.il

Dr. Sergei B. Popov
Sternberg Astronomical Institute
Universitetskii pr. 13
119899 Moscow
Russia
polar@xray.sai.msu.ru

Dr. Vladimir Ptuskin
Inst. for Terrestrial Magnetism
Russian Academy of Sciences
(IZMIRAN)
142092 Troitsk, Moscow Region
Russia
vptuskin@izmiran.troitsk.ru

Dr. Mohamed Abd El Aziz Rassem
Astronomy Department
Cairo University
Cairo
Egypt
rassem2000@hotmail.com

Ms. Katherine Rawlins
University of Wisconsin - Madison
Department of Physics
1150 University Ave.
Madison, WI 53706
USA
Kath@alizarin.physics.wisc.edu

Dr. Yoel Rephaeli
School of Physics & Astronomy
Tel Aviv University
Tel Aviv, 69978
Israel
yoelr@wise.tau.ac.il

Prof. Maurice M. Shapiro
205 Yoakum Parkway
#1514
Alexandria, VA 22304
USA
Shapiro@Estart.com

Mr. Giorgio Riccobene
University of Catania and INFN
LNS: Via S. Sofia, 44
I-95123 Catania
Italy
giorgio.riccobene@romal.infn.it
riccobene@lns.infn.it

Ms. Agnieszka Sierpowska
Dept. of Experimental Physics
University of Lodz
Pomorska 149/153
90-236 Lodz
Poland
asierp@kfd2.fic.uni.lodz.pl

Prof. Norma Sanchez
Observatoire de Paris Demirm
61 Ave. de l'Observatoire
75014 Paris
France
norma.sanchez@obspm.fr

Mr. Valeri Smirichinski
Bogoliubov Lab. of Theoretical Physics
Joint Institute for Nuclear Research
Dubna, Moscow Region, 141980
Russia
smirvi@thsun1.jinr.ru

Prof. Livio Scarsi
D. di Energeticae Applicazione della
Fisica
Universita di Palermo
Viale Delle Scienze
90128 Palermo
Italy
scarsi@ifcai.pa.cnr.it

Dr. Arthur E. Smith
University of Oxford
Clarendon Laboratory
Park Road
Oxford
United Kingdom
a.smith1@physics.ox.ac.uk

Mr. Thomas P. Schweizer
IFAE, Edifici Cn.
Universitat Autonoma de Barcelona
08193 Bellaterra,
Barcelona
Spain
tschweiz@ifae.es

Ms. Iglika F. Spassovska
Predio D, sala 49
CPG-IFGW-UNICAMP
C.P. 6165
13083-970
CAMPINAS - SP
Brazil
iglika@ifi.unicamp.br

Prof. Todor Stanev
Bartol Research Institute
University of Delaware
Newark, DE 19716
USA
stanev@bartol.udel.edu

Dr. Francesco Sylos-Labini
Dept. de Physique Theorique
Universite de Geneve
Quai E. Ansermet 24
CH-1211 Geneve
Switzerland
sylos@amorgos.unige.ch

Dr. Gabor Szecsenyi-Nagy
Department of Astronomy
Eotvos University of Budapest
ELTE Csillagaszati Tanszek
Pazmany P. setany 1./A.
Budapest H-1117
Hungary
szena@ludens.elte.hu

Mr. Martin Tluczykont
University of Hamburg
II. Institute for Experimental Physics
Luruper Chaussee 149
D-22761 Hamburg
Germany
tluczym@mail.desy.de

Ms. Maria Diaz Trigo
Max Planck Institute for Physics
Foehringer Ring 6
D-80805 Munich
Germany
mdiaz@mppmu.mpg.de

Prof. Joachim Trümper
MPE Garching
Postfach 1603
85748 Garching
Germany
jtrumper@mpe.mpg.de

Dr. Oleg Udovyk
National Academy of Sciences
13 Chokolivsky Blvd.
03680 Kyiv
Ukraine
udovyk@erriu.ukrpack.net

Dr. Yury A. Uvarov
Dept. of Theoretical Astrophysics
Ioffe Physical Technical Institute
Polytekhnicheskaya 26
194021 St. Petersburg
Russia
uv@astro.ioffe.rssi.ru

Ms. Silvia Valchierotti
Dipartimento di fisica generale
v. Pietro Giuria 1
10125 Torino
Italy
valchier@to.infn.it

Ms. Giada Valle
Istituto Nazionale di Fisica Nucleare INFN
via Vecchia Livornese, 1291
56010, S. Piero a Grado, Pisa
Italy
giada.valle@pi.infn.it

Ms. Elena Vannuccini
Universitii de Firenze
Dipartimento di Fisics
Largo E.Fermi, 2
50125 Firenze
Italy
vannucci@fi.infn.it

Mr. Nikita R. Vavilov
Moscow State Engineering
Physics Institute
Kashirskoe shosse, 31
115409 Moscow
Russia
vavilov@space.exp.mephi.ru

Prof. John P. Wefel
Department of Physics & Astronomy
Louisiana State University
Baton Rouge, LA 70803
USA
wefel@phunds.phys.lsu.edu

Mr. Grzegorz Wieczorek
Department of Physics
University of Lodz
Pomorska 149/153
90-236 Lodz
Poland
gjw@kfd2.fic.uni.lodz.pl

Dr. Ralf Wischnewski
Deutsches Elektronen-Synchrotron
DESY
Platanenalle 6
D-15738 Zeuthen
Germany
wischnew@ifh.de

Mr. Michael Wood-Vasey
Lawrence Berkeley Laboratory
1 Cyclotron Road
MS 50-232
Berkeley, CA 94720
USA
wmwood-vasey@lbl.gov

Dr. Vladimir Zirakashvili
Astrophysical Research Laboratory
Institute for Terrestrial Magnetism
IZMIRAN
142190 Troitsk, Moscow region
Russia
zirak@izmiran.rssi.ru